Platelets in Health and Medicine

Platelets in Health and Medicine

Editor: Luis Tucker

AMERICAN
MEDICAL PUBLISHERS
www.americanmedicalpublishers.com

AMERICAN
MEDICAL PUBLISHERS
www.americanmedicalpublishers.com

Cataloging-in-Publication Data

Platelets in health and medicine / edited by Luis Tucker.
p. cm.
Includes bibliographical references and index.
ISBN 978-1-63927-787-2
1. Blood platelets. 2. Blood platelets--Health aspects. 3. Blood platelets--Transfusion.
4. Regenerative medicine. 5. Blood cells. I. Tucker, Luis.
QP97 .P53 2023
612.117--dc23

American Medical Publishers,
41 Flatbush Avenue,
1st Floor, New York,
NY 11217, USA

ISBN 978-1-63927-787-2 (Hardback)

Contents

Preface

Platelets are tiny blood cells that help in the formation of clots which helps to stop bleeding. They rush to the damaged site in the body, where they form a clot as a plug to stop the damage. They are formed in the bone marrow, and white and red blood cells. The body requires a specific number of platelet count to function properly. A normal platelet count in human body is considered to be 150,000 to 450,000 platelets per microliter of blood. The risk of bleeding develops if the platelet count falls below 10,000 to 20,000. Abnormal platelet count may result in several health conditions, such as thrombocytopenia, essential thrombocythemia, secondary thrombocytosis and platelet dysfunction. Thrombocytopenia is a condition in which bone marrow makes too few platelets. The common causes of this disease include cancer, liver disease, pregnancy, infections, and an abnormal immune system. Diagnosis of conditions related to platelets involves a simple blood test to determine the platelet count. This book provides significant information of platelets to help develop a good understanding of their role in health and medicine. Those in search of information to further their knowledge will be greatly assisted by it.

Significant researches are present in this book. Intensive efforts have been employed by authors to make this book an outstanding discourse. This book contains the enlightening chapters which have been written on the basis of significant researches done by the experts.

Finally, I would also like to thank all the members involved in this book for being a team and meeting all the deadlines for the submission of their respective works. I would also like to thank my friends and family for being supportive in my efforts.

Editor

Platelet-Activating Factor-Receptor Signaling Mediates Targeted Therapies-Induced Microvesicle Particles Release in Lung Cancer Cells

Shreepa J. Chauhan [1], Anita Thyagarajan [1,*], Yanfang Chen [1], Jeffrey B. Travers [1,2] and Ravi P. Sahu [1,*]

1 Department of Pharmacology and Toxicology, Boonshoft School of Medicine Wright State University, Dayton, OH 45345, USA; Chauhan.22@wright.edu (S.J.C.); yanfang.chen@wright.edu (Y.C.); jeffrey.travers@wright.edu (J.B.T.)
2 Department of Dermatology, Wright State Physicians, Wright State University, Dayton, OH 45345, USA
* Correspondence: anita.thyagarajan@wright.edu (A.T.); ravi.sahu@wright.edu (R.P.S.)

Abstract: Microvesicle particles (MVP) secreted by a variety of cell types in response to reactive oxygen species (ROS)-generating pro-oxidative stressors have been implicated in modifying the cellular responses including the sensitivity to therapeutic agents. Our previous studies have shown that expression of a G-protein coupled, platelet-activating factor-receptor (PAFR) pathway plays critical roles in pro-oxidative stressors-mediated cancer growth and MVP release. As most therapeutic agents act as pro-oxidative stressors, the current studies were designed to determine the role of the PAFR signaling in targeted therapies (i.e., gefitinib and erlotinib)-mediated MVP release and underlying mechanisms using PAFR-expressing human A549 and H1299 non-small cell lung cancer (NSCLC) cell lines. Our studies demonstrate that both gefitinib and erlotinib generate ROS in a dose-dependent manner in a process blocked by antioxidant and PAFR antagonist, verifying their pro-oxidative stressor's ability, and the role of the PAFR in this effect. We observed that these targeted therapies induce MVP release in a dose- and time-dependent manner, similar to a PAFR-agonist, carbamoyl-PAF (CPAF), and PAFR-independent agonist, phorbol myristate acetate (PMA), used as positive controls. To confirm the PAFR dependency, we demonstrate that siRNA-mediated PAFR knockdown or PAFR antagonist significantly blocked only targeted therapies- and CPAF-mediated but not PMA-induced MVP release. The use of pharmacologic inhibitor strategy suggested the involvement of the lipid ceramide-generating enzyme, acid sphingomyelinase (aSMase) in MVP biogenesis, and observed that regardless of the stimuli used, aSMase inhibition significantly blocked MVP release. As mitogen-activated protein kinase (MAPK; ERK1/2 and p38) pathways crosstalk with PAFR, their inhibition also significantly attenuated targeted therapies-mediated MVP release. These findings indicate that PAFR signaling could be targeted to modify cellular responses of targeted therapies in lung cancer cells.

Keywords: lung cancer; targeted therapies; platelet-activating factor-receptor; microvesicle particles

---

1. Introduction

Lung cancer is currently the second most prevalent type of cancer in both men and women in the United States, which continues to be a major health and socioeconomic issues [1–3]. Several risk factors have been known to be associated with lung carcinogenesis, including environmental factors and pollutants such as smoking and arsenic exposure, etc., as well as genetic factors [4–6]. While lung cancer cases have also been documented in non-smoking patients, smoking remains the most common contributing factor as long-term exposure to smoking has been shown to induce dysregulation and disruption in healthy lungs [7–9]. Of two main types, non-small cell lung carcinoma (NSCLC) accounts for the majority of the cases diagnosed compared to a less prevent small cell lung carcinoma (SCLC) [10,11]. Though significant achievements have been made in molecular pathogenesis and cellular signaling pathways involved in the development of lung cancer, the mechanisms leading to immunoevasion or the induction of tumor resistance to the known therapeutic regimens are yet to be fully explored [12–14]. Depending upon the lung cancer stages, the treatment options include surgical resection, chemotherapy, radiation therapy, targeted and immune-based approaches, as well as a combination of two or more therapeutic agents [15–19].

Among targeted therapies, receptor tyrosine kinase inhibitors (TKIs) including gefitinib and erlotinib have been used to treat NSCLC patients [20–22]. The primary mechanism of TKIs is to target tumor cells harboring activating mutations in the epidermal growth factor receptor (EGFR), a receptor tyrosine kinase of the ErbB family, which plays important roles in sustained tumor cell proliferation [20–22]. However, EGFR-independent mechanisms of action of these TKIs have also been documented, of which the most notable is the induction of oxidative stress, attributable to their pro-oxidative stressors ability to generate reactive oxygen species (ROS) [23–25]. While ROS generation remains one of the common mechanisms of cancer therapies including TKIs to induce cytotoxic effects, this ROS-mediated induction of oxidative stress has also been implicated in the development of tumor resistance mechanisms to such therapeutic options [26–28].

Studies, including ours, have shown that ROS-generating pro-oxidative stressors such as chemotherapeutic agents and radiation therapy produce oxidized lipid mediator glycerophosphocholines (ox-GPCs) as a bystander effect [29–31]. Many of these ox-GPCs exhibit platelet-activating factor (PAF)-like agonistic activity and bind to a G-protein coupled, PAF-receptor (*Ptafr*; PAFR), expressed on a variety of cell types including tumor cells [29–31]. Relevant to lung cancer, our previous studies have shown that exposure to cigarette smoke generates PAF agonists in a PAFR-dependent manner that induce systemic immunosuppression in a process blocked by antioxidants, cyclooxygenase types 2 (COX-2) inhibitors, and PAF-metabolizing enzyme PAF-acetyl hydrolase (PAF-AH), as well as depleting antibodies against immunosuppressive regulatory T cells (Tregs) [32]. We have also shown that the systemic administration of a known PAFR-agonist, carbamoyl-PAF (CPAF) can augment the growth of Lewis lung carcinoma (LLC1) tumors and its metastatic ability in a PAFR-dependent manner [33]. However, the role of the PAFR signaling in modulating targeted therapy effects in lung cancer models has not been studied. Given that these PAF agonists are metabolically labile and upon generation get readily metabolized by PAF-AH, our recent studies have defined the critical roles of a subpopulation of extracellular vesicles known as microvesicles or microvesicle particles as a novel mechanism by which these potent lipids are not only protected but circulated to exert local as well as delayed systemic effects [34–36].

Extracellular vesicles (EVs) are released from a wide variety of cell types and serve to mediate cell-to-cell communications, both in physiological and pathological conditions. These EVs mainly comprise of three types, which differ particularly in their sizes, for example, exosomes (size ~30–100 nm), microvesicle particles (MVP, size ~100–1000 nm), and apoptotic bodies (size >1000 nm) [35,36], their contents, and mechanism of formation. In as much as PAFR signaling plays critical roles ranging from acute pro-inflammatory to delayed systemic immunosuppression and augmentation of tumor growth as well as limiting the efficacy of therapeutic agents [29–33], the current studies were sought to determine its relevance in targeted therapies-mediated MVP release and define the underlying mechanisms in lung cancer cells. Overall, our studies demonstrate that targeted therapies act as pro-oxidative stressors, and induce MVP release in a PAFR-dependent manner via mechanisms involving acid sphingomyelinase (aSMase) enzyme, and mitogen-activated protein kinase (MAPK) pathway.

2. Results

2.1. Exposure to Gefitinib and Erlotinib Generates ROS

As several human malignancies including lung cancer express functional PAFR [29,37–39], which upon the exposure to pro-oxidative stressors including chemotherapeutic agents and radiation therapy can produce PAF agonists [29–31], our first studies verified the pro-oxidative stressors ability of targeted therapies to generate ROS. For this, we took advantage of a recent study, which identified that some NSCLC cell lines including A549 and H1299 express functional PAFR [38]. Besides, these cell lines also harbor wild-type EGFR, and have been used in studies with gefitinib and/or erlotinib [28,40–43].

To measure ROS generation, we treated A549 cells (used as a model) with or without gefitinib and erlotinib at various doses (25, 50, and 75 μM), similar to as previously reported [28,40–42]. Since ROS generation is an earlier event, we assessed its generation after 30 min of incubation along with assay buffer and negative and positive controls via measuring DCFDA fluorescence as per the kit's protocol. We observed that both these agents generate ROS in a dose-dependent manner, which verified that these targeted therapies act as pro-oxidative stressors (Figure 1A), as also reported previously [24,25]. To confirm the roles of ROS, and PAFR on targeted therapies-induced ROS generation, A549 cells were pretreated with antioxidant, N-acetylcysteine (NAC), and PAFR antagonist, WEB2086 compounds. Following 1 h of incubation, cells were treated with or without gefitinib and erlotinib at an optimum dose of 50 μM, along with PAFR agonist, CPAF, and PAFR-independent agonist, phorbol myristate acetate (PMA) used as positive controls. We also used two different solvents (i.e., DMSO and ethanol [EOH]) as negative controls to precisely rule out their effects on gefitinib and erlotinib (dissolved in DMSO) versus CPAF and PMA (dissolved in EOH)-induced ROS generation. After 30 min of incubation, ROS generation was assessed. Our studies demonstrate that targeted therapies-induced ROS generation is significantly reduced by NAC and WEB2086 compounds, similar to as observed by CPAF treatment (Figure 1B). However, PMA-induced ROS generation is only significantly attenuated by NAC but not by WEB2086 (Figure 1B). These studies confirmed that targeted therapies induce ROS generation, and suggest the role of the PAFR signaling in mediating this effect.

2.2. Gefitinib and Erlotinib Treatments Induce MVP Release from NSCLC Cell Lines in a Time- and Dose-Dependent Manner

Given that gefitinib and erlotinib generate ROS in a process blocked by antioxidant and PAFR antagonist (Figure 1), and our previous reports demonstrating that pro-oxidative stressors induce MVP release in a PAFR-dependent manner [34–36], we tested our working hypothesis if these targeted therapies can induce MVP release in lung cancer cells. It should be noted that depending upon the nature of the stimuli or the cell types, the secretion of MVPs has been shown to be time and/or concentration-dependent [44]. Thus, our first studies evaluated the time-dependent response of gefitinib (used as a drug model) on MVP release from the A549 cell line using CPAF and PMA as positive controls, as per our previous reports [34–36]. We observed that gefitinib induces MVP release in a time-dependent manner that significantly peaks between 4 to 8 h, yet no differences in MVP release were noted between these time points (Figure 2A). Thus, we used a 4-hours' time point to evaluate the dose-response effects of these targeted therapies along with appropriate controls on MVP release from A549 and H1299 cell lines. Our studies demonstrate that gefitinib and erlotinib induce MVP release from both A549 (Figure 2B,C) and H1299 (Figure 2D,E) cell lines in a dose-dependent manner.

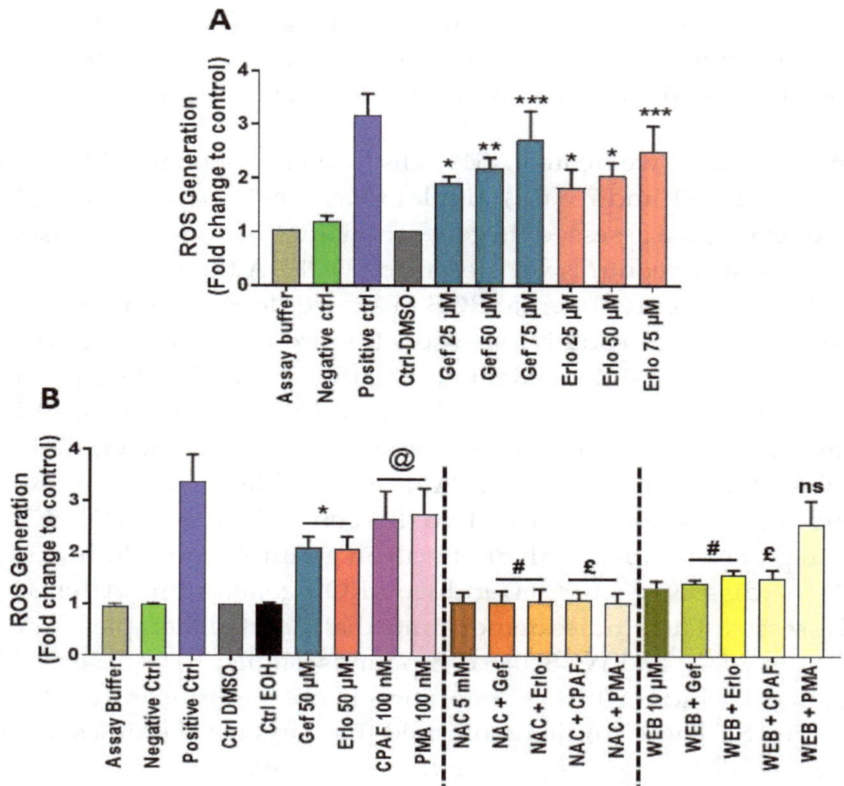

Figure 1. Effects of gefitinib and erlotinib on reactive oxygen species (ROS) generation. (**A**) A549 cells were treated with 0.1% dimethyl sulfoxide (DMSO) as vehicle control, and gefitinib (Gef) or erlotinib (Erlo) at 25, 50, and 75 μM concentrations. The cells were also treated with assay buffer, negative and positive controls from the manufacturer's kit. After 30 min of incubation, ROS generation was measured using a Synergy microplate reader. Data are expressed as mean ± SE of three independent experiments and presented as ROS generation (Fold change to control) versus treatment groups. Statistically significant differences were observed between Ctrl-DMSO and Gef or Erlo at various doses (* = $p < 0.05$, ** = $p < 0.01$, *** = $p < 0.001$). (**B**) A549 cells were pretreated with N-acetylcysteine (NAC; 5 mM) or WEB2086 (10 μM) for 1 h followed by the treatments with or without Gef, Erlo, carbamoyl-PAF (CPAF), or phorbol myristate acetate (PMA), and after 30 min of incubation, ROS generation was measured. Data are expressed as mean ± SE of five independent experiments and presented similar to Figure 1A. Statistically significant differences were observed between Ctrl-DMSO and Gef or Erlo (* $p < 0.05$); Ctrl-ethanol (EOH) vs. CPAF or PMA (@ = $p < 0.05$); Gef vs. NAC + Gef or Erlo vs. NAC + Erlo (# = $p < 0.01$); Gef vs. WEB + Gef or Erlo vs. WEB + Erlo (£ = $p < 0.05$); CPAF vs. NAC + CPAF or PMA vs. NAC + PMA (# = $p < 0.05$); and CPAF vs. WEB + CPAF (£ = $p < 0.05$). ns denotes non-significant change between PMA and WEB + PMA.

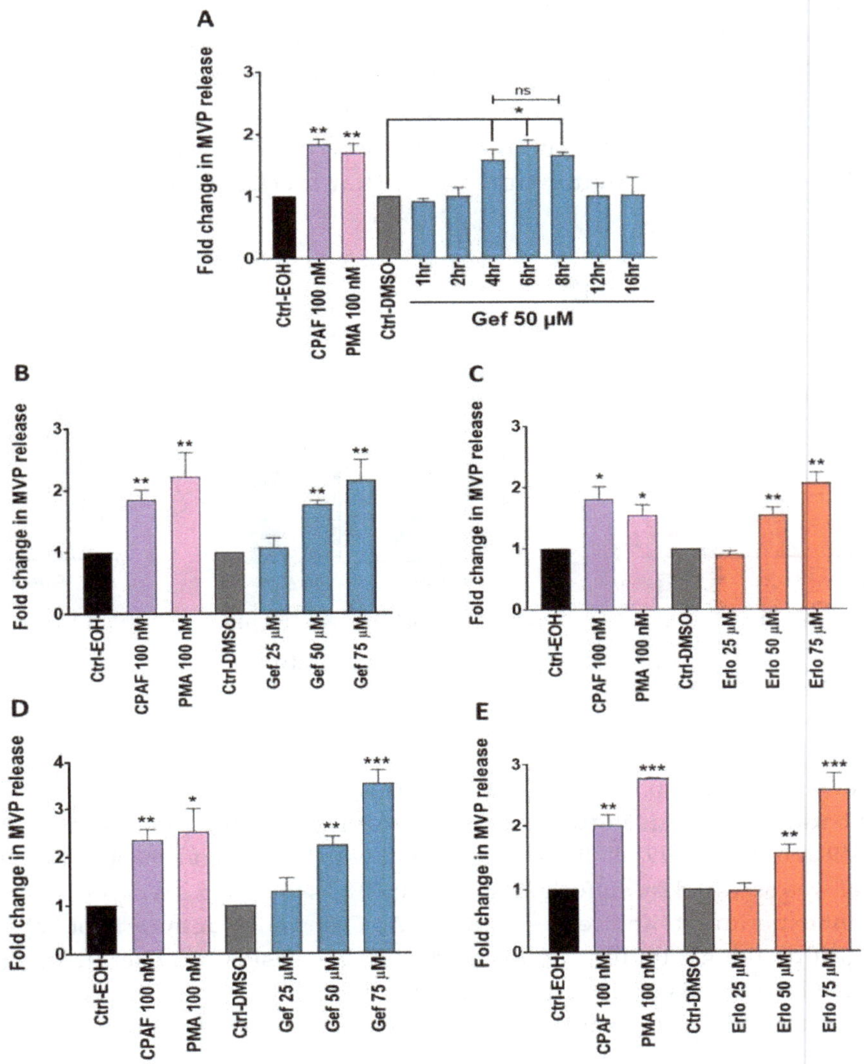

Figure 2. Time and dose-response assessments of gefitinib and erlotinib on microvesicle particles (MVP) release. (**A**) A549 cells were treated with 0.1% EOH or DMSO as vehicle controls, 100 nM CPAF or PMA as positive controls, or 50 μM Gef. After incubation at the given time points, MVPs were isolated and analyzed. (**B**,**C**) Similarly, A549 cells were treated with vehicle controls, CPAF, PMA, and Gef or Erlo at 25, 50, and 75 μM doses. (**D**,**E**) Similar experiments as mentioned for A549 cells were performed with H1299 cells. After 4 h of incubation, MVPs were isolated and analyzed. Data are representative of mean ± SE of three independent experiments, normalized per 1×10^6 cells. Statistically significant differences were observed between Ctrl-EOH and CPAF or PMA (* = $p < 0.05$, ** = $p < 0.01$); Ctrl-DMSO and Gef or Erlo at various doses (* = $p < 0.05$, ** = $p < 0.01$, *** = $p < 0.001$); ns denotes non-significant changes between the analyzed groups.

2.3. *Effect of Gefitinib and Erlotinib Treatments on Apoptosis Induction*

To address if gefitinib and erlotinib specifically induce MVP, or can also stimulate the secretion of other extracellular vesicles such as apoptotic bodies, we assessed apoptosis induction at the same experimental settings that were used to measure MVP release. For this, A549 cells were treated with or without gefitinib and erlotinib at an optimal dose of 50 μM, and after 4 h of incubation, apoptosis induction was assessed by quantitative caspase 3/7 activity assay. We did not observe apoptosis induction by these targeted therapies as compared to vehicle-treated cells at 4 hours' time point (Figure 3A). However, significantly increased apoptosis was detected following gefitinib and erlotinib treatments compared to the vehicle-treated cells at 72 hours' time point (Figure 3A). We also noticed significantly decreased % cell survival by gefitinib and erlotinib treatments compared to the vehicle-treated A549 and H1299 cell lines at 72 hours' time point (Figure 3B,C), as also reported previously [28,40,43]. These data indicate that targeted therapies-mediated MVP release is an earlier event as also confirmed by the findings that MVP release peaks between 4 to 8 hours' time point, and then declines at 12 hours' time point (Figure 2A). These data also indicate that apoptosis induction or cell growth inhibition occurs at a later time point that mediates the cytotoxic effects of gefitinib and erlotinib.

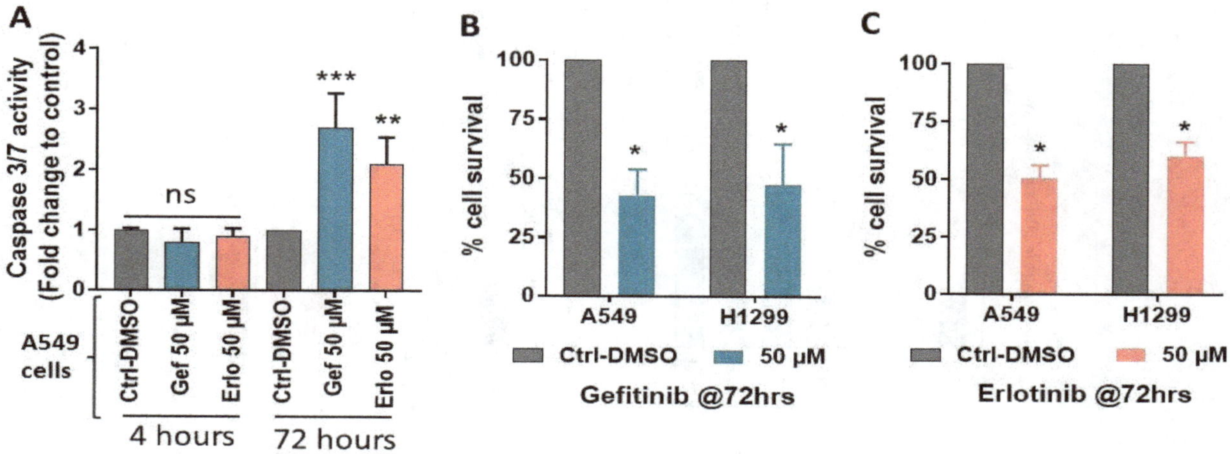

Figure 3. Effects of gefitinib and erlotinib on apoptosis induction and survival of NSCLC cell lines. (**A**) A549 cells were treated with 0.1% DMSO as vehicle control and Gef or Erlo at 50 µM dose. After 4 h and 72 h of incubation, apoptosis induction was assessed by quantitative caspase 3/7 activity assay. Similarly, (**B**) A549 and (**C**) H1299 cell lines were treated with vehicle control, and Gef or Erlo and after 72 h, cell survival was assessed by sulforhodamine B (SRB) assay. Data are expressed as mean ± SE of three independent experiments and represented as (**A**) Caspase 3/7 activity (Fold change to control) over treatment groups along with time points, and (**B**,**C**) % cell survival over treatment groups with their respective cell lines. Statistically significant differences were observed between Ctrl-DMSO and Gef or Erlo with their respective time points or cell lines used (* = $p < 0.05$, ** = $p < 0.01$, *** = $p < 0.001$); ns indicates non-significant changes between the analyzed groups.

2.4. Blockade of the PAFR Attenuates Erlotinib and Gefitinib-Induced MVP Release

To confirm if PAFR signaling is essential in gefitinib and erlotinib-induced MVP release, we adopted two separate approaches: (1) knocking down PAFR expression via PAFR specific siRNA; and (2) using a specific PAFR antagonist, WEB2086. Given the relatively same expression of the PAFR protein in A549 and H1299 cell lines [38] with similar dose-response effects of gefitinib and erlotinib on MVP release (~2–3 folds; Figure 2B,E), or inhibition of % cell survival (~50%; Figure 3B,C), we used A549 cells for our next studies. To optimize PAFR gene silencing, A549 cells were transfected with scrambled siRNA for control (i.e., Scr-ctrl) or three separate clones of PAFR siRNA, and the knockdown efficiency was evaluated by quantitative RT-PCR (qPCR) assay. We found that all PAFR siRNA clones decreased its mRNA expression with significant knockdown was detected with clone 3 (i.e., siRNA3) compared to Scr-ctrl (Figure 4A). In the next studies, A549 cells were transfected with either Scr-ctrl or siRNA3 followed by the treatments with or without CPAF, PMA, gefitinib, or erlotinib, and MVP release was quantified. In separate experiments, we tested the effect of WEB2086 pretreatment on CPAF, PMA, gefitinib, or erlotinib-mediated MVP secretion. Our studies demonstrate that both the knockdown or blockade of the PAFR resulted in significantly decreased MVP release induced by CPAF, gefitinib, and erlotinib as compared to these treatments alone (Figure 4B,C). However, neither PAFR siRNA3 nor WEB2086 exerted any effects on PMA-induced MVP release when compared to PMA-alone or the vehicle control treatments (Figure 4B,C). These findings indicate the necessity of the PAFR signaling in targeted therapies-induced MVP release.

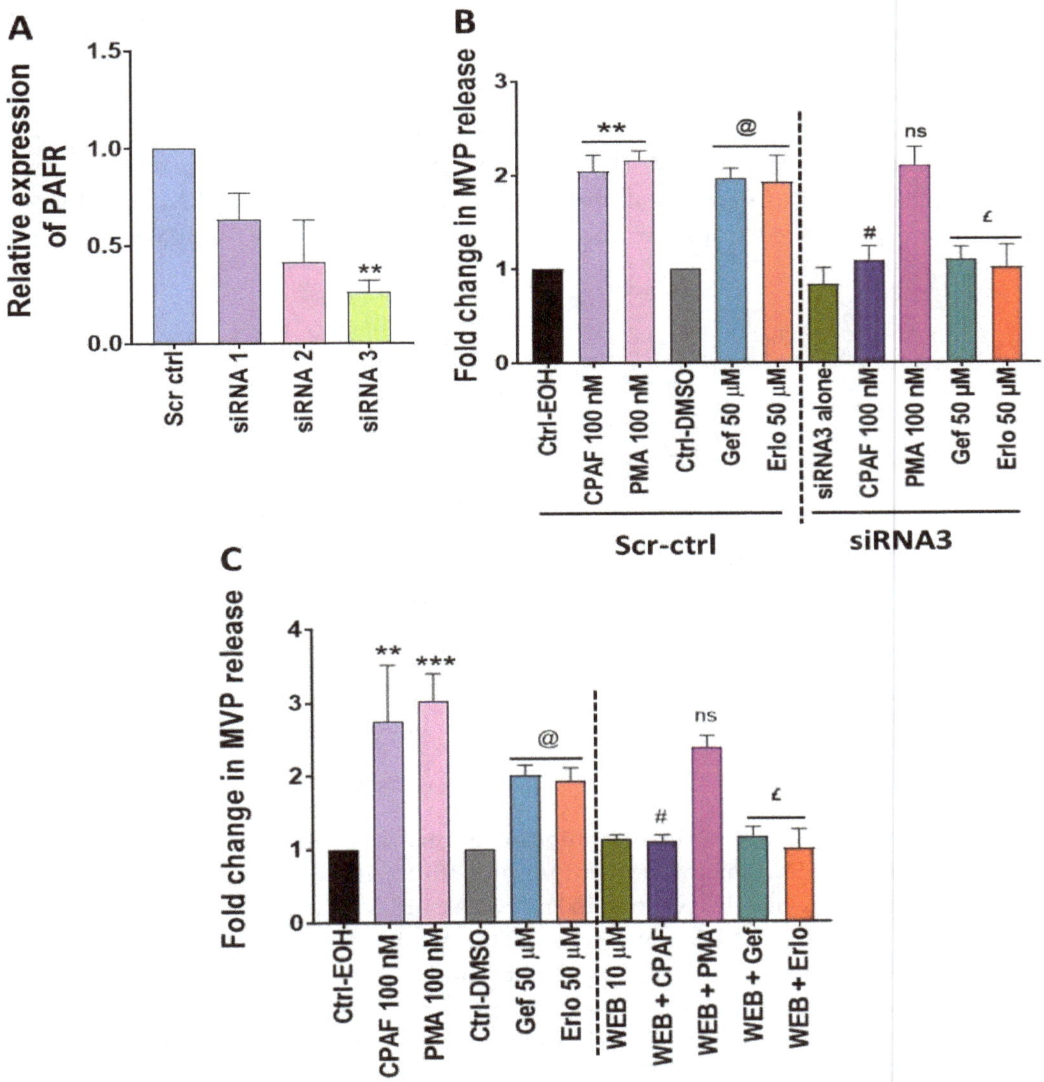

Figure 4. Effects of the PAFR-specific siRNA and PAFR antagonist on gefitinib and erlotinib-induced MVP release. (**A**) A549 cells were transfected with scrambled siRNA (i.e., Scr-ctrl) for control or three separate clones of PAFR siRNA and after 48 h, the PAFR knockdown efficiency was evaluated by qPCR assay. (**B**) A549 cells were transfected with Scr-ctrl or PAFR siRNA clone 3 (i.e, siRNA3) and after 48 h treated with 0.1% EOH or DMSO as vehicle controls, 100 nM CPAF or PMA as positive controls, and 50 μM Gef or Erlo. (**C**) A549 cells were pretreated with PAFR antagonist, WEB2086 (10 μM, 1 h) followed by the treatments with or without CPAF, PMA, Gef, or Erlo at the given doses. After 4 h of incubation, MVPs were isolated and analyzed. Data are mean ± SE of three independent experiments, normalized per 1×10^6 cells, and represented as (**A**) Relative expression of PAFR over Scr-ctrl and 3 separate clones of PAFR siRNA, and (**B**,**C**) Fold change in MVP release over various treatment groups. The statistically significant differences were observed between (**A**) Scr-ctrl and siRNA3 (** = $p < 0.01$), (**B**) Ctrl-EOH and CPAF or PMA (** = $p < 0.01$); ctrl-DMSO and Gef or Erlo (@ = $p < 0.01$); CPAF vs. siRNA3 + CPAF (# = $p < 0.01$); PMA vs. siRNA3 + PMA (ns); Gef vs. siRNA3 + Gef, and Erlo vs. siRNA3 + Erlo (£ = $p < 0.01$), (**C**) Ctrl-EOH and CPAF or PMA (** = $p < 0.01$, *** = $p < 0.001$); Ctrl-DMSO and Gef or Erlo (@ = $p < 0.01$); CPAF vs. WEB + CPAF (# = $p < 0.01$); PMA vs. WEB + PMA (ns); Gef vs. WEB + Gef or Erlo vs. WEB + Erlo (£ = $p < 0.01$); ns indicates non-significant changes between the analyzed groups.

2.5. Inhibition of aSMase Blocks MVP Release

The biogenesis of MVP involves activation of the aSMase pathway, and previous reports including ours have shown that the inhibition of aSMase blocks MVP release regardless of the stimuli used [34–36,45–47]. To that end, our next studies tested the effect of a specific pharmacologic inhibitor of aSMase (i.e., imipramine) on targeted therapies-induced MVP release. The A549 cells were pretreated with imipramine for 1 h followed by the treatments with or without CPAF, PMA, gefitinib, or erlotinib and MVP release was quantified. We observed that imipramine (Imip) treatment not only significantly blocked CPAF and PMA-mediated but also gefitinib and erlotinib-induced MVP release compared to these treatments alone (Figure 5), confirming the role of aSMase in MVP release.

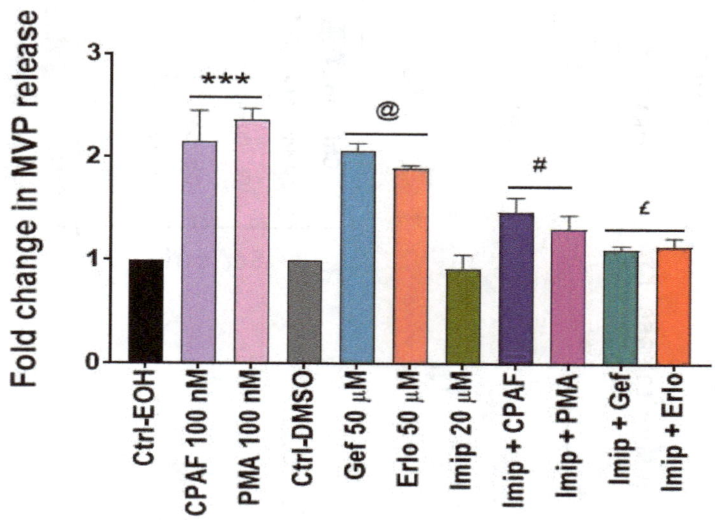

Figure 5. Acid sphingomyelinase (aSMase) inhibition abrogates gefitinib-induced MVP release. A549 cells were pretreated with aSMase inhibitor, imipramine (Imip; 20 µM, 1 h) followed by the treatments with 0.1% EOH or DMSO as vehicle controls, 100 nM CPAF or PMA as positive controls, and 50 µM Gef or Erlo. After 4 h of incubation, MVPs were isolated and quantified. Data are mean ± SE from three independent experiments, normalized per 1×10^6 cells, and represented as Fold change in MVP release over various treatment groups. Statistically significant differences were observed between Ctrl-EOH and CPAF or PMA (*** = $p < 0.001$); ctrl-DMSO and Gef or Erlo (@ = $p < 0.001$); CPAF vs. Imip + CPAF or PMA vs. Imip + PMA (# = $p < 0.05$); Gef vs. Imip + Gef or Erlo vs. Imip + Erlo (£ = $p < 0.01$).

2.6. MAPK Pathway Mediates Erlotinib and Gefitinib-Mediated MVP Release

Studies, including ours, have shown that the MAPK pathway, particularly extracellular signal-regulated kinase (ERK1/2) and p38-MAPK mediate not only PAFR and chemotherapy-induced effects, but also aSMase-mediated MVP biogenesis and secretion [34,48]. To that end, our next studies examined the effects of blocking ERK1/2 and p38-MAPK pathways using their specific inhibitors (i.e., PD90859 [PD] and SB202190 [SB] compounds), respectively. For this, A549 cells were pretreated with PD or SB compounds for 1 h followed by the treatments with or without CPAF, PMA, gefitinib, or erlotinib, and MVP release was quantified. We demonstrate that both ERK1/2 and p38-MAPK inhibition significantly attenuated CPAF, gefitinib, and erlotinib-induced MVP secretion compared to these treatments alone (Figure 6). Interestingly, PMA-induced MVP release was also significantly blocked by PD and SB compounds (Figure 6). These findings are not entirely surprising given that the PMA-induced phospholipase C (PLC) activation interacts with the phosphoinositide 3-kinase (PI3K) pathway, which indicates the crosstalk of the MAPK signaling with both PI3K and PAFR pathways [49,50]. The schematic representation of our working model is shown in Figure 7.

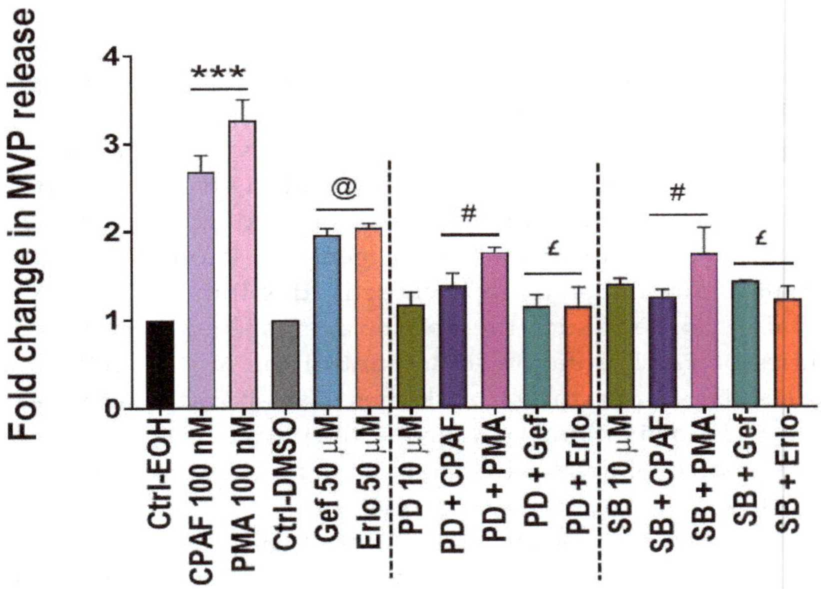

Figure 6. Effects of the extracellular signal-regulated kinase (ERK1/2) and p38-MAPK inhibitors on gefitinib and erlotinib-induced MVP release. A549 cells were pretreated with PD98059 and SB202190 (10 μM, 1 h) followed by the treatments with 0.1% EOH or DMSO as negative controls, 100 nM CPAF or PMA as positive controls, and 50 μM Gef or Erlo. After 4 h of incubation, we isolated and analyzed MVP secretion. Data are mean ± SE from three independent experiments, normalized per 1×10^6 cells, and represented as Fold change in MVP release over various treatment groups. Statistically significant differences were observed between Ctrl-EOH and CPAF or PMA (*** = $p < 0.001$); Ctrl-DMSO and Gef or Erlo (@ = $p < 0.01$); CPAF vs. PD + CPAF or SB + CPAF (# = $p < 0.001$); PMA vs. PD + PMA or SB + PMA (# = $p < 0.001$); Gef vs. PD + Gef or SB + Gef, and Erlo vs. PD + Erlo or SB + Erlo (£ = $p < 0.05$).

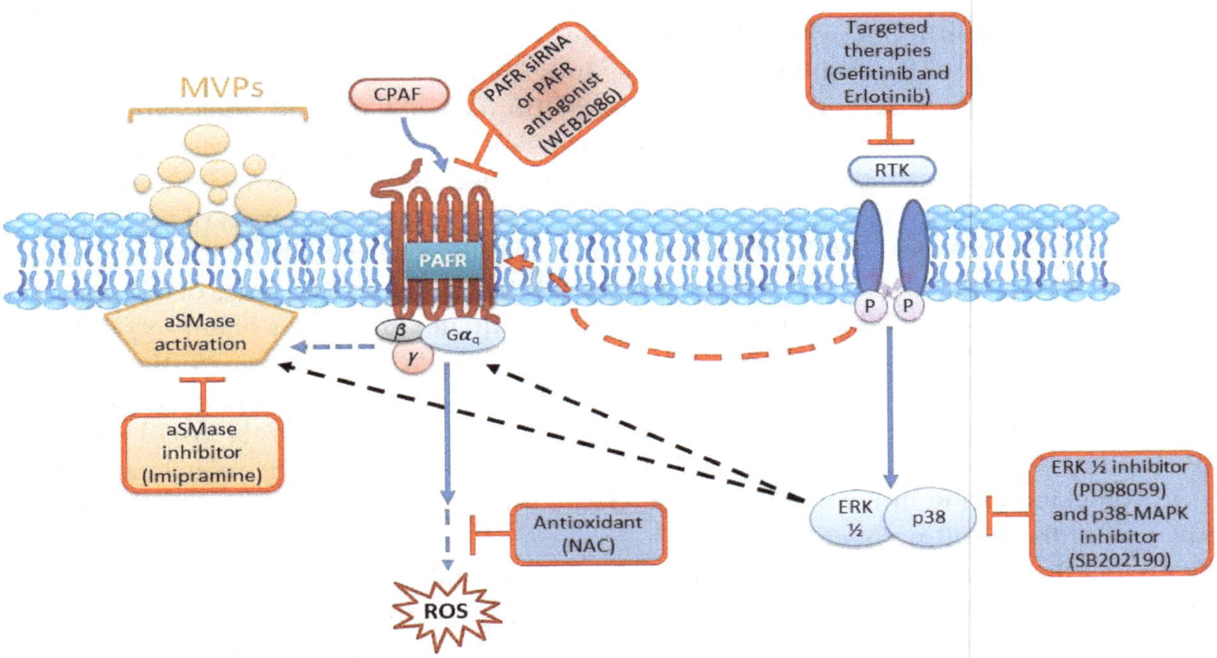

Figure 7. Schematic representation of a working model of PAFR-dependent targeted therapies-induced MVP release. In this model, targeted therapies activate the PAFR likely via the production of PAF agonists, similar to CPAF, which generates ROS in a process blocked by PAFR antagonist and antioxidant. The PAFR activation also stimulates aSMase, which involves ERK1/2 and p38-MAPK pathways resulting in MVPs formation and release in a process blocked by PAFR siRNA and PAFR antagonist, aSMase inhibitor, and ERK1/2 and p38-MAPK inhibitors.

3. Discussion

PAF is a potent phospholipid mediator, implicated in several pathological as well as inflammatory conditions, which mediates its effects via binding to a widely-expressed single seven-transmembrane G protein-coupled PAFR [32,51–53]. Relevant to cancer, growing evidence has demonstrated the potential role of the PAFR signaling in modulating the in vitro and in vivo growth of various tumor types in response to ROS-generating pro-oxidative stressors including therapeutic agents [29–31,54–56]. In in vitro models, PAFR-expressing tumor cells generate more PAF agonists in response to chemotherapeutic agents and radiation therapy via activating tumoral PAFR, which has been shown to induce a prosurvival response to such therapeutic agents [29–31]. To that end, multiple studies have tested the abilities of PAFR antagonists to control the in vitro cell proliferation, or in vivo tumor progression [29,54,55]. Importantly, our studies have shown that in the experimental murine melanoma model, host PAFR activation can augment the growth of tumor xenografts or impede the anti-tumoral immune response of therapeutic agents via mechanisms involving COX-2-dependent upregulation of Tregs [30,31,56].

Particularly to the lung cancer model, we have shown that activation of the host PAFR (via studies using PAFR-expressing wild type and PAFR-deficient mice) not only augments the growth but also the metastatic ability of murine Lewis lung carcinoma (LLC1) cells [33]. Notably, Chen and colleagues have also demonstrated that PAFR activation enhances the in vitro cell proliferation, as well as in vivo growth of NSCLC via involving STAT3 signaling [38]. Importantly, high tumoral PAFR expression has also been found to be positively correlated with increased tumor invasiveness and tumor stages as well as the decreased survival probability of NSCLC patients [38]. As lung cancer remains one of the difficult malignancies to treat, despite the advancements in the treatment modalities including the targeted therapy approaches, studies have been directed towards deciphering the mechanisms associated with modulating the sensitivity or efficacy of therapeutic agents. Given that ROS-generating pro-oxidative stressors can produce oxidized PAF agonists from a variety of PAFR-expressing cell types, which can then travel via the MVP to exert local, as well as systemic effects [34–36], the current studies were designed to test our hypothesis if PAFR signaling can modulate targeted therapy-induced MVP release from lung cancer cells, and then define the underlying mechanisms.

Given a recent report demonstrating that some human NSCLC cell lines express PAFR and that A549 and H1299 cells have similar protein expression of PAFR [38], and also harbor wild type EGFR [28], we used them as models. Our first studies verified that gefitinib and erlotinib in addition to their primary mechanism of targeting tumor cells harboring mutated forms of EGFR [20–22] also act as pro-oxidative stressors [23–25]. We demonstrate that these targeted therapies generate ROS in a dose-dependent manner, as also previously reported [24,25]. To confirm the roles of ROS, and PAFR on targeted therapies-induced ROS generation, our studies evaluated the effects of antioxidant, NAC and PAFR antagonist, WEB2086 and demonstrated that both these compounds significantly attenuate gefitinib and erlotinib-induced ROS generation, similar to as observed by CPAF, used as a positive control [57]. However, PMA-induced ROS generation, used as another control [58–60] is blocked only by NAC but not by the WEB2086 compound. While NAC is a known ROS quencher, the WEB2086 effect is via blocking PAFR, indicating that targeted therapies-induced ROS generation is mediated via the activation of the PAFR pathway.

Given that these targeted therapies generate ROS, its potential role in exosomes and MVP release has been directly or indirectly evidenced in modifying the cellular responses, or the efficacy of targeted and conventional anticancer agents in model systems including cancer cells. While ROS are involved in the production of MVPs via its ability to induce oxidative stress, MVPs shed by cell types including tumor cells have also been shown to exhibit altered redox balances with elevated ROS levels, and that the number of MVPs are further affected by oxidative conditions such as those induced by ROS-generating pro-oxidative stressors [34,35,61]. In addition, Li and colleagues have demonstrated that exosomes-derived from gefitinib-treated EGFR-mutant human NSCLC PC9 cell line decrease the antitumor effects of cisplatin, and that pharmacological inhibition of exosomes secretion resulted in a modest synergistic effect of cisplatin and gefitinib combination [62]. Along similar lines, exosomes and MVPs inhibitors have been shown to augment the cytotoxic efficacy of a chemotherapy drug 5-fluorouracil (5-FU) against human prostate cancer PC3 and breast cancer MCF-7 cell lines compared to 5-FU alone treatment [63]. However, there is no direct evidence of the PAFR involvement in targeted therapies-induced MVP release. To that end, our next studies evaluated the time and dose responses of gefitinib on MVP release from NSCLC cell lines using CPAF and PMA as positive controls. We demonstrate that exposure to gefitinib induces MVP release in a time and dose-dependent manner. Similar dose-dependent effects of erlotinib were also observed on MVP release.

While several comprehensive studies have shown that the crosstalk between various signaling pathways mediates the cytotoxic (e.g., apoptotic) effects of anti-cancer drugs, the cell-to-cell communication by extracellular vesicles such as MVP and exosomes plays vital roles in diverse biological, and pathological processes including cancer growth and metastasis, as well as modulating the sensitivity of therapeutic agents [44,45,64,65]. To evaluate if gefitinib and erlotinib at experimental settings used to measure MVP release, can also induce apoptosis, we analyzed caspase 3/7 activity. We observed that treatments of these targeted therapies did not induce apoptosis as compared to the vehicle control treatment. However, at a later time point, we detected significantly increased apoptosis and decreased cell survival by gefitinib and erlotinib as compared to the vehicle control treatment, similar to as previously reported [28]. These findings indicate that MVP release is an earlier event, which has been implicated in modulating cancer growth, sensitivity, or the responses of anti-cancer agents [64–66] and that cancer growth inhibition or apoptosis induction occurs at a later time point(s) which mediate their cytotoxic effects [28,40,41,43]. Notably, MVP release during the pathological states including from tumor cells has also been implicated as potential biomarkers to predict disease stages/conditions [66,67]. Importantly, since tumor-derived exosomes and MVPs exacerbate the tumor microenvironment, multiple studies have also suggested that removal of such vesicles could be exploited to sensitize tumor cells to chemotherapy or immunotherapy approaches, or could be used as a vehicle for drug delivery [68–70].

The current studies are significant given the findings of another report demonstrating that circulating MVP derived from the lung cancer patients enhanced tumor angiogenesis in in vitro HUVEC cell model [71]. Besides, these results are compatible with our group's previous reports demonstrating that exposure to UVB radiation or thermal burn injury induces MVP release from PAFR-expressing epidermal HaCaT, and KBP cell lines in a dose-dependent manner [35,36]. Moreover, another study demonstrated that gemcitabine treatment induces MVP release significantly from PAFR-expressing cell line-PANC1 but not from PAFR-deficient Hs766T cell line [34]. The translational relevance of these studies is supported by our published findings demonstrating that MVP derived from chemotherapy-treated PAFR-expressing cells contain PAF agonists [34]. These separate lines of evidence are in agreement with our current studies indicating that ROS-generating pro-oxidative stressors stimulate MVP secretion in a PAFR-dependent manner. Moreover, the data demonstrating that the blockade of PAFR activation either by PAFR antagonist or PAFR specific siRNA approach only attenuated MVP release via CPAF and targeted therapy but not by PMA confirmed the absolute requirement of the PAFR in mediating targeted therapy-induced MVP release from NSCLC cell lines.

Since the biosynthesis and release of MVP in response to multiple stimuli are aSMase-dependent [34–36,45–48], and that aSMase inhibition by imipramine has been used as a tool to block aSMase-mediated MVP generation [46,47]. Consistent with this notion, our studies have observed that imipramine not only blocked CPAF-induced, but also PMA, as well as gefitinib and erlotinib-mediated MVP release, confirming the role of aSMase in MVP release. In addition to the primary mechanism (i.e., targeting activating mutations in EGFR), several other targets of tyrosine kinase inhibitors including MAPK, and PI3K/AKT have been identified in NSCLC models [24–28,40]. Notably, the MAPK (i.e., ERK1/2 and p38) pathways have also been shown to play important roles in aSMase-dependent MVP generation, and mediating the effects of PAFR signaling in pro-oxidative stressors-induced local and systemic responses [34,36,48,72]. To verify the roles of ERK1/2 and p38 pathways in MVP release, we used their specific inhibitors and found that the inhibition of both ERK1/2 and p38 pathways resulted in significantly reduced MVP secretion mediated by CPAF, erlotinib, and gefitinib compared to these treatments alone. However, these inhibitors also blocked the PMA-induced MVP release. These findings indicate that the activation of both the PAFR and MAPK pathways are involved in mediating gefitinib and erlotinib-induced aSMase-dependent MVP release, and also supported by the studies demonstrating that MAPK crosstalks with PAFR pathway [48]. Consistent with this notion that tyrosine kinase inhibitors interact with PAFR, we have found that treatment of PAFR-positive KBP cells with gefitinib and erlotinib and results in IL-8 release, which is not found in PAFR-negative KBM cells (a separate project, data not shown).

Overall, these studies indicate the potential role of the PAFR signaling in targeted therapies-mediated MVP release, and that these observed effects are mediated via an interplay between the PAFR and MAPK pathways. Importantly, our studies are of high significance as these provide the rationale of exploring the translational relevance of this PAFR signaling targeted therapies-induced MVP release, which could have cellular and systemic effects in lung cancer.

4. Materials and Methods

4.1. Reagents

The culture media was purchased from GE Healthcare Biosciences (Marlborough, MA, USA). The PAF-R agonist (CPAF), PAF-R antagonist (WEB2086), PAF-R siRNA, erlotinib, gefitinib, ERK1/2, and p38 inhibitors, and imipramine were purchased from Cayman Chemicals Co. (Ann Arbor, MI, USA). Cell culture media, fetal bovine serum (FBS) was from Corning (Corning, NY, USA), antibiotic-antimycotic was from Gibco (Gaithersburg, MD, USA), and penicillin-streptomycin was purchased from Hyclone (Logan, UT, USA). The RNA extraction kit was purchased from Invitrogen Life Technologies (Carlsbad, CA, USA). PAFR siRNAs (FlexiTube) were from Qiagen (Germantown, MD, USA) and lipofectamine 3000 was from Invitrogen (Thermo Fisher Scientific, Carlsbad, CA, USA). The PAF-R, GAPDH primers, high-capacity cDNA Reverse Transcription kit, SYBR green, and PCR reagents were purchased from Applied Biosystems (Carlsbad, CA, USA). The ROS detection assay kit was from BioVision (BioVision Incorporated (Milpitas, CA, USA), and the Caspase-Glo 3/7 Assay kit for the measurement of apoptosis induction was from Promega Corporation (Madison, WI, USA).

4.2. Cell Culture

Human non-small cell lung cancer (NSCLC) cell lines A549 and H1299 (a kind gift from Dr. Weiwen Long at WSU) were used for these studies as these cell lines harbor wild type EGFR and express relatively similar protein expression of PAFR [28,38]. The A549 cell line was cultured in F-12K medium with 10% FBS, 2.5 mL antibiotic-antimycotic, 2.5 mL penicillin-streptomycin, and 15 μL of 2M magnesium chloride. H1299 cells were cultured in RPMI-1640 medium with 10% fetal bovine serum, 2.5 mL antibiotic-antimycotic, 2.5 mL penicillin-streptomycin, 2.25 mL of 40% glucose, and 5 mL of 100 mM sodium pyruvate. These cell lines were maintained at 37 °C with 5% CO2 and 95% humidity.

4.3. Measurement of ROS Generation

The assessment of ROS generation was done by the ROS detection assay kit, which is based on widely-used H2DCFDA methodology. According to the kit's manual protocol, A549 cells were seeded in 96 well plates, and cultured overnight. The media was removed and cells were washed with 100 μL ROS assay buffer followed by the addition of 100 μL of 1X ROS label per well, diluted in ROS assay buffer, and incubated for 45 min at 37 °C in the dark. After that, the ROS label was removed, and a 100 μL ROS assay buffer was added, and cells were then treated with 0.1% DMSO or various doses of gefitinib and erlotinib (25, 50, and 75 μM in 100 μL), and incubated for 30 min. Cells containing only 100 μL ROS assay buffer were used as an additional control, cells with only 100 μL of 1X ROS label were used as a negative control, and cells with 100 μL of 1X ROS inducer served as a positive control. In separate experiments, the effects of antioxidant, N-acetylcysteine (NAC), and PAFR antagonist, WEB2086 were evaluated on targeted therapies, and CPAF and PMA-induced ROS generation. The fluorescence was measured at Ex/Em = 495/529 using a Synergy microplate reader (BioTek, Winooski, VT, USA) at Proteome Analysis Laboratory (PAL) core facilities.

4.4. Cell Survival Assay

Cell survival was measured by the sulforhodamine-B (SRB) assay as per our previous report [54]. The A549 and/or H1299 cell lines were seeded in 96-well plate at 5×10^3 cell density per well with 200 µL of 1% serum media and following treatments with 0.1% DMSO as vehicle control, and 50 µM concentration of erlotinib and gefitinib in 1% serum media. After 72 h, cells were fixed by adding 100 µL of 20% trichloroacetic acid (TCA) followed by 1-h incubation at 4 °C. Then cells were washed thrice with distilled water and stained with 100 µL 0.4% (weight/volume) SRB (dissolved in 1% acetic acid) followed by 15 min incubation in dark at room temperature. Unbound dye was removed by three washes with 1% acetic acid followed by air drying. Finally, the protein-bound dye was dissolved by 150 µL of 10 mM unbuffered Tris base [tris (hydroxymethyl) aminomethane] for 10 min on a shaker, and absorbance was read at 570 nm with the Synergy H1Mf plate reader. Treatment groups were normalized with the control (0.1% DMSO) group.

4.5. Apoptosis Assay

The apoptosis induction was assessed by quantitative caspase 3/7 activity assay using kit's manual protocol, as described by us previously [73]. For this, A549 cells were seeded overnight into 6-well plates followed by treatments with 0.1% DMSO or erlotinib or gefitinib (50 µM). After 4 and 72 h of incubation, cells were homogenized in hypotonic extraction buffer (HEB) and 0.5 mg of total extracted protein was incubated with Caspase-3/7 Glo reagent. Finally, caspase 3/7 activity was measured using Synergy H1 Luminescence microplate reader (BioTek, Winooski, VT, USA) at Proteome Analysis Laboratory (PAL) core facilities.

4.6. siRNA Transfection, RNA Extraction and Quantitative Real-Time PCR (qRT-PCR) Analysis

The A549 cells were transfected using lipofectamine 3000 with three separate clones (i.e., transcripts) of PAFR siRNA as per the manufacturer's protocol. To optimize the condition, following transfection PAFR knockdown was evaluated at 24 and 48 hours' time points by analyzing the relative PAFR mRNA expression. For this, the total RNA was extracted from A549 cells by TRIzol extraction method, and extracted RNA was quantified with Nanodrop One spectrophotometer (Thermo Fisher Scientific, Waltham, MA, USA). High-Capacity cDNA Reverse Transcription kit was used to transcribed RNA samples to cDNA for the analysis of the PAFR mRNA expression using an SYBR green-based, quantitative fluorescent PCR method as per our previous reports [34,66]. The fluorescence was detected using a StepOne Real-Time PCR machine (Applied Biosystems, Foster City, CA, USA). The primers used in this experiment were specific for PAFR, and GAPDH was used as an endogenous control to normalize PAFR expression. The quantification of each PCR product was analyzed using the $2^{-\Delta\Delta Ct}$ method.

4.7. Microvesicle Particles (MVP) Extraction and Analysis

The extraction and analysis of MVP were done from the culture supernatants of NSCLC cell lines as per our previous reports [34–36]. In brief, A549 and H1299 cell lines were cultured overnight or until cells reach 80% to 90% confluency followed by washing the cells thrice with HBSS (no phenol red) only. Then cells were treated with 0.1% EOH and DMSO for negative controls, 100 nM CPAF and PMA for positive controls, and gefitinib and erlotinib at various concentrations (25, 50, and 75 µM) in HBSS containing 1% BSA. After incubations at various time points (1, 2, 4, 6, 8, 12, and 16 h) for initial experiments, and for 4 h for the rest of the experiments, the supernatants were collected and centrifuged at 2000× g for 20 min at 4 °C to remove cells, and cell debris. Then, the supernatants were collected and centrifuged the cell-free solution again at 20,000× g for 70 min at 4 °C. The pelleted MVPs were resuspended with 100 μL of filtered PBS and used for the nanoparticle tracking analysis (NTA). The size and concentration of these MVPs were detected by using a NanoSight 300 (NS300) instrument (Malvern Instruments, UK). MVPs counts were normalized with the cell count of each treatment [34–36]. Similarly, the effects of PAFR knockdown, aSMase inhibition, and MAPK pathways inhibitors on targeted therapies-induced MVP release were analyzed.

4.8. Statistical Analysis

Statistical analysis was assessed by GraphPad Prism software version 7.0 (GraphPad Software, San Diego, CA, USA). All in vitro experiments were repeated, independently, at least three times. Data were analyzed by Student's t-test (to compare between two groups) or one-way ANOVA (for more than two groups) with post hoc Tukey or Bonferroni multiple comparison tests. The value of $p < 0.05$ was considered to indicate a statistically significant difference between the tested groups.

Author Contributions: A.T. and R.P.S. were involved in the conceptualization of the project. S.J.C. and A.T. were involved in generating all the data. A.T., Y.C., J.B.T. and R.P.S. were involved in critical discussion and providing inputs. A.T. and R.P.S. wrote the manuscript. S.J.C., A.T., Y.C., J.B.T. and R.P.S. were involved in editing, and approving the final version of the manuscript. All authors have read and agreed to the published version of the manuscript.

Abbreviations

PAF	Platelet-activating factor
PAFR	Platelet-activating factor-receptor
Ox-GPCs	Oxidized glycerophosphocholines
MVP	Microvesicle particles
ROS	Reactive oxygen species
MAPK	Mitogen-activated protein kinase
STAT3	Signal transduction and activator of transcription
NSCLC	Non-small cell lung cancer
SCLC	Small-cell lung cancer
aSMase	Acid sphingomyelinase
Imip	Imipramine
PMA	Phorbol myristate acetate
CPAF	Carbamoyl-PAF
COX-2	Cyclooxygenase type 2
PAF-AH	PAF-acetyl hydrolase
Tregs	Regulatory T cells
EGFR	Epidermal growth factor receptor
Gef	Gefitinib
Erlo	Erlotinib
ERK	Extracellular signal-regulated kinase
PLC	Phospholipase C
PI3K	Phosphoinositide 3-kinase

References

1. Siegel, R.L.; Miller, K.D.; Jemal, A. Cancer statistics, 2020. *CA Cancer J. Clin.* **2020**, *70*, 7–30. [CrossRef]
2. McIntyre, A.; Ganti, A.K. Lung cancer-A global perspective. *J. Surg. Oncol.* **2017**, *115*, 550–554. [CrossRef]
3. Mao, Y.; Yang, D.; He, J.; Krasna, M.J. Epidemiology of Lung Cancer. *Surg. Oncol. Clin. N. Am.* **2016**, *25*, 439–445. [CrossRef]

4. Soza-Ried, C.; Bustamante, E.; Caglevic, C.; Rolfo, C.; Sirera, R.; Marsiglia, H. Oncogenic role of arsenic exposure in lung cancer: A forgotten risk factor. *Crit. Rev. Oncol. Hematol.* **2019**, *139*, 128–133. [CrossRef]
5. Malhotra, J.; Malvezzi, M.; Negri, E.; La Vecchia, C.; Boffetta, P. Risk factors for lung cancer worldwide. *Eur. Respir. J.* **2016**, *48*, 889–902. [CrossRef]
6. Bastian, L.A.; Gray, K.E.; DeRycke, E.; Mirza, S.; Gierisch, J.M.; Haskell, S.G.; Magruder, K.M.; Wakelee, H.A.; Wang, A.; Ho, G.Y.; et al. Differences in Active and Passive Smoking Exposures and Lung Cancer Incidence Between Veterans and Non-Veterans in the Women's Health Initiative. *Gerontologist* **2016**, *56* (Suppl. 1), S102–S111. [CrossRef]
7. Corrales, L.; Rosell, R.; Cardona, A.F.; Martín, C.; Zatarain-Barrón, Z.L.; Arrieta, O. Lung cancer in never smokers: The role of different risk factors other than tobacco smoking. *Crit. Rev. Oncol. Hematol.* **2020**, *148*, 102895. [CrossRef]
8. Akhtar, N.; Bansal, J.G. Risk factors of Lung Cancer in nonsmoker. *Curr. Probl. Cancer* **2017**, *41*, 328–339. [CrossRef] [PubMed]
9. Rivera, G.A.; Wakelee, H. Lung Cancer in Never Smokers. *Adv. Exp. Med. Biol.* **2016**, *893*, 43–57. [CrossRef]
10. Nasim, F.; Sabath, B.F.; Eapen, G.A. Lung cancer. *Med. Clin. N. Am.* **2019**, *103*, 463–473. [CrossRef]
11. Latimer, K.M.; Mott, T.F. Lung cancer: Diagnosis, treatment principles, and screening. *Am. Fam. Physician* **2015**, *91*, 250–256. [PubMed]
12. Rodriguez-Canales, J.; Parra-Cuentas, E.; Wistuba, I.I. Diagnosis and Molecular Classification of Lung Cancer. *Cancer Treat. Res.* **2016**, *170*, 25–46. [CrossRef] [PubMed]
13. Varella-Garcia, M. Chromosomal and genomic changes in lung cancer. *Cell Adh. Migr.* **2010**, *4*, 100–106. [CrossRef] [PubMed]
14. Zugazagoitia, J.; Guedes, C.; Ponce, S.; Ferrer, I.; Molina-Pinelo, S.; Paz-Ares, L. Current Challenges in Cancer Treatment. *Clin. Ther.* **2016**, *38*, 1551–1566. [CrossRef] [PubMed]
15. Lemjabbar-Alaoui, H.; Hassan, O.U.; Yang, Y.W.; Buchanan, P. Lung cancer: Biology and treatment options. *Biochim. Biophys. Acta* **2015**, *1856*, 189–210. [CrossRef]
16. Pallis, A.G.; Gridelli, C.; Wedding, U.; Faivre-Finn, C.; Veronesi, G.; Jaklitsch, M.; Luciani, A.; O'Brien, M. Management of elderly patients with NSCLC; updated expert's opinion paper: EORTC Elderly Task Force, Lung Cancer Group and International Society for Geriatric Oncology. *Ann. Oncol.* **2014**, *25*, 1270–1283. [CrossRef]
17. Nagasaka, M.; Gadgeel, S.M. Role of chemotherapy and targeted therapy in early-stage non-small cell lung cancer. *Expert Rev. Anticancer Ther.* **2018**, *18*, 63–70. [CrossRef]
18. Shroff, G.S.; de Groot, P.M.; Papadimitrakopoulou, V.A.; Truong, M.T.; Carter, B.W. Targeted Therapy and Immunotherapy in the Treatment of Non-Small Cell Lung Cancer. *Radiol. Clin. N. Am.* **2018**, *56*, 485–495. [CrossRef]
19. Reck, M.; Heigener, D.; Reinmuth, N. Immunotherapy for small-cell lung cancer: Emerging evidence. *Future Oncol.* **2016**, *12*, 931–943. [CrossRef]
20. Wang, S.; Song, Y.; Liu, D. EAI045: The fourth-generation EGFR inhibitor overcoming T790M and C797S resistance. *Cancer Lett.* **2017**, *385*, 51–54. [CrossRef]
21. Rosell, R.; Dafni, U.; Felip, E.; Curioni-Fontecedro, A.; Gautschi, O.; Peters, S.; Massutí, B.; Palmero, R.; Aix, S.P.; Carcereny, E.; et al. Erlotinib and bevacizumab in patients with advanced non-small-cell lung cancer and activating EGFR mutations (BELIEF): An international, multicentre, single-arm, phase 2 trial. *Lancet Respir. Med.* **2017**, *5*, 435–444. [CrossRef]
22. Chung, C. Tyrosine kinase inhibitors for epidermal growth factor receptor gene mutation-positive non-small cell lung cancers: An update for recent advances in therapeutics. *J. Oncol. Pharm. Pract.* **2016**, *22*, 461–476. [CrossRef] [PubMed]
23. Teppo, H.R.; Soini, Y.; Karihtala, P. Reactive Oxygen Species-Mediated Mechanisms of Action of Targeted Cancer Therapy. *Oxid. Med. Cell Longev.* **2017**, *2017*, 1485283. [CrossRef] [PubMed]
24. Okon, I.S.; Coughlan, K.A.; Zhang, M.; Wang, Q.; Zou, M.-H. Gefitinib-mediated reactive oxygen specie (ROS) instigates mitochondrial dysfunction and drug resistance in lung cancer cells. *J. Biol. Chem.* **2015**, *290*, 9101–9110. [CrossRef] [PubMed]
25. Shan, F.; Shao, Z.; Jiang, S.; Cheng, Z. Erlotinib induces the human non–small-cell lung cancer cells apoptosis via activating ROS-dependent JNK pathways. *Cancer Med.* **2016**, *5*, 3166–3175. [CrossRef]

26. Marcar, L.; Bardhan, K.; Gheorghiu, L.; Dinkelborg, P.; Pfäffle, H.; Liu, Q.; Wang, M.; Piotrowska, Z.; Sequist, L.V.; Borgmann, K.; et al. Acquired Resistance of EGFR-Mutated Lung Cancer to Tyrosine Kinase Inhibitor Treatment Promotes PARP Inhibitor Sensitivity. *Cell Rep.* **2019**, *27*, 3422–3432. [CrossRef]
27. Krall, E.B.; Wang, B.; Munoz, D.M.; Ilic, N.; Raghavan, S.; Niederst, M.J.; Yu, K.; Ruddy, D.A.; Aguirre, A.J.; Kim, J.W.; et al. KEAP1 loss modulates sensitivity to kinase targeted therapy in lung cancer. *eLife* **2017**, *6*, e18970. [CrossRef]
28. Li, H.; Schmid-Bindert, G.; Wang, D.; Zhao, Y.; Yang, X.; Su, B.; Zhou, C. Blocking the PI3K/AKT and MEK/ERK signaling pathways can overcome gefitinib-resistance in non-small cell lung cancer cell lines. *Adv. Med. Sci.* **2011**, *56*, 275–284. [CrossRef]
29. da Silva, A., Jr.; Chammas, R.; Lepique, A.P.; Jancar, S. Platelet-activating factor (PAF) receptor as a promising target for cancer cell repopulation after radiotherapy. *Oncogenesis* **2017**, *6*, e296. [CrossRef]
30. Sahu, R.P.; Ocana, J.A.; Harrison, K.A.; Ferracini, M.; Touloukian, C.E.; Al-Hassani, M.; Sun, L.; Loesch, M.; Murphy, R.C.; Althouse, S.K.; et al. Chemotherapeutic agents subvert tumor immunity by generating agonists of platelet-activating factor. *Cancer Res.* **2014**, *74*, 7069–7078. [CrossRef]
31. Sahu, R.P.; Harrison, K.A.; Weyerbacher, J.; Murphy, R.C.; Konger, R.L.; Garrett, J.E.; Chin-Sinex, H.J.; Johnston, M.E., 2nd; Dynlacht, J.R.; Mendonca, M.; et al. Radiation therapy generates platelet-activating factor agonists. *Oncotarget* **2016**, *7*, 20788–20800. [CrossRef] [PubMed]
32. Sahu, R.P.; Petrache, I.; Van Demark, M.J.; Rashid, B.M.; Ocana, J.A.; Tang, Y.; Yi, Q.; Turner, M.J.; Konger, R.L.; Travers, J.B. Cigarette smoke exposure inhibits contact hypersensitivity via the generation of platelet-activating factor agonists. *J. Immunol.* **2013**, *190*, 2447–2454. [CrossRef] [PubMed]
33. Hackler, P.C.; Reuss, S.; Konger, R.L.; Travers, J.B.; Sahu, R.P. Systemic Platelet-activating Factor Receptor Activation Augments Experimental Lung Tumor Growth and Metastasis. *Cancer Growth Metastasis* **2014**, *7*, 27–32. [CrossRef] [PubMed]
34. Thyagarajan, A.; Kadam, S.M.; Liu, L.; Kelly, L.E.; Rapp, C.M.; Chen, Y.; Sahu, R.P. Gemcitabine Induces Microvesicle Particle Release in a Platelet-Activating Factor-Receptor-Dependent Manner via Modulation of the MAPK Pathway in Pancreatic Cancer Cells. *Int. J. Mol. Sci.* **2018**, *20*, 32. [CrossRef]
35. Fahy, K.; Liu, L.; Rapp, C.M.; Borchers, C.; Bihl, J.C.; Chen, Y.; Simman, R.; Travers, J.B. UVB-generated Microvesicle Particles: A Novel Pathway by Which a Skin-specific Stimulus Could Exert Systemic Effects. *Photochem. Photobiol.* **2017**, *93*, 937–942. [CrossRef]
36. Liu, L.; Fahy, K.E.; Awoyemi, A.A.; Thapa, P.; Kelly, L.E.; Chen, J.; Bihl, J.C.; Cool, D.R.; Chen, Y.; Rapp, C.M.; et al. Thermal Burn Injury Generates Bioactive Microvesicles: Evidence for a Novel Transport Mechanism for the Lipid Mediator Platelet-Activating Factor (PAF) That Involves Subcellular Particles and the PAF Receptor. *J. Immunol.* **2020**, *205*, 193–201. [CrossRef]
37. Yu, Y.; Zhang, M.; Zhang, X.; Cai, Q.; Zhu, Z.; Jiang, W.; Xu, C. Transactivation of epidermal growth factor receptor through platelet-activating factor/receptor in ovarian cancer cells. *J. Exp. Clin. Cancer Res.* **2014**, *33*, 85. [CrossRef]
38. Chen, J.; Lan, T.; Zhang, W.; Dong, L.; Kang, N.; Zhang, S.; Fu, M.; Liu, B.; Liu, K.; Zhan, Q. Feed-Forward Reciprocal Activation of PAFR and STAT3 Regulates Epithelial-Mesenchymal Transition in Non-Small Cell Lung Cancer. *Cancer Res.* **2015**, *75*, 4198–4210. [CrossRef]
39. Marques, S.A.; Dy, L.C.; Southall, M.D.; Yi, Q.; Smietana, E.; Kapur, R.; Marques, M.; Travers, J.B.; Spandau, D.F. The Platelet-Activating Factor Receptor Activates the Extracellular Signal-Regulated Kinase Mitogen-Activated Protein Kinase and Induces Proliferation of Epidermal Cells through an Epidermal Growth Factor-Receptor-Dependent Pathway. *J. Pharmacol. Exp. Ther.* **2002**, *300*, 1026–1035. [CrossRef]
40. Steen, N.V.D.; Potze, L.; Giovannetti, E.; Cavazzoni, A.; Ruijtenbeek, R.; Rolfo, C.; Pauwels, P.; Peters, G.J. Molecular mechanism underlying the pharmacological interactions of the protein kinase C-β inhibitor enzastaurin and erlotinib in non-small cell lung cancer cells. *Am. J. Cancer Res.* **2017**, *7*, 816–830.
41. Howe, G.A.; Xiao, B.; Zhao, H.; Al-Zahrani, K.N.; Hasin, M.S.; Villeneuve, J.; Sekhon, H.S.; Goss, G.D.; Sabourin, L.A.; Dimitroulakos, J.; et al. Focal Adhesion Kinase Inhibitors in Combination with Erlotinib Demonstrate Enhanced Anti-Tumor Activity in Non-Small Cell Lung Cancer. *PLoS ONE* **2016**, *11*, e0150567. [CrossRef] [PubMed]
42. Lin, C.; Qin, Y.; Zhang, H.; Gao, M.Y.; Wang, Y.F. EGF upregulates RFPL3 and hTERT via the MEK signaling pathway in non-small cell lung cancer cells. *Oncol. Rep.* **2018**, *40*, 29–38. [CrossRef]

43. Choi, J.; Kang, M.; Nam, S.H.; Lee, G.H.; Kim, H.J.; Ryu, J.; Cheong, J.G.; Jung, J.W.; Kim, T.Y.; Lee, H.Y.; et al. Bidirectional signaling between TM4SF5 and IGF1R promotes resistance to EGFR kinase inhibitors. *Lung Cancer* **2015**, *90*, 22–31. [CrossRef]
44. Neri, T.; Pergoli, L.; Petrini, S.; Gravendonk, L.; Balia, C.; Scalise, V.; Amoruso, A.; Pedrinelli, R.; Paggiaro, P.; Bollati, V.; et al. Particulate matter induces prothrombotic microparticle shedding by human mononuclear and endothelial cells. *Toxicol. In Vitro* **2016**, *32*, 333–338. [CrossRef] [PubMed]
45. Tricarico, C.; Clancy, J.; D'Souza-Schorey, C. Biology and biogenesis of shed microvesicles. *Small GTPases* **2017**, *8*, 220–232. [CrossRef] [PubMed]
46. Albouz, S.; Hauw, J.J.; Berwald-Netter, Y.; Boutry, J.M.; Bourdon, R.; Baumann, N. Tricyclic antidepressants induce sphingomyelinase deficiency in fibroblast and neuroblastoma cell cultures. *Biomedicine* **1981**, *35*, 218–220.
47. Catalano, M.; O'Driscoll, L. Inhibiting extracellular vesicles formation and release: A review of EV inhibitors. *J. Extracell. Vesicles* **2019**, *9*, 1703244. [CrossRef]
48. Landis, M.; Yi, Q.; Hyatt, A.M.; Travers, A.R.; Lewis, D.A.; Travers, J.B. Involvement of P38 MAP kinase in the augmentation of UVB-mediated apoptosis via the epidermal platelet-activating factor receptor. *Arch. Dermatol. Res.* **2007**, *299*, 263–266. [CrossRef]
49. Chao, W.; Deng, J.S.; Li, P.Y.; Kuo, Y.H.; Huang, G.J. Inotilone from Inonotus linteus suppresses lung cancer metastasis in vitro and in vivo through ROS-mediated PI3K/AKT/MAPK signaling pathways. *Sci. Rep.* **2019**, *9*, 2344. [CrossRef]
50. Tang, H.; Xue, G. Major Physiological Signaling Pathways in the Regulation of Cell Proliferation and Survival. *Handb. Exp. Pharmacol.* **2018**, *249*, 13–30. [CrossRef]
51. Braquet, P.; Touqui, L.; Shen, T.Y.; Vargaftig, B.B. Perspectives in platelet-activating factor research. *Pharmacol. Rev.* **1987**, *39*, 97–145.
52. Walterscheid, J.P.; Ullrich, S.E.; Nghiem, D.X. Platelet-activating factor, a molecular sensor for cellular damage, activates systemic immune suppression. *J. Exp. Med.* **2002**, *195*, 171–179. [CrossRef]
53. Konger, R.L.; Marathe, G.K.; Yao, Y.; Zhang, Q.; Travers, J.B. Oxidized glycerophosphocholines as biologically active mediators for ultraviolet radiation-mediated effects. *Prostaglandins Other Lipid Mediat.* **2008**, *87*, 1–8. [CrossRef]
54. Sahu, R.P. Expression of the platelet-activating factor receptor enhances benzyl isothiocyanate-induced apoptosis in murine and human melanoma cells. *Mol. Med. Rep.* **2015**, *12*, 394–400. [CrossRef]
55. Onuchic, A.C.; Machado, C.M.L.; Saito, R.F.; Rios, F.J.; Jancar, S.; Chammas, R. Expression of PAFR as part of a prosurvival response to chemotherapy: A novel target for combination therapy in melanoma. *Mediat. Inflamm.* **2012**, *2012*, 175408. [CrossRef]
56. Sahu, R.P.; Turner, M.J.; DaSilva, S.C.; Rashid, B.M.; Ocana, J.A.; Perkins, S.M.; Konger, R.L.; Touloukian, C.E.; Kaplan, M.H.; Travers, J.B. The environmental stressor ultraviolet B radiation inhibits murine antitumor immunity through its ability to generate platelet-activating factor agonists. *Carcinogenesis* **2012**, *33*, 1360–1367. [CrossRef]
57. Perry, S.W.; Norman, J.P.; Litzburg, A.; Zhang, D.; Dewhurst, S.; Gelbard, H.A. HIV-1 transactivator of transcription protein induces mitochondrial hyperpolarization and synaptic stress leading to apoptosis. *J. Immunol.* **2005**, *74*, 4333–4344. [CrossRef]
58. Lee, H.B.; Yu, M.R.; Song, J.S.; Ha, H. Reactive oxygen species amplify protein kinase C signaling in high glucose-induced fibronectin expression by human peritoneal mesothelial cells. *Kidney Int.* **2004**, *65*, 1170–1179. [CrossRef]
59. Kim, A.; Im, M.; Yim, N.-H.; Jung, Y.P.; Ma, J.Y. Aqueous extract of Bambusae Caulis in Taeniam inhibits PMA-induced tumor cell invasion and pulmonary metastasis: Suppression of NF-κB activation through ROS signaling. *PLoS ONE* **2013**, *8*, e78061. [CrossRef]
60. Ponath, V.; Kaina, B. Death of Monocytes through Oxidative Burst of Macrophages and Neutrophils: Killing in Trans. *PLoS ONE* **2017**, *12*, e0170347. [CrossRef]
61. Bodega, G.; Alique, M.; Puebla, L.; Carracedo, J.; Ramírez, R.M. Microvesicles: ROS scavengers and ROS producers. *J. Extracell. Vesicles* **2019**, *8*, 1626654. [CrossRef]
62. Li, X.-Q.; Liu, J.-T.; Fan, L.-L.; Liu, Y.; Cheng, L.; Wang, F.; Yu, H.-Q.; Gao, J.; Wei, W.; Wang, H.; et al. Exosomes derived from gefitinib-treated EGFR-mutant lung cancer cells alter cisplatin sensitivity via up-regulating autophagy. *Oncotarget* **2016**, *7*, 24585–24595. [CrossRef]

63. Kosgodage, U.S.; Trindade, R.P.; Thompson, P.R.; Inal, J.M.; Lange, S. Chloramidine/Bisindolylmaleimide-I-Mediated Inhibition of Exosome and Microvesicle Release and Enhanced Efficacy of Cancer Chemotherapy. *Int. J. Mol. Sci.* **2017**, *18*, 1007. [CrossRef]
64. Becker, A.; Thakur, B.K.; Weiss, J.M.; Kim, H.S.; Peinado, H.; Lyden, D. Extracellular Vesicles in Cancer: Cell-to-Cell Mediators of Metastasis. *Cancer Cell* **2016**, *30*, 836–848. [CrossRef]
65. Willms, E.; Cabañas, C.; Mäger, I.; Wood, M.; Vader, P. Extracellular Vesicle Heterogeneity: Subpopulations, Isolation Techniques, and Diverse Functions in Cancer Progression. *Front. Immunol.* **2018**, *9*, 738. [CrossRef]
66. D'Souza-Schorey, C.; Clancy, J.W. Tumor-derived microvesicles: Shedding light on novel microenvironment modulators and prospective cancer biomarkers. *Genes Dev.* **2012**, *26*, 1287–1299. [CrossRef]
67. Barreiro, K.; Holthofer, H. Urinary extracellular vesicles. A promising shortcut to novel biomarker discoveries. *Cell Tissue Res.* **2017**, *369*, 217–227. [CrossRef]
68. Moore, C.; Kosgodage, U.; Lange, S.; Inal, S.M. The emerging role of exosome and microvesicle- (EMV-) based cancer therapeutics and immunotherapy. *Int. J. Cancer* **2017**, *141*, 428–436. [CrossRef]
69. Balachandran, B.; Yuana, Y. Extracellular vesicles-based drug delivery system for cancer treatment. *Cogent Med.* **2019**, *6*, 1635806. [CrossRef]
70. Mentkowski, K.I.; Snitzer, J.D.; Rusnak, S.; Lang, J.K. Therapeutic Potential of Engineered Extracellular Vesicles. *AAPS J.* **2018**, *20*, 50. [CrossRef]
71. Sheu, J.; Lee, F.; Wallace, C.G.; Tsai, T.H.; Leu, S.; Chen, Y.L.; Chai, H.T.; Lu, H.I.; Sun, C.K.; Yip, H.K. Administered circulating microparticles derived from lung cancer patients markedly improved angiogenesis, blood flow and ischemic recovery in rat critical limb ischemia. *J. Transl. Med.* **2015**, *13*, 59. [CrossRef] [PubMed]
72. Chen, Z.; Schubert, P.; Bakkour, S.; Culibrk, B.; Busch, M.P.; Devine, D.V. p38 mitogen-activated protein kinase regulates mitochondrial function and microvesicle release in riboflavin- and ultraviolet light-treated apheresis platelet concentrates. *Transfusion* **2017**, *57*, 1199–1207. [CrossRef] [PubMed]
73. Thyagarajan, A.; Saylae, J.; Sahu, R.P. Acetylsalicylic acid inhibits the growth of melanoma tumors via SOX2-dependent-PAF-R-independent signaling pathway. *Oncotarget* **2017**, *8*, 49959–49972. [CrossRef] [PubMed]

2

Molecular Drivers of Platelet Activation: Unraveling Novel Targets for Anti-Thrombotic and Anti-Thrombo-Inflammatory Therapy

Madhumita Chatterjee [1], Agnes Ehrenberg [2], Laura Mara Toska [2], Lisa Maria Metz [2], Meike Klier [2], Irena Krueger [2], Friedrich Reusswig [2] and Margitta Elvers [2,*]

1 Department of Cardiology and Angiology, Universitätsklinikum Tübingen, Medizinische Klinik III, 72076 Tübingen, Germany; madhumita.chatterjee@med.uni-tuebingen.de

2 Department of Vascular and Endovascular Surgery, Experimental Vascular Medicine, Heinrich-Heine University Medical Center, 40225 Düsseldorf, Germany; agnes.ehrenberg@hhu.de (A.E.); laura.toska@hhu.de (L.M.T.); lisa.metz@hhu.de (L.M.M.); mekli100@uni-duesseldorf.de (M.K.); irena.krueger@uni-duesseldorf.de (I.K.); Friedrich.reusswig@uni-duesseldorf.de (F.R.)

* Correspondence: elvers@uni-duesseldorf.de

Abstract: Cardiovascular diseases (CVDs) are the leading cause of death globally—partly a consequence of increased population size and ageing—and are major contributors to reduced quality of life. Platelets play a major role in hemostasis and thrombosis. While platelet activation and aggregation are essential for hemostasis at sites of vascular injury, uncontrolled platelet activation leads to pathological thrombus formation and provokes thrombosis leading to myocardial infarction or stroke. Platelet activation and thrombus formation is a multistage process with different signaling pathways involved to trigger platelet shape change, integrin activation, stable platelet adhesion, aggregation, and degranulation. Apart from thrombotic events, thrombo-inflammation contributes to organ damage and dysfunction in CVDs and is mediated by platelets and inflammatory cells. Therefore, in the past, many efforts have been made to investigate specific signaling pathways in platelets to identify innovative and promising approaches for novel antithrombotic and anti-thrombo-inflammatory strategies that do not interfere with hemostasis. In this review, we focus on some of the most recent data reported on different platelet receptors, including GPIb-vWF interactions, GPVI activation, platelet chemokine receptors, regulation of integrin signaling, and channel homeostasis of NMDAR and PANX1.

Keywords: platelets; integrins; glycoprotein Ib; glycoprotein VI; reelin; phospholipase D; CXCR; pannexin 1; NMDAR

---

1. Introduction

Cardiovascular diseases (CVDs) include ischemic heart disease, stroke, heart failure, peripheral arterial disease, and a number of other cardiac and vascular conditions. They are the leading cause of death globally. Every year, more people die from cardiovascular disease than any other cause, with 31% of deaths following cardiovascular diseases worldwide [1–3].

Platelets are small anucleate cells of the hematopoietic system and are formed by megakaryocytes (MKs) in the bone marrow. At sites of vascular injury, platelets adhere to the injured vessel wall, forming a hemostatic plug to avoid excessive blood loss, and therefore play a fundamental role in hemostasis. In contrast, platelets trigger thrombotic events because uncontrolled platelet activation can induce acute vessel occlusion, leading to myocardial infarction or stroke at areas of atherosclerotic plaque rupture [4,5]. Ischemic CVDs such as myocardial infarction and stroke as well as infectious diseases are characterized by thrombotic and inflammatory events that contribute to both cell death and organ failure. This thrombo-inflammation is mediated by platelets and immune cells such as T cells, macrophages and neutrophils and triggers organ dysfunction [6–8].

Injury of the endothelium leads to exposure of extracellular matrix proteins that serve as substrates for adhesion and activation of circulating platelets. Platelet activation and thrombus formation include different signaling pathways leading to platelet shape change, integrin activation and degranulation. Platelet recruitment to the vessel wall and initial adhesion is dependent on the interaction between glycoprotein (GP) Ib-IX-V and von Willebrand factor (vWF), that binds to subendothelial collagen [9]. In the second step, platelets become activated characterized by shape change, activation of integrins and secretion [10]. In the third step, platelet adhesion to the vessel wall is stabilized followed by platelet aggregation and thrombus formation. These latter processes are mainly mediated by integrins [11], but the mechanisms involved are not fully understood. Thus, many efforts have been made to identify new molecular targets for anti-thrombotic and anti-thrombo-inflammatory therapy in the last years.

2. GPIb-vWF-Axis

During vascular injury, subendothelial matrix proteins such as collagen, fibronectin, thrombospondin and endothelial released vWF are exposed to the blood stream. Platelet surface exposed receptors are able to bind to these sub-endothelial matrix components to avoid massive blood loss. A first, transient binding of platelets to the injured vessel wall is achieved by binding of GPIb, exclusively expressed on the platelet surface, to collagen-adherent vWF, thus tethering platelets to the vessel wall. However, binding of GPIb to vWF is characterized by a rapid on-off rate that facilitates binding of the collagen receptor GPVI to fibrillar collagen in the injured vessel wall to initiate platelet activation, firm adhesion and thrombus formation [12]. Under high shear conditions, GPIb-vWF interaction is essential for the recruitment of platelets to the injured vessel wall [5] and initiates intracellular signaling cascades involving tyrosine phosphorylation of SYK, the Src family kinases Lyn and Fyn, and phospholipase (PL) Cγ2 to induce $\alpha_{IIb}\beta_3$ integrin activation and firm adhesion [13–15]. Thus it is not surprising that inhibition of the GPIb-vWF axis could be a promising target for therapy in ischemic stroke to improve cerebral reperfusion rates and reduce cerebral injury after ischemia and reperfusion as demonstrated by Denorme et al. [16].

Reelin and PLD1 Are New Regulators of GPIb Signaling

In recent years, different modulators of GPIb activation and signaling have been identified. The glycoprotein reelin has been shown to be important for GPIb-dependent thrombus formation [17]. Reelin was first identified in the central nervous system where it plays a prominent role in brain development by controlling neuronal positioning and migration to mediate the formation of laminated brain structures. It is produced by Cajal-Retzius neurons and interacts with the lipoprotein receptors apolipoprotein E receptor 2 (ApoER2), mediating the formation of filopodia and lamellipodia, and with the very-low-density lipoprotein receptor (VLDLR) [18,19]. Reelin is found in blood plasma and is expressed in endothelia cells, in the liver [20,21] and in megakaryocytes and platelets, where it co-localizes to F-actin and regulates platelet spreading on fibrinogen [17,22]. Recent study identified reelin as a new promising therapeutic target to prevent occlusive thrombus formation. Reelin-deficient mice are protected against arterial thrombosis and focal cerebral ischemia but show normal hemostasis. Reelin deficient platelets display reduced adhesion to recombinant vWF under high shear conditions but exhibit no defects in vWF binding under static conditions, when binding of vWF to GPIb is induced by the snake venom botrocetin. After GPIb engagement, integrin activation and the phosphorylation of Erk and Akt are reduced in reelin deficient platelets. Reelin-induced GPIb activation is mediated by the interaction of GPIb and the major platelet reelin receptor amyloid precursor protein (APP) suggesting that reelin-induced effects on GPIb signaling are mediated by APP–GPIb interactions (Figure 1) [17].

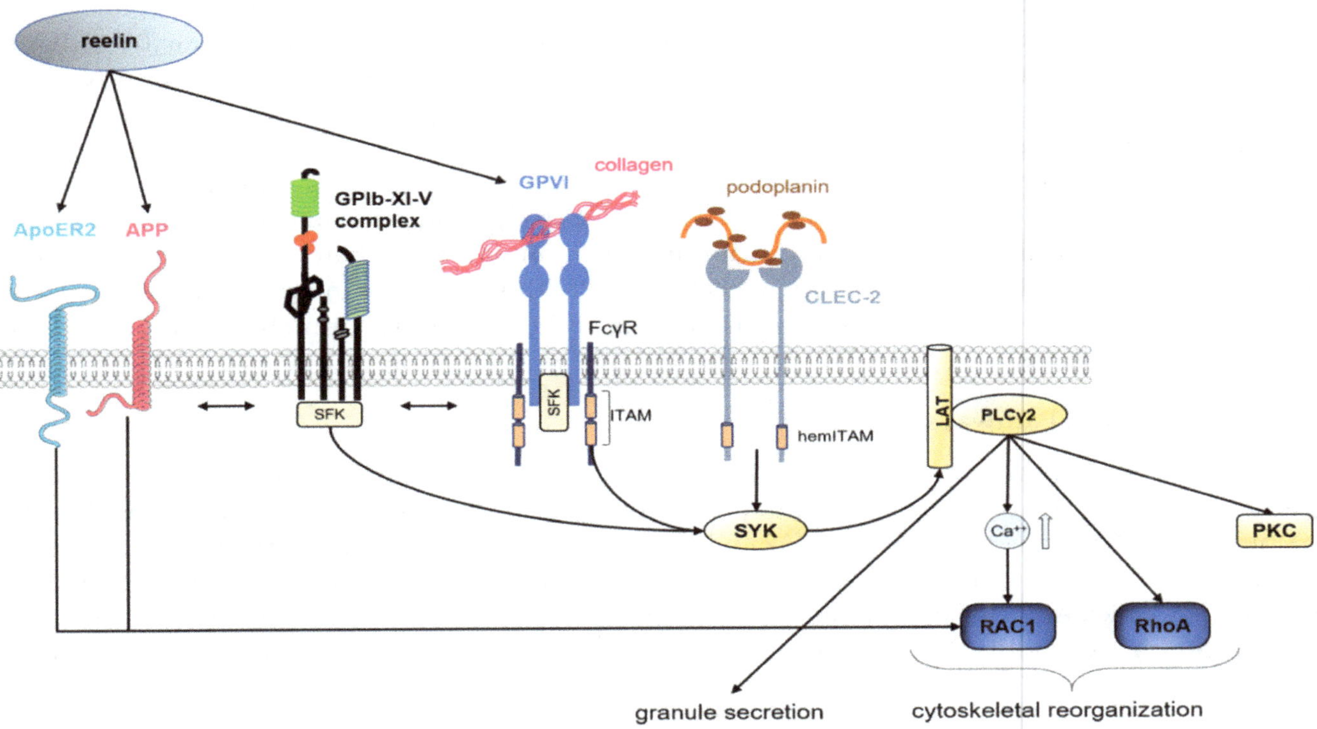

Figure 1. Impact of reelin in platelet activation and ITAM signaling. Extracellular reelin binds to ApoER2 and APP receptors at the platelet surface, resulting in platelet activation through RAC-1 activation. Additionally, reelin binding indirectly modulates GPIb signaling through APP and GPVI. Ligand binding to GPVI leads to dimerization of GPVI monomers, inducing auto-transphosphorylation of the GPVI-associated FcγR-chain ITAM motif through activation of Src-family kinases (SFKs). Besides, podoplanin binding to the CLEC-2 receptor causes activation of the CLEC-2-integrated hemITAM motif. Phosphorylated ITAM motifs lead to phosphorylation of SYK, which activates LAT kinases, being in direct proximity to PLCγ2. Activated PLCγ2 causes granule release, activation of protein kinase C (PKC), and activation of RAC1 and RhoA through calcium influx leading to cytoskeletal reorganization. This figure was created using images from Servier Medical Art Commons Attribution 3.0 Unported License.

The family of PLD, including the most prominent isoforms PLD1 and PLD2 [23], catalyzes the degradation of phosphatidylcholine into choline and phosphatidic acid (PA) [24], representing a very important messenger in many cellular processes [25]. An important role for PLD1 in GPIb-dependent thrombus formation under flow conditions has been shown using PLD1-deficient mice. Loss of PLD1 in platelets impairs integrin $\alpha_{IIb}\beta_3$ activation and GPIb-dependent aggregate formation, thereby protecting against thrombosis and ischemic brain infarction without altering hemostasis in PLD1-deficient mice [26]. In contrast, the loss of PLD2 does not affect platelet activation [27,28]. Recent studies provided evidence for yet another important role of PLD1 in platelet-mediated inflammation [29]. According to the prominent role of GPIb in the adhesion of platelets to endothelial cells (ECs) via binding to selectins under high shear rates [30], PLD modulates the up-regulation of endothelial adhesion molecules and regulates the adhesion of platelets to ECs. In addition, PLD1 contributes to platelet-leukocyte recruitment under inflammatory conditions. Thus, PLD1 plays a role in thrombo-inflammatory processes known to aggravate organ damage following ischemia/reperfusion injury after ischemic stroke and myocardial infarction [26,31]. Indeed, in an experimental model of myocardial infarction, loss of PLD1 leads to defective cell adhesion and migration of inflammatory cells into the infarct border zone, and to altered scar formation resulting in enhanced infarct size and declined myocardial function [31]. Recently, a role for PLD2 in platelet-mediated inflammation was identified. PLD2-deficient mice exhibit enhanced

IL-6 plasma levels concomitant with enhanced migration of inflammatory cells into the infarct border zone 24 h after acute myocardial infarction. This was due to enhanced integrin $\alpha_{IIb}\beta_3$ activation of PLD2-deficient platelets under inflammatory conditions, resulting in enhanced IL-6 release of ECs to accelerate inflammation after myocardial infarction, suggesting that PLD2 is an effector of thrombo-inflammation [28].

These results emphasize the important role of the vWF-GPIb axis not only in hemostasis and thrombosis but also in thrombo-inflammatory processes following ischemia and reperfusion injury after stroke and myocardial infarction contributing to organ damage. Therefore, the identification of novel modulators of GPIb signaling such as reelin and PLD1 is of utmost interest and might represent promising therapeutic targets to prevent thrombosis and thrombo-inflammation without inducing bleeding complications.

3. ITAM-Signaling Pathways

Loss of the immunoreceptor tyrosine-based activation motif (ITAM)-coupled receptors glycoprotein (GP) VI or C-type lectin-like receptor (CLEC)-2 is linked to only a mild bleeding diathesis in patients as well as in mice [32–34]. Both receptors share the conserved immunoreceptor tyrosine-based activation motif (ITAM) sequence containing four amino acids. A tyrosine separated from an (iso)-leucine by two other amino acids, called the YXXL motif. The GPVI/Fcγ chain consists of two cytosolic YXXL motifs. In contrast, CLEC-2 consists of only a singular cytosolic YXXL motif, thus named hemITAM [35–37].

3.1. GPVI Signaling

GPVI is the major receptor for collagen and is expressed on megakaryocytes and platelets. Surface plasmon resonance studies revealed that fibrous collagen binds to dimeric but not to monomeric GPVI via the glycine-proline-hydroxyproline (GPO) sequence [38]. GPVI activation leads to Src family kinase-mediated phosphorylation of the two ITAMs and induces the binding and phosphorylation of SYK. Phosphorylated SYK in turn induces the phosphorylation of the adaptor protein LAT and the recruitment of different pathway effectors resulting in the activation of PLCγ2. The activation of PLCγ2 by phosphorylation triggers the mobilization of intracellular Ca^{2+} stores as well as the activation of PKC. These events lead to the release of intracellular α-granules and dense granules and inside-out activation of platelet integrins, resulting in platelet aggregation [37].

In recent years, several GPVI ligands beside collagen have been identified including fibrin, diesel exhaust particles (DEP) and large polysaccharides such as fucoidan and dextran sulfate [39–41]. Interestingly, GPVI was shown to bind to polymerized fibrin to amplify thrombin generation and to recruit additional platelets to the thrombus. Thus, GPVI is a receptor for fibrin, supports phosphatidylserine (PS) exposure and promotes thrombus growth and stability [42].

Since GPVI is a key receptor involved in the pro-thrombotic stage of acute coronary syndrome (ACS) it is presumed to be a useful biomarker for the early detection of atherosclerotic diseases and may contribute to risk stratification and prediction of clinical outcome in patients with ACS [43–47]. Furthermore, soluble GPVI may serve as a target for molecular imaging to identify vulnerable plaques [48]. In addition to the property of GPVI as a diagnostic tool, therapeutic implications have been developed such as the blockage of collagen binding sides of GPVI by a soluble recombinant GPVI-Fc protein. This protein binds to collagen upon vessel injury and avoids binding of platelets via membrane bound GPVI to the injured vessel. Furthermore, anti-GPVI antibodies are useful to block platelet aggregation in static blood [49–54]. Treatment of mice with GPVI-Fc or the bifunctional protein consisting of an stromal cell-derived factor 1 (SDF-1) domain and a GPVI domain (SDF-1-GPVI) preserved cardiac function in a mouse model of myocardial infarction [55,56]. Thus, therapeutic anti-GPVI strategies could serve as a promising strategy for anti-thrombotic and anti-atherosclerotic therapy.

Reelin Amplifies GPVI Signaling in Platelets

Recently, reelin was identified as a novel regulator of GPVI signaling beside its regulatory role in GPIb signaling and shear-dependent thrombus formation.

Reelin is released after platelet activation with collagen and binds to GPVI with sub-nanomolar affinity in a concentration-dependent manner to amplify GPVI-dependent signaling in platelets [57]. In detail, reelin binding to GPVI induces tyrosine phosphorylation of GPVI target proteins to support platelet binding to collagen and GPVI-dependent RAC1 activation, PLCγ2 phosphorylation and platelet aggregation (Figure 1). Antibody-mediated deletion of GPVI from the platelet surface in reelin-deficient mice completely abolishes thrombus formation in vivo. Thrombus formation is only partly reduced in either GPVI-depleted or reelin-deficient mice. Thus, interfering with reelin-GPVI interaction might be a novel strategy to avoid arterial thrombosis [57].

3.2. CLEC-2 Signaling

The ability of CLEC-2 monomers to form homodimers is essential for signal transduction. To date podoplanin expressed on lymphatic endothelium and on tumor cells but not on vascular endothelial cells, is identified as the endogenous ligand found for CLEC-2. During podoplanin-mediated platelet aggregation, glycosylated podoplanin interacts with CLEC-2, leading to receptor clustering and phosphorylation of the hemITAM motifs. Moreover, although it is well known that GPVI phosphorylation of the ITAM motif is induced by the Src family kinases (SFK) Fyn and Lyn, it is still controversial how CLEC-2 is initially phosphorylated. Until now, several models of CLEC-2 hemITAM phosphorylation are published. Severin et al., demonstrated that—in contrast to GPVI activation—the tyrosine kinase SYK mediates CLEC-2 phosphorylation independently of the SFKs Fyn, Lyn or Src [58]. In contrast, others postulated that a phosphoinositide 3-kinase (PI3K)/tyrosine-protein kinase Tec (Tec)-axis regulates SYK activation downstream of CLEC-2 [59] and that CLEC-2 phosphorylation is initiated by SFKs. However, once recruited, phosphorylated SYK induces the engagement and phosphorylation of the adaptor protein LAT similar to GPVI. From this point, GPVI and CLEC-2 share the same signaling cascade leading to platelet aggregation and secretion. To date, podoplanin is the only known natural ligand for CLEC-2. However, some studies report increased tail bleeding times and reduced thrombus formation in CLEC-2-deficient mice, suggesting the presence of an additional ligand in the vasculature [60,61].

In the past, different reports provided evidence for GPVI or CLEC-2 to be a major player in platelet-mediated processes of inflammation. Transfer of GPVI or CLEC-2-deficient platelets to thrombocytopenic mice revealed an important role for ITAM-coupled receptors in inflammatory hemostasis of the skin and the lung [36]. In a mouse model of stroke, GPVI was identified as a key player in the processes of neuronal damage following cerebral reperfusion injury [49,62]. Similarly, antibody-mediated deletion of GPVI or treatment of mice with recombinant SDF-1-GPVI reduced inflammation and infarct size demonstrating a pivotal role of GPVI in ischemia reperfusion injury following ligation of the left anterior descending artery (LAD) in a mouse model of myocardial infarction [56,63]. More recently, it was shown that the inhibition of GPVI reduced the adhesion of amyloid beta (Aβ)-activated platelets to injured carotid arteries in mice suggesting a role of GPVI in Aβ-mediated inflammation in Alzheimer's disease [64].

Only a few studies investigated the role of CLEC-2 in inflammatory processes suggesting that CLEC-2 is a major player in thrombo-inflammatory diseases. Inflammatory events in the vessel wall characterize deep vein thrombosis (DVT) where reduced thrombosis was observed in CLEC-2 deficient mice or after treatment of mice with an anti-podoplanin antibody [65]. So far, there is only experimental evidence for a role of CLEC-2 in thrombo-inflammation in mice. However, Nicolson et al. detected up-regulated podoplanin in the venous valves near to a femoral vein thrombus in a patient with DVT. Interestingly, no alterations in podoplanin expression have been detected in the unaffected valves of the same vein or the equivalent valves in the contralateral leg [66].

4. G-Protein Coupled Receptor Activation—CXCRs as Emerging Modulators

G-Protein-coupled-receptors (GPCRs) are seven-pass transmembrane receptors that transduce intracellular signals through physical interaction with heterotrimeric G-proteins [67–69], e.g., G_s, G_i, G_q, and $G_{12/13}$, located at the cytoplasmic face of the plasma membrane [70,71]. Platelets express a number of GPCRs that are directly engaged by activating stimuli, protease-activated-receptors (PAR) for thrombin, purinergic receptors-P2Y12, P2Y1 for adenosine diphosphate (ADP), thromboxane receptor (TP) for thromboxane A_2 (TxA_2) or by physiological inhibitors (prostaglandin I_2 receptor (IP) for prostaglandin I_2 (PGI_2) [72]) released from the vasculature [67,73]. PAR1 and PAR4 transduce signals through $G\alpha_q$ and $G\alpha_{13}$; $P2Y_1$ is coupled to $G\alpha_q$; $P2Y_{12}$ is coupled to $G\alpha_i$ [74–77], while the TxA_2 interacting $TP\alpha$ receptor is coupled to $G\alpha_q$ and $G\alpha_{13}$ in platelets [78]. Therefore, these GPCR are top candidates for therapeutic intervention [71,73,79]. On the other end of the spectrum arachidonic acid (AA)-derived PGI_2, also called prostacyclin, acts as a physiological anti-platelet agent by engaging $G\alpha_s$-coupled IP receptor. Prostaglandin D_2 (PGD_2) activates $G\alpha_s$-coupled prostaglandin D_2 receptor 1 (DP_1), which triggers the platelet inhibitory adenylyl cyclase-protein kinase A cascade [69,72,80]. Prostaglandin E2 (PGE_2) exhibits a biphasic, dose-dependent effect through EP_1, EP_2, EP_3 and EP_4 receptors either activating or inactivating platelet response [67,69,80]. Recent research in search of novel anti-platelet targets to overcome the challenge of bleeding complications associated with some of the GPCR antagonist in clinical practice (vorapaxar) have revealed some unconventional targets that may fine-tune platelet reactivity.

CXCR Chemokine Receptors on Platelets Fine-Tune Thrombosis

Platelet chemokine receptors (e.g., C-C chemokine receptor type 1 (CCR1), CCR3, CCR4, CXC receptor 2 (CXCR2), CXCR4, CXCR6 and CXCR7) are a category of GPCRs that aggravate or reduce thrombotic response [81–88] (Figure 2). Platelets derived chemokines, e.g., C-X-C motif chemokine ligand 12 (CXCL12), CXCL16, and the cytokine macrophage migration inhibitory factor (MIF), released upon activation, engage with their cognate receptors and modulate thrombotic attributes, while in a paracrine manner elicit thrombo-inflammatory response from CXCR4, CXCR6, CXCR7 expressing immune and vascular cells [88–93]. Soluble CXCL16, acting through CXCR6 on platelets, triggers the PI3K-Akt activating signaling cascade to substantiate aggregation, degranulation, integrin $\alpha_{IIb}\beta_3$ activation and platelet shape change [88] (Figure 2). Serum levels of CXCL16 are significantly elevated in ACS patients as compared to those with stable angina pectoris (SAP) [94]. They correlate with inflammatory and metabolic risk factors [95], are shown to be prognostically unfavorable [96,97] and independently associated with cardiovascular death and morbidity in ACS patients [98] enrolled in the PLATO (platelet inhibition and patient outcome) trial. Surface expression of scavenger receptor (SR-PSOX/CXCL16) for the atherogenic mediator oxidized low density lipoprotein (oxLDL) on platelets is significantly enhanced in ACS patients and correlates with the circulatory inflammatory marker C-reactive protein (CRP) and creatinine kinase (CK) [99]. CXCL16 is abundantly deposited on inflammatory atherosclerotic plaques and carotid endarterectomy specimens and promotes CXCR6-dependent platelet adhesion to the inflamed endothelial cells and vWF [100], suggesting its potential to propagate thrombo-inflammation.

The Gi-coupled canonical CXCR4 and atypical/non-canonical CXCR7 are platelet receptors for the pro-inflammatory chemokine SDF-1α/CXCL12 and the chemokine-like cytokine MIF, whereas interferon inducible T cell alpha chemoattractant (ITAC, also named CXCL11) exclusively ligates CXCR7 [101,102] (Figure 2). Moreover, the presence of ligands in the immediate microenvironment influences a dynamic alteration in CXCR4/CXCR7 surface expression in platelets [101,102]. Platelet CXCR4/CXCR7 surface expression is significantly elevated in coronary artery disease (CAD) patients and influences prognosis [87] following myocardial infarction (MI) [86,87,103]. Mechanistic insights reveal the functional dichotomy of CXCR4 and CXCR7 in substantiating thrombotic response through CXCR4, whereas a pro-survival and anti-thrombotic impact was observed through CXCR7 [101,103]. MIF does not alter degranulation, ADP- or TxA2-analog (U46619)-induced aggregation or spreading on fibrinogen [91,92,101], either alone or synergistically in combination with other agonists. However, the pro-survival effect of MIF-CXCR7, significantly attenuates the exposure of thrombogenic phospholipid phosphatidylserine and counteracts thrombus build up [101] (Figure 2). On the contrary, CXCL12/SDF-1 acting through Gi-coupled CXCR4 clearly exerts a pro-thrombotic impact (Figure 2). Platelets exhibit flow directed migration towards CXCL12/SDF-1 [104] and also transmigrate through endothelium following a CXCL12 gradient, attributed to CXCR4 [105]. Once in a CXCL12/SDF-1 enriched microenvironment, e.g., atherosclerotic plaques, the CXCL12/CXCR4-Gi axis in platelets triggers intracellular calcium mobilization, PI3K activation, while dampening platelet inhibitory cAMP levels, leading to augmented aggregation [7,13–17,106–109] and thrombotic potential (Figure 2). SDF-1α-CXCR4 localized to lipid rafts in the platelet membrane trigger platelet reactivity [110], while decreased surface expression of CXCR4 in patients with inherited thrombocytopenia corroborates with impaired CXCL12/SDF1α-triggered platelet aggregation [111]. CXCL12 induces the primary phase of aggregation at lower concentrations, but dose dependently instigates a biphasic aggregation profile involving PI3K and prostanoids [107]. CXCL12 also synergistically substantiates aggregation induced by sub-threshold concentration of ADP and thrombin [108] and relatively weaker agonist serotonin (5HT), an effect substantiated by granular release (ADP, adenosine triphosphate (ATP)) [109], P-selectin exposure [108], PLC activation, and release of TxA_2 [109]. CXCL12 increases platelet adhesion to collagen type IV and fibrinogen, an effect counteracted by apyrase [108] and mediated through ATP release, TxA_2 production, promoting aggregation and thrombus formation [109,112], in contrast to the anti-thrombotic influence of MIF [101] (Figure 2).

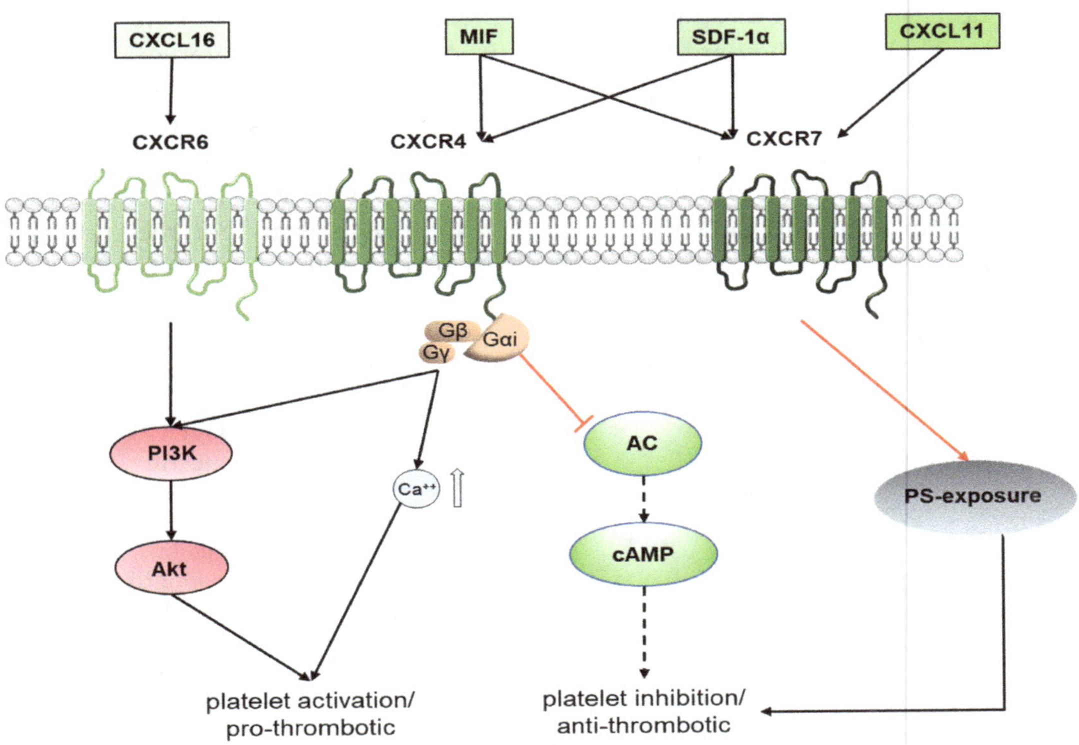

Figure 2. Regulation of platelet activation by CXCR chemokine receptors. Different chemokine receptors like CXCR6, CXCR4 and CXCR7 induce GPCR signaling in platelets. The chemokines CXCL16, and CXCL12/SDF-1α activate the phosphatidylinositol 3-kinases (PI3K) pathway leading to the activation of protein kinase B (Akt) and calcium mobilization resulting in platelet activation. Simultaneously, AC is inhibited through Gαi downstream of CXCR4 ligation by CXCL12/SDF-1α. MIF binding to CXCR7 substantiates platelet survival through PI3K-Akt pathway and downregulates phosphatidyl serine (PS) exposure. This figure was created using images from Servier Medical Art Commons Attribution 3.0 Unported License.

Plasma levels of CXCL12/SDF1α [113] and MIF [114] are elevated in ACS patients and associated with progressive disease severity, which may influence platelet responsiveness differentially through CXCR4/CXCR7. Platelet CXCR4 surface expression correlates inversely whereas CXCR7 correlates positively with inflammatory and atherogenic oxLDL in platelets from CAD patients. CXCR4 and CXCR7 also mediate the synergistic effects of pro-inflammatory CXCL12 on LDL and oxLDL induced changes to platelet morphodynamics and functional response [112], highlighting the potential of CXCL12/CXCR4/CXCR7 in executing thrombo-inflammatory platelet functions in a hyperlipidemic environment to promote atheroprogression. Differences in release kinetics and specific trigger for secretion may direct the course and pattern of functional impact exerted by different chemokines on platelet response as they engage CXCR6, CXCR4, or CXCR7. These GPCRs may provide promising therapeutic alternatives to check thrombotic functions without compromising the hemostatic and regenerative capacity of platelets.

5. Integrin Signaling in Platelets

5.1. Integrin Structure

Integrins are heterodimeric transmembrane proteins expressed on the cell surface of different cells. They serve as adhesion receptors that trigger intracellular signaling pathways by binding extracellular ligands [115]. Integrins are composed of two non-covalently bound subunits [115,116]. Overall, there are eight β- and 18 α-chains known. Through different combinations of these various subunits, 24 integrins have been identified in mammals that can bind a variety of adhesion ligands [117,118]. Due to their function as adhesion receptors, integrins mediate the adhesion to the extracellular matrix and between cells. Therefore, they are involved in processes like cell differentiation, proliferation, and migration in organisms, as well as in platelet adhesion and aggregation [115,117,119].

Five different integrins are described at the platelet surface. The classification of these integrins is based on the different β-subunits. There are three β_1- and two β_3-class integrin receptors exposed at the platelet surface. Members of the β_1-class integrin receptors are the collagen receptor $\alpha_2\beta_1$, the fibronectin receptor $\alpha_5\beta_1$ and the laminin receptor $\alpha_6\beta_1$. The fibrinogen receptor $\alpha_{IIb}\beta_3$ and the vitronectin receptor $\alpha_v\beta_3$ correspond to the β_3-class integrin receptors (Table 1) [118].

Table 1. Platelet integrins with classification, glycoprotein (GP) nomenclature and corresponding ligands.

Classification	Integrin	GP Nomenclature	Ligands
β1-class	$\alpha_2\beta_1$	GPIa/IIa	collagen
	$\alpha_5\beta_1$	GPIc/IIa	fibronectin
	$\alpha_6\beta_1$	GPIc'/IIa	laminin
β3-class	$\alpha_{IIb}\beta_3$	GPIIb/IIIa	fibrinogen, fibronectin, thrombospondin, vitronectin, von Willebrand factor
	$\alpha_v\beta_3$	GPαv/IIIa	vitronectin, fibrinogen, fibronectin, collagen, osteopontin, thrombospondin, von Willebrand factor

The most redundant and prominent integrin receptor on platelets is the fibrinogen receptor integrin $\alpha_{IIb}\beta_3$ (also known as GPIIb/IIIa). About 80,000 of these glycoprotein complexes can be found in their inactive state on the surface of resting platelets [120]. Furthermore, the membrane of α-granules expresses $\alpha_{IIb}\beta_3$ integrins. Activation-induced secretion of α-granules leads to externalization of $\alpha_{IIb}\beta_3$

integrins at the platelet surface and serves as a marker of platelet activation [121]. Rare mutations in the integrin subunits have been shown to modify the interaction between the subunits of the integrin. This connection is formed by a salt bridge that links the intracytoplasmic part of αIIb to the β3 unit of the $\alpha_{IIb}\beta_3$ integrin. Mutations in the salt bridge cause abnormal proplatelet formation with abnormal large α-granules and reduced but not absent platelet aggregation [122].

5.2. *Bidirectional Signaling of Integrin $\alpha_{IIb}\beta_3$: Role of Paxillin, Reelin and PLD1*

In the plasma membrane of platelets, integrin $\alpha_{IIb}\beta_3$ serves as bidirectional receptor for inside-out and outside-in signaling [115]. The inside-out signaling is induced by binding of different soluble or immobilized agonists to platelets (e.g., thrombin, ADP, TxA_2, epinephrine, vWF, collagen, etc.) via GPCRs or GPs such as GPIb or GPVI at the platelet membrane. Platelet activation triggers intracellular signaling cascades that lead to a conformational change of integrin $\alpha_{IIb}\beta_3$ [117]. The binding of the intracellular protein talin to the cytoplasmic part of integrin $\alpha_{IIb}\beta_3$ results in an unclasping of the cytoplasmic and transmembrane domain of α_{IIb} and β_3. This unclasping finally triggers a conformational change of the extracellular domain. The talin-triggered conformational change of integrin $\alpha_{IIb}\beta_3$ is supported by the interaction of kindlin and the C-terminal region of the β3 cytoplasmic domain [123–125]. Investigating the mechanistic details, Gao et al. recently showed that kindlin directly interacts with paxillin. The disruption of this binding through mutations in the binding site significantly impaired the activation of integrin $\alpha_{IIb}\beta_3$. This suggests that the interaction between kindlin and paxillin supports talin-mediated integrin $\alpha_{IIb}\beta_3$ activation that could be a target for anti-thrombotic therapies [126]. The conformational change of the $\alpha_{IIb}\beta_3$ integrin activates the receptor from its inactive into its active state [115]. This activation drives binding of ligands with higher affinity (Figure 3) [117]. Fibrinogen as major ligand binds to $\alpha_{IIb}\beta_3$ via its HHLGGAKQAGV sequence that is located in the C-terminus of the γ-chain of fibrinogen and the RGD sequence in the α-chain of integrin $\alpha_{IIb}\beta_3$. Since the RGD sequence can be found in other proteins, such as vitronectin, fibronectin, and von Willebrand factor, it is not surprising that these proteins are ligands that can be bound to $\alpha_{IIb}\beta_3$ integrin as well [125].

The binding of ligands to the extracellular domain of integrin $\alpha_{IIb}\beta_3$ induces integrin clustering and promotes integrin outside-in signalling [115]. Integrin outside-in signaling triggers a set of intracellular processes like platelet spreading, cytoskeleton reorganization, clot retraction, granule secretion as well as platelet adhesion and aggregation that finally leads to thrombus formation and stabilization [127]. During integrin $\alpha_{IIb}\beta_3$ ($\alpha_5\beta_1$ and $\alpha_2\beta_1$)-mediated platelet adhesion, the protein disulfide isomerase (PDI) inherits an important role as enzymatic mediator for disulfide exchange [128]. Lahav et al. could show that specific blockers of PDI inhibit the aggregation of platelets [129]. In mice, quercetin-3-rutinoside block platelet accumulation as well as fibrin generation at the vascular injury site [130]. In patients who are at risk of thrombosis, quercetin flavonoids inhibit PDI activity in plasma and reduce platelet-dependent thrombin generation [131].

Regulation of integrin $\alpha_{IIb}\beta_3$ outside-in signaling involves different interacting proteins, which associate with the cytoplasmic tails of $\alpha_{IIb}\beta_3$. The tyrosine kinase c-Src comes into proximity to the active cytoplasmic tail of the β_3 integrin subunit to induce full catalytic activation by trans-autophosphorylation to initiate outside-in signaling [127,132]. This is followed by coordinated interactions of Csk, Src, and SYK kinases with integrin $\alpha_{IIb}\beta_3$ to induce integrin signaling finally resulting in cytoskeletal reorganization [133]. Furthermore, a Src-kinase-dependent activation of PLCγ2 and phosphorylation of MLC mediates clot retraction downstream of integrin $\alpha_{IIb}\beta_3$ [134]. Enhanced integrin $\alpha_{IIb}\beta_3$ outside-in signaling was observed in mice with a gain-of-function mutation in PLCγ2 [135]. Platelets from these mice show accelerated spreading on different matrices and elevated clot retraction suggesting a major role for PLCγ2 in integrin outside-in signaling.

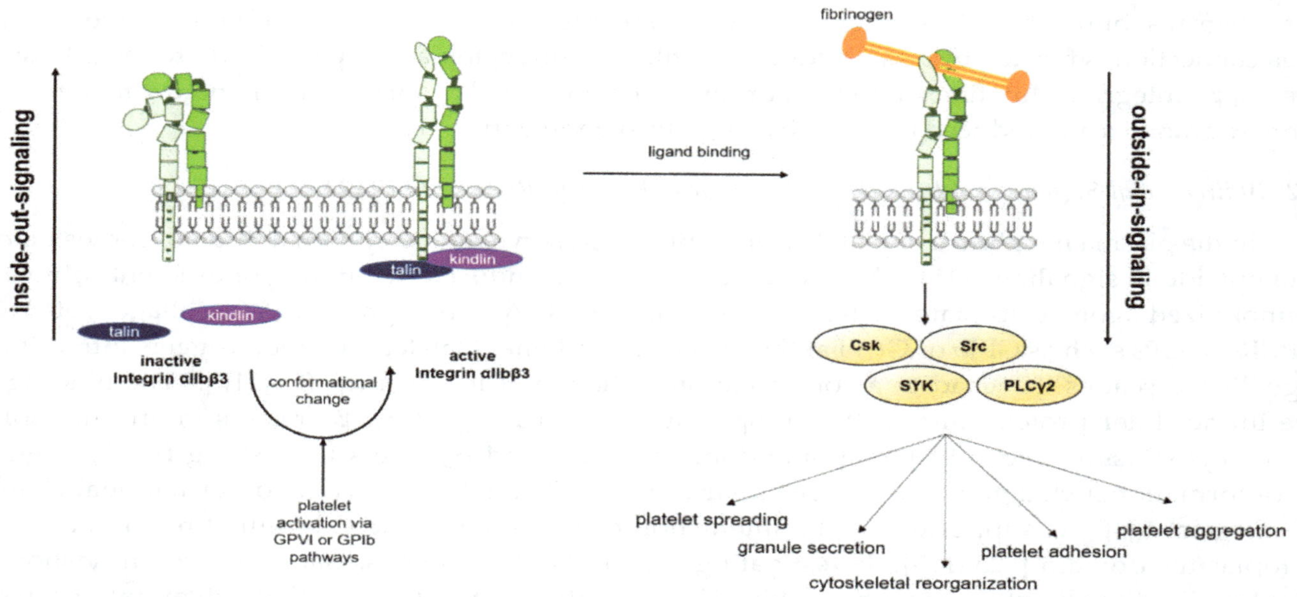

Figure 3. Inside-out- and outside-in-signaling of integrin $\alpha_{IIb}\beta_3$. In resting platelets, integrin $\alpha_{IIb}\beta_3$ is in an inactive conformation that prevents ligand binding. However, an initial activation of platelets (e.g., via GPIb or GPVI signaling) leads to binding of cytosolic kindlin and talin. This binding induces a conformational change of integrin $\alpha_{IIb}\beta_3$ that allows the binding of fibrinogen and other ligands. The process which induces the conformational change of integrin $\alpha_{IIb}\beta_3$ is called inside-out signaling. Binding of a ligand like fibrinogen to the extracellular domain of integrin $\alpha_{IIb}\beta_3$ triggers a signaling cascade through various kinases. This amplifies different platelet processes such as spreading, granule secretion and aggregation. The process of extracellular ligand-induced activation of integrin $\alpha_{IIb}\beta_3$ is called outside-in signaling. This figure was created using images from Servier Medical Art Commons Attribution 3.0 Unported License.

Based on the finding that ibrutinib can inhibit platelet integrin $\alpha_{IIb}\beta_3$ outside-in signaling, Dobie and colleagues investigated the effects of this inhibitor on platelet activation and glycoprotein expression in more detail by [136]. Ibrutinib induced the shedding of integrin $\alpha_{IIb}\beta_3$ from the platelet surface in a time- and dose-dependent manner as detected by reduced integrin $\alpha_{IIb}\beta_3$ surface expression of platelets isolated from patients with chronic lymphocytic leukemia, who were treated with ibrutinib. The underlying molecular mechanisms including the responsible sheddases are still unknown [137]. PLD1 modulates platelet $\alpha_{IIb}\beta_3$ integrin activation; therefore reduced shear-dependent thrombus formation in PLD1-deficient mice was observed [26]. More recently, Klier et al., could show reduced platelet adhesion to endothelial cells caused by defective integrin activation confirming a role for PLD1 not only in hemostasis and thrombosis but also in platelet-mediated inflammation [29]. Analyzing the impact of reelin signaling in platelets and its role for integrin activation and thrombus formation, Gowert et al. showed that reelin is a mediator of GPIb-dependent integrin activation as well [17]. In contrast to cytoplasmic PLD1 that affects integrin signaling by the activation of Src [29], reelin binds to its main receptor APP at the platelet surface leading to the co-localization of APP and GPIb to induce integrin activation. However, the molecular mechanisms behind the impact of reelin on integrin activation are not entirely clear to date.

6. Polymorphonuclear Leukocyte Released Platelet Agonists

Activated platelets release pro-inflammatory chemokines and cytokines and show an increased expression of P-selectin and CD40-ligand on their surface. These platelets are able to interact with leukocytes and endothelial cells leading to the recruitment of neutrophils into the inflamed tissue [138].

Interaction of polymorphonuclear leukocytes (PMNLs) with platelets is important for host defense and is associated with an increased risk of thrombosis induced by inflammatory pathways. The ability of PMNLs to release antimicrobial peptides, so called human neutrophil peptides (HNPs) or α-defensins, is an important mechanism of the immune response. Once released, HNPs are able to destroy microorganism membranes to exhibit their antimicrobial activity [139]. Besides their role in host defense, it was shown that HNPs act as platelet agonists, leading to platelet degranulation and even shedding of microparticles. Horn et al. showed that HNPs are able to induce platelet apoptosis by the formation of amyloid-like structures. Moreover, HNPs induce the formation of polymeric fibrinogen and thrombospondin-1 structures where platelets adhere and form aggregates. Furthermore, microorganisms are captured by these amyloid-like structures thus linking thrombosis and infection. Blocking of integrin $\alpha_{IIb}\beta_3$ (GPIIb/IIIa) strongly inhibits HNP-induced activation of platelets [140].

Recently, it was shown that transgenic mice expressing human α-Def-1 (Def^{++}) developed larger occlusive neutrophil-rich clots after ligation of the inferior vena cava (IVC), characterized by abnormal fibrin networks. Moreover, these mice were resistant to thrombo-prophylactic treatment with heparin. The inhibition of HNP synthesis or release was able to rescue this phenotype showing smaller thrombi and restoration of the responsiveness to heparin [141].

These observations explain an increased thrombotic risk at sites of inflammation and provide further insights into mechanisms linking inflammation to thrombosis. Glycosaminoglycans and serpins, which are potent inhibitors of platelet activation by HNPs, should be evaluated for therapeutic use in diseases characterized by activated hemostasis and inflammation [140]. Furthermore, drugs that inhibit neutrophil degranulation could be useful to treat or prevent thrombosis that develop during systemic inflammation with enhanced neutrophil activation [141].

7. Ion Channels: N-methyl-d-aspartate Glutamate Receptor (NMDAR) and Pannexin-1 (PANX1) as Modulators of Platelet Function

Ion homeostasis in each cell is a fundamental regulator of cellular functions and protection against apoptosis. In the last decades, the role of different ions was extensively investigated and Ca^{2+} was identified as a critical second wave mediator in different cell types [142]. Platelet activation in response to different agonists trigger different signaling pathways to mobilize cytosolic Ca^{2+} [143]. This process is essential for the reorganization of the cytoskeleton, the so called shape change [144]. Rapid increase in cytosolic Ca^{2+} is mediated by two main sources: (i) release of Ca^{2+} by an endomembrane system called dense tubular system (DTS) or (ii) by entry of Ca^{2+} through the plasma membrane [145]. Moreover, recent studies provide evidence that acidic organelles such as lysosomes are involved in the uptake and release of Ca^{2+}, proposing a role for acidic organelles in platelet-mediated Ca^{2+} signaling [146]. Here, we focus (i) on the N-methyl-d-aspartate glutamate receptor (NMDAR) on the platelet membrane as well as (ii) on pannexin-1 (PANX1), an ion channel that mediates platelet activation and thrombus formation and how they potentially interact to mediate platelet responses.

7.1. N-methyl-d-aspartate Glutamate Receptor (NMDAR)

The N-methyl-d-aspartate glutamate receptor (NMDAR) belongs to the ionotropic glutamate receptor family and functionally consists of a heterotetramer or heteropentamer. Each receptor is composed of the essential NMDA receptor 1 (GluN1) subunit, a various number of NR2 subunits: GluN2A, GluN2B, GluN2C and GluN2D [147] and rarely GluN3 (A and B) subunits [148]. NMDAR plays a critical role in neuronal signaling and are activated by binding of its main agonists L-glutamate and glycine to the GluN1 and GluN2 subunit. Ligand-gated opening removes the Mg^{2+} plug which finally leads to Ca^{2+} exchange to maintain Ca^{2+} homeostasis [149].

Platelets store about 400 μM glutamate in their dense granules which is released upon platelet activation and aggregation [150]. High extracellular glutamate concentrations lead to Na^+ influx mediated by α-amino-3-hydroxy-5-methyl-4-isoxazolepropionic acid (AMPA) and kainite receptors contributing to membrane depolarization [150]. This process might enable L-glutamate binding to the

GluN1 and GluN2 subunit to remove the Mg^{2+} plug leading to NMDAR activation and increased Ca^{2+} conduction in platelets (Figure 4).

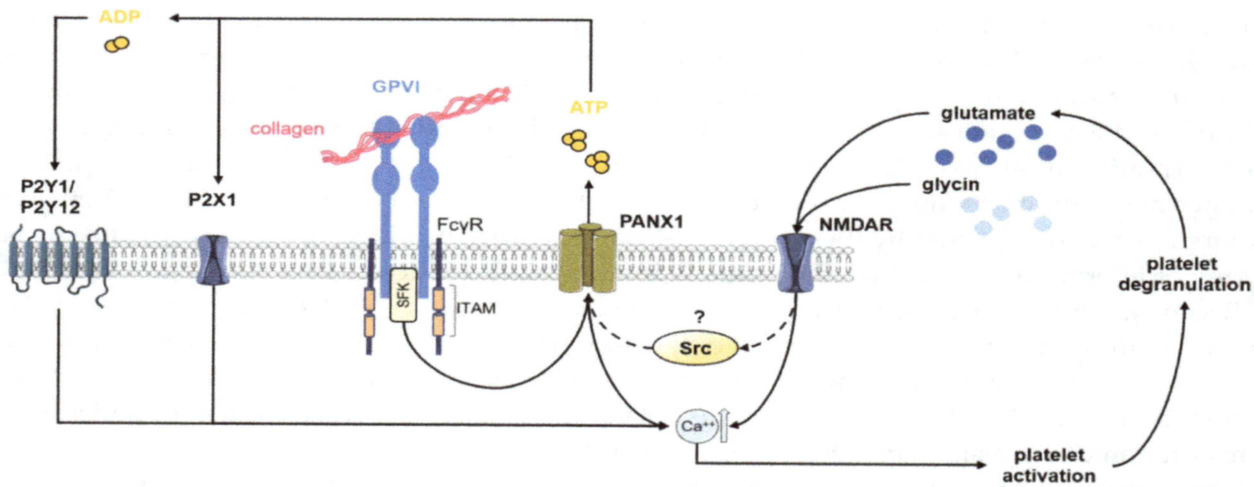

Figure 4. PANX1 and NMDAR activation upon platelet activation and signaling. Collagen binding induces activation of Src-family kinases (SFKs) which in turn activates PANX1 channels. Activation of PANX1 leads to an exchange of calcium and ATP, both modulating platelet activation through intra-and extracellular pathways. ATP directly binds to P2X1 or is degraded to ADP and binds to the purineric receptors P2Y1/P2Y12, all inducing platelet activation. Additionally, NMDAR might influence PANX1 activation through Src kinases, but this remains unclear in platelets to date. Besides, NMDAR activation induces calcium mobilization resulting in platelet activation. In a feedback loop, activated platelets release glutamate, an essential ligand for NMDAR activation. This figure was created using images from Servier Medical Art Commons Attribution 3.0 Unported License.

Using immunocytochemistry, the group of Genever first described the expression of the initial GluN1 subunit on the platelet surface, but their function remains unclear [151]. In 2004, Kalev-Zylinska and colleagues identified the four subunits: GluN1, GluN2A, GluN2D and GluN3A that form the NMDARs on the platelet surface. Interestingly, the fully functional receptor complex is transported to the plasma membrane after platelet activation. This supports the hypothesis of a potential role of NMDARs in platelet mediated aggregation and thrombus formation [152]. However, human platelet aggregation was only slightly reduced in the presence of the NMDAR antagonist MK-801. A few years later, the same group demonstrated a role of NMDARs in promoting dense granule release, thrombus formation and stabilization, using an inhibitory antibody against the GluN1 subunit [153]. These data provide evidence that NMDARs might be involved in Ca^{2+} homeostasis upon platelet aggregation and thrombus formation. However, Ca^{2+} measurements with platelets from GluN1 knock-out mice have to be performed in the near future to clarify the role of NMDARs in intraplatelet calcium homeostasis and thrombotic responses.

7.2. *Pannexin-1 (PANX1)*

20 years ago a new subgroup of transmembrane proteins, called pannexins (PANX1, PANX2 and PANX3), were identified in many mammalian tissues [154]. PANX1 is ubiquitously expressed, whereas PANX2 is mostly found in the central nervous system (CNS) and PANX3 in bone and skin [155]. All three isoforms function as ion channels for small molecules with a high affinity for ATP. In 2014, Taylor and colleagues identified transcripts for PANX1, but not PANX2 and PANX3 in human platelets. Immunochemistry analysis revealed that PANX1 is predominantly located on the surface of platelets. Targeting PANX1 channels by probenecid (Prb) in vitro revealed that platelet activation upon stimulation with low concentrations of collagen induces opening of PANX1

channels and thereby amplifies Ca^{2+} influx and aggregation by activation of the P2X1 channel [156,157]. The activation of PANX1 channels via collagen is mediated by GPVI, leading to the phosphorylation of SFKs (Figure 4) [157]. Inhibition of SFKs reduces PANX1-dependent platelet aggregation as well as intracellular phosphorylation of PANX1 at Tyr308 [158]. Ex vivo thrombus formation on collagen under arterial shear rates is reduced when whole blood is treated with Prb [156]. More recently, the impact of platelet PANX1 in thrombosis and hemostasis has been confirmed using platelet-specific PANX1 knockout mice (Panx1(fl/fl)/PF4-cre+) that display extended bleeding times and defective arterial thrombosis in vivo [158].

In neuronal tissue, activated NMDARs cross-activate PANX1 through phosphorylation of SFKs [159]. Moreover, inhibiting Src kinase-mediated phosphorylation of PANX1 at Tyr308 blocks metabotropic NMDAR signaling in brain slices. Furthermore, disturbing the NMDAR-Src-PANX1 complex has been found to be neuroprotective after ischemia or stroke, suggesting that the NMDAR–PANX1 axis might be involved in thrombo-inflammation [160].

8. Conclusions

The high burden of CVDs and the impact of platelets in thrombotic and thrombo-inflammatory processes indicate a high medical need for effective and novel anti-thrombotic and anti-thrombo-inflammatory therapies that avoid thrombotic and inflammatory events while preserving hemostasis.

Platelet receptors have been exploited as therapeutic targets since decades. However, residual platelet reactivity and incidences of bleeding among susceptible individuals often lead to fatal consequences as evident from the outcome of several clinical trials conducted with FDA-approved antiplatelet agents, like P2Y12 antagonists [161] clopidogrel (in CAPRIE, CURE, CREDO, CLARITY, CHARISMA), prasugrel (in TRITON, TRILOGY), ticagrelor (in PLATO and PEGASUS), and competitive inhibitors of PAR-1 against thrombin, i.e., vorapaxar (TRACER [162] and TRA2P [163–165]). Currently, the choice of anti-platelet therapeutics in clinical practice is fairly limited to irreversible COX-1 inhibitor aspirin (ASA) for primary prevention of cardiovascular disease and dual anti-platelet therapy (ASA in combination with reversible (ticagrelor, cangrelor) and irreversible (clopidogrel, prasugrel) inhibitors of P2Y12) for secondary prevention of recurrent thrombotic events following acute coronary syndrome (ACS) or in patients with coronary stent implantation. Clinical trials, like PEGASUS-TIMI54 [166,167], DAPT [168–170], OPTIDUAL [171,172], COMPASS [173–175], and GEMINI-ACS1 [176], reflect their drawbacks in increasing non-fatal/fatal bleeding and falling short of expectations. These trials have validated the potential of single (aspirin), dual and triple anti-platelet therapy or anti-platelet therapy administered in combination with anti-coagulants (rivaroxaban) [176,177]. The therapeutic goal is to ascertain the cornerstone of antithrombotic treatment following ACS.

It is essential to prevent recurrent thrombo-ischemic events with efficient thromboprophylactic strategies and reassess the benefit-to-risk ratio [168–172,178,179], dosage regimens, and treatment duration. Novel anti-thrombotic mediators which can discriminate between the physiological processes of thrombosis and hemostasis may potentially achieve this fine balance between anti-thrombotic efficacy and bleeding risk [180]. Of these, specific inhibitors of PI3Kp110β [181] and PKC [182–184], besides tyrosine kinase inhibitors widely recommended against cancer, like SYK, BTK, those of the Src family [185,186], agents antagonizing integrin $\alpha_{IIb}\beta_3$ by specifically targeting the active form of the receptor [187,188] or the outside-in signaling [189,190], a fusion protein combining the extracellular domain of CD39, and a single-chain antibody (Targ-CD39) that specifically detects activated integrin $\alpha_{IIb}\beta_3$ [191] have emerged in recent years. Therapeutics utilizing CD39 [192–194], GPVI [195], and GPIb-IX-V [180] as targets/mediators have been of prime interest. These novel therapeutic targets have been effectively validated in animal models of thrombosis where they were administered before an experimental induction of thrombotic/thrombo-ischemic events, while PI3Kβ-inhibitor [196] and soluble GPVI (Revacept®) are in clinical trials. Some of these strategies like those against tyrosine kinases may exhibit off-target adverse effects on other cells or organs requiring restricted use in limited dose and over a limited time period. The potential anti-thrombotic mediators discussed in this review may exert significant influence on thrombotic and thrombo-inflammatory platelet function as these cells interact with the inflamed vascular bed or inflammatory cells, but essentially, these mediators are not prime drivers of platelet function. They may offer a fine-tuning adjustment in modulating important signaling events in platelets to retain physiological hemostasis but check pathological thrombosis. Additionally, potential anti-thrombo-inflammatory benefits of emerging anti-platelet approaches, like those targeting platelet secretion, e.g., inhibitors of platelet lysosome-derived protein disulfide isomerase (PDI) [197,198], recombinant ectonucleases that degrade the platelet agonist ADP [199], platelet-derived soluble P-selectin antagonist as validated in the SELECT-ACS trial [200] and those discussed in this review may be of vital significance in the prevention of thrombo-inflammatory complications associated with ischemic stroke [8], atherosclerosis [201], and other CVDs [202,203] in combating target organ damage following a thrombo-ischemic episode, recurrent thromboembolic events, and in-stent thrombosis. Anti-inflammatory effects of aspirin alone or in combination with P2Y12 antagonists, also rivaroxaban have been seen in clinical settings and in animal models [201]. Most of these beneficial effects stem from reduction in levels of inflammatory mediators like C-reactive protein, IL-6, TNF-α, MCP-1, IL-1β. The anti-inflammatory benefits of IL-1β targeting monoclonal antibody canakinumab, in the CANTOS-trial [204], in significantly reducing the rate of recurrent cardiovascular events affirm a promising future. Platelets are an active source of thrombo-inflammatory mediators [82,205]. The significant association between plasma IL-1β levels and circulatory platelet count, also platelet single-nucleotide polymorphisms (SNPs) assessed in GWAS, while inverse association between platelet count and plasma IL-1β antagonist α-1-anti-trypsin, signify the thrombo-inflammatory potential of platelets in CVD [206]. In this light novel therapeutics with additional impact on thrombo-inflammatory processes will certainly have an edge over conventional anti-platelet therapies in clinical practice. However, several important aspects need to be contemplated and verified for a translational implication. The differences in structure and functional response of animal (most often murine) and human platelets, the type and surface availability of receptors, and interspecies differences in the physiological processes of thrombosis-hemostasis are of particular significance. Nevertheless, constant preclinical and clinical research endeavors in delineating the molecular mechanisms dissecting hemostatic and thrombotic or thrombo-inflammatory platelet functions will present more options for therapeutic interventions and broaden the currently limited choice of anti-platelet strategies.

Author Contributions: Conceptualization, M.E.; Writing—original draft preparation, M.C., A.E., L.M.T., L.M.M., M.K., I.K., and M.E.; writing—review and editing, M.K., F.R., I.K., and M.E.; format analysis and figure preparation, F.R.; funding acquisition, M.E. and M.C. All authors have read and agreed to the published version of the manuscript.

Abbreviations

AA	arachidonic acid
ACS	acute coronary syndrome
ADP	adenosine diphosphate
ATP	adenosine triphosphate
ApoER2	apolipoprotein E receptor 2
APP	amyloid precursor protein
Aβ	amyloid beta
CAD	coronary artery diseases
CCR1 (3, 4)	C-C chemokine receptor type 1 (3, 4)
CLEC	C-type lectin-like receptor

CNS	central nervous system
CVD	cardiovascular diseases
CXCL11 (12, 16)	C-X-C motif chemokine ligand 11 (12, 16)
DEP	diesel exhaust particles
DP_1	prostaglandin D_2 receptor 1
DVT	deep vein thrombosis
DTS	dense tubular system
EC	endothelial cells
EP_{1-4}	prostaglandin E_2 receptors 1-4
Erk	extracellular stimuli-responsive kinase
GPCR	G-protein-coupled-receptors
GPIb	glycoprotein Ib
GPO	glycine-proline-hydroxyproline
GPVI	glycoprotein VI
IL-6	interleukin-6
ITAM	immunoreceptor tyrosine-based activation motif
LAD	left anterior descending artery
LAT	linker for the activation of T cells
MI	myocardial infarction
MIF	macrophage migration inhibitory factor
MK	megakaryocytes
NMDAR	N-methyl-D-aspartate receptor
oxLDL	oxidized low density lipoprotein
PA	phosphatidic acid
PANX1	pannexin-1
PDI	protein disulfide isomerase
PGD_2	prostaglandin D_2
PGE_2	prostaglandin E_2
PGI_2	prostaglandin I_2
PKC	protein kinase C
PLATO	Platelet Inhibition and Patient Outcome
PLCγ2	phospholipase Cγ2
PLD	phospholipase D
Prb	probenecid
PS	phosphatidylserine
SAP	stable angina pectoris
SDF-1	stromal cell-derived factor 1, CXCL12
SFK	Src family kinases
SR	scavenger receptor
TP	TXA_2 receptor
TxA2	Thromboxane A_2
VLDLR	very low density lipoprotein receptor
vWF	von Willebrand factor

References

1. GBD 2017 Causes of Death Collaborators. Global, regional, and national age-sex-specific mortality for 282 causes of death in 195 countries and territories, 1980–2017: A systematic analysis for the Global Burden of Disease Study 2017. *Lancet* **2018**, *392*, 1736–1788. [CrossRef]
2. GBD 2017 DALYs and HALE Collaborators. Global, regional, and national disability-adjusted life-years (dalys) for 359 diseases and injuries and healthy life expectancy (HALE) for 195 countries and territories, 1990–2017: A systematic analysis for the Global Burden of Disease Study 2017. *Lancet* **2018**, *392*, 1859–1922. [CrossRef]
3. Read, S.H.; Wild, S.H. Prevention of premature cardiovascular death worldwide. *Lancet* **2020**, *395*, 758–760. [CrossRef]

4. Gawaz, M. Role of platelets in coronary thrombosis and reperfusion of ischemic myocardium. *Cardiovasc. Res.* **2004**, *61*, 498–511. [CrossRef]
5. Ruggeri, Z.M. Platelets in atherothrombosis. *Nat. Med.* **2002**, *8*, 1227–1234. [CrossRef]
6. Guo, L.; Rondina, M.T. The era of thromboinflammation: Platelets Are dynamic sensors and effector cells during infectious diseases. *Front. Immunol.* **2019**, *10*, 2204. [CrossRef]
7. Schanze, N.; Bode, C.; Duerschmied, D. Platelet Contributions to Myocardial Ischemia/Reperfusion Injury. *Front. Immunol.* **2019**, *10*, 1260. [CrossRef]
8. Stoll, G.; Nieswandt, B. Thrombo-inflammation in acute ischaemic stroke—Implications for treatment. Nature reviews. *Neurology* **2019**, *15*, 473–481. [CrossRef]
9. Savage, B.; Almus-Jacobs, F.; Ruggeri, Z.M. Specific synergy of multiple substrate-receptor interactions in platelet thrombus formation under flow. *Cell* **1998**, *94*, 657–666. [CrossRef]
10. Varga-Szabo, D.; Pleines, I.; Nieswandt, B. Cell adhesion mechanisms in platelets. *Arterioscler. Thromb. Vasc. Biol.* **2008**, *28*, 403–412. [CrossRef]
11. Nieswandt, B.; Varga-Szabo, D.; Elvers, M. Integrins in platelet activation. *J. Thromb. Haemost. JTH* **2009**, *7*, 206–209. [CrossRef] [PubMed]
12. Jackson, S.P. Arterial thrombosis—Insidious, unpredictable and deadly. *Nat. Med.* **2011**, *17*, 1423–1436. [CrossRef] [PubMed]
13. Asazuma, N.; Ozaki, Y.; Satoh, K.; Yatomi, Y.; Handa, M.; Fujimura, Y.; Miura, S.; Kume, S. Glycoprotein Ib-von Willebrand factor interactions activate tyrosine kinases in human platelets. *Blood* **1997**, *90*, 4789–4798. [CrossRef] [PubMed]
14. Falati, S.; Edmead, C.E.; Poole, A.W. Glycoprotein Ib-V-IX, a receptor for von Willebrand factor, couples physically and functionally to the Fc receptor gamma-chain, Fyn, and Lyn to activate human platelets. *Blood* **1999**, *94*, 1648–1656. [CrossRef] [PubMed]
15. Ozaki, Y.; Suzuki-Inoue, K.; Inoue, O. Platelet receptors activated via mulitmerization: Glycoprotein VI, GPIb-IX-V, and CLEC-2. *J. Thromb. Haemost. JTH* **2013**, *11*, 330–339. [CrossRef]
16. Denorme, F.; De Meyer, S.F. The VWF-GPIb axis in ischaemic stroke: Lessons from animal models. *Thromb. Haemost.* **2016**, *116*, 597–604. [CrossRef]
17. Gowert, N.S.; Kruger, I.; Klier, M.; Donner, L.; Kipkeew, F.; Gliem, M.; Bradshaw, N.J.; Lutz, D.; Kober, S.; Langer, H.; et al. Loss of Reelin protects mice against arterial thrombosis by impairing integrin activation and thrombus formation under high shear conditions. *Cell. Signal.* **2017**, *40*, 210–221. [CrossRef]
18. Herz, J.; Chen, Y. Reelin, lipoprotein receptors and synaptic plasticity. *Nat. Rev. Neurosci.* **2006**, *7*, 850–859. [CrossRef]
19. Leemhuis, J.; Bouche, E.; Frotscher, M.; Henle, F.; Hein, L.; Herz, J.; Meyer, D.K.; Pichler, M.; Roth, G.; Schwan, C.; et al. Reelin signals through apolipoprotein E receptor 2 and Cdc42 to increase growth cone motility and filopodia formation. *J. Neurosci. Off. J. Soc. Neurosci.* **2010**, *30*, 14759–14772. [CrossRef]
20. Perez-Costas, E.; Fenton, E.Y.; Caruncho, H.J. Reelin expression in brain endothelial cells: An electron microscopy study. *BMC Neurosci.* **2015**, *16*, 16. [CrossRef]
21. Smalheiser, N.R.; Costa, E.; Guidotti, A.; Impagnatiello, F.; Auta, J.; Lacor, P.; Kriho, V.; Pappas, G.D. Expression of reelin in adult mammalian blood, liver, pituitary pars intermedia, and adrenal chromaffin cells. *Proc. Natl. Acad. Sci. USA* **2000**, *97*, 1281–1286. [CrossRef] [PubMed]
22. Tseng, W.L.; Huang, C.L.; Chong, K.Y.; Liao, C.H.; Stern, A.; Cheng, J.C.; Tseng, C.P. Reelin is a platelet protein and functions as a positive regulator of platelet spreading on fibrinogen. *Cell. Mol. Life Sci. CMLS* **2010**, *67*, 641–653. [CrossRef] [PubMed]
23. Kodaki, T.; Yamashita, S. Cloning, expression, and characterization of a novel phospholipase D complementary DNA from rat brain. *J. Biol. Chem.* **1997**, *272*, 11408–11413. [CrossRef] [PubMed]
24. McDermott, M.; Wakelam, M.J.; Morris, A.J. Phospholipase D. *Biochem. Cell Biol.* **2004**, *82*, 225–253. [CrossRef]
25. Oude Weernink, P.A.; Lopez de Jesus, M.; Schmidt, M. Phospholipase D signaling: Orchestration by PIP2 and small GTPases. *Naunyn Schmiedebergs Arch. Pharmacol.* **2007**, *374*, 399–411. [CrossRef]
26. Elvers, M.; Stegner, D.; Hagedorn, I.; Kleinschnitz, C.; Braun, A.; Kuijpers, M.E.; Boesl, M.; Chen, Q.; Heemskerk, J.W.; Stoll, G.; et al. Impaired alpha(IIb)beta(3) integrin activation and shear-dependent thrombus formation in mice lacking phospholipase D1. *Sci. Signal.* **2010**, *3*, ra1. [CrossRef]

27. Thielmann, I.; Stegner, D.; Kraft, P.; Hagedorn, I.; Krohne, G.; Kleinschnitz, C.; Stoll, G.; Nieswandt, B. Redundant functions of phospholipases D1 and D2 in platelet alpha-granule release. *J. Thromb. Haemost. JTH* **2012**, *10*, 2361–2372. [CrossRef]
28. Klose, A.M.; Klier, M.; Gorressen, S.; Elvers, M. Enhanced Integrin Activation of PLD2-Deficient Platelets Accelerates Inflammation after Myocardial Infarction. *Int. J. Mol. Sci.* **2020**, *21*. [CrossRef]
29. Klier, M.; Gowert, N.S.; Jackel, S.; Reinhardt, C.; Elvers, M. Phospholipase D1 is a regulator of platelet-mediated inflammation. *Cell. Signal.* **2017**, *38*, 171–181. [CrossRef]
30. Simon, D.I.; Chen, Z.; Xu, H.; Li, C.Q.; Dong, J.; McIntire, L.V.; Ballantyne, C.M.; Zhang, L.; Furman, M.I.; Berndt, M.C.; et al. Platelet glycoprotein ibalpha is a counterreceptor for the leukocyte integrin Mac-1 (CD11b/CD18). *J. Exp. Med.* **2000**, *192*, 193–204. [CrossRef]
31. Schönberger, T.; Jürgens, T.; Müller, J.; Armbruster, N.; Niermann, C.; Gorressen, S.; Sommer, J.; Tian, H.; di Paolo, G.; Scheller, J.; et al. Pivotal role of phospholipase D1 in tumor necrosis factor-α-mediated inflammation and scar formation after myocardial ischemia and reperfusion in mice. *Am. J. Pathol.* **2014**, *184*, 2450–2464. [CrossRef]
32. Bender, M.; May, F.; Lorenz, V.; Thielmann, I.; Hagedorn, I.; Finney, B.A.; Vogtle, T.; Remer, K.; Braun, A.; Bosl, M.; et al. Combined in vivo depletion of glycoprotein VI and C-type lectin-like receptor 2 severely compromises hemostasis and abrogates arterial thrombosis in mice. *Arterioscler. Thromb. Vasc. Biol.* **2013**, *33*, 926–934. [CrossRef]
33. Dumont, B.; Lasne, D.; Rothschild, C.; Bouabdelli, M.; Ollivier, V.; Oudin, C.; Ajzenberg, N.; Grandchamp, B.; Jandrot-Perrus, M. Absence of collagen-induced platelet activation caused by compound heterozygous GPVI mutations. *Blood* **2009**, *114*, 1900–1903. [CrossRef] [PubMed]
34. Moroi, M.; Jung, S.M.; Okuma, M.; Shinmyozu, K. A patient with platelets deficient in glycoprotein VI that lack both collagen-induced aggregation and adhesion. *J. Clin. Investig.* **1989**, *84*, 1440–1445. [CrossRef] [PubMed]
35. Bergmeier, W.; Stefanini, L. Platelet ITAM signaling. *Curr. Opin. Hematol.* **2013**, *20*, 445–450. [CrossRef] [PubMed]
36. Boulaftali, Y.; Hess, P.R.; Getz, T.M.; Cholka, A.; Stolla, M.; Mackman, N.; Owens, A.P., III; Ware, J.; Kahn, M.L.; Bergmeier, W. Platelet ITAM signaling is critical for vascular integrity in inflammation. *J. Clin. Investig.* **2013**, *123*, 908–916. [CrossRef]
37. Rayes, J.; Watson, S.P.; Nieswandt, B. Functional significance of the platelet immune receptors GPVI and CLEC-2. *J. Clin. Investig.* **2019**, *129*, 12–23. [CrossRef]
38. Miura, Y.; Takahashi, T.; Jung, S.M.; Moroi, M. Analysis of the interaction of platelet collagen receptor glycoprotein VI (GPVI) with collagen. A dimeric form of GPVI, but not the monomeric form, shows affinity to fibrous collagen. *J. Biol. Chem.* **2002**, *277*, 46197–46204. [CrossRef]
39. Alshehri, O.M.; Hughes, C.E.; Montague, S.; Watson, S.K.; Frampton, J.; Bender, M.; Watson, S.P. Fibrin activates GPVI in human and mouse platelets. *Blood* **2015**, *126*, 1601–1608. [CrossRef]
40. Alshehri, O.M.; Montague, S.; Watson, S.; Carter, P.; Sarker, N.; Manne, B.K.; Miller, J.L.; Herr, A.B.; Pollitt, A.Y.; O'Callaghan, C.A.; et al. Activation of glycoprotein VI (GPVI) and C-type lectin-like receptor-2 (CLEC-2) underlies platelet activation by diesel exhaust particles and other charged/hydrophobic ligands. *Biochem. J.* **2015**, *468*, 459–473. [CrossRef]
41. Onselaer, M.B.; Hardy, A.T.; Wilson, C.; Sanchez, X.; Babar, A.K.; Miller, J.L.C.; Watson, C.N.; Watson, S.K.; Bonna, A.; Philippou, H.; et al. Fibrin and D-dimer bind to monomeric GPVI. *Blood Adv.* **2017**, *1*, 1495–1504. [CrossRef]
42. Mammadova-Bach, E.; Ollivier, V.; Loyau, S.; Schaff, M.; Dumont, B.; Favier, R.; Freyburger, G.; Latger-Cannard, V.; Nieswandt, B.; Gachet, C.; et al. Platelet glycoprotein VI binds to polymerized fibrin and promotes thrombin generation. *Blood* **2015**, *126*, 683–691. [CrossRef]
43. Bigalke, B.; Elvers, M.; Schönberger, T.; Gawaz, M. Platelet and soluble glycoprotein VI-novel applications in diagnosis and therapy. *Curr. Drug Targets* **2011**, *12*, 1821–1830. [CrossRef] [PubMed]
44. Bigalke, B.; Lindemann, S.; Ehlers, R.; Seizer, P.; Daub, K.; Langer, H.; Schonberger, T.; Kremmer, E.; Siegel-Axel, D.; May, A.E.; et al. Expression of platelet collagen receptor glycoprotein VI is associated with acute coronary syndrome. *Eur. Heart J.* **2006**, *27*, 2165–2169. [CrossRef]
45. Chatterjee, M.; Gawaz, M. Clinical significance of receptor shedding-platelet GPVI as an emerging diagnostic and therapeutic tool. *Platelets* **2017**, *28*, 362–371. [CrossRef]

46. Ollikainen, E.; Mikkelsson, J.; Perola, M.; Penttilä, A.; Karhunen, P.J. Platelet membrane collagen receptor glycoprotein VI polymorphism is associated with coronary thrombosis and fatal myocardial infarction in middle-aged men. *Atherosclerosis* **2004**, *176*, 95–99. [CrossRef] [PubMed]
47. Samaha, F.F.; Hibbard, C.; Sacks, J.; Chen, H.; Varello, M.A.; George, T.; Kahn, M.L. Density of platelet collagen receptors glycoprotein VI and alpha2beta1 and prior myocardial infarction in human subjects, a pilot study. *Med. Sci. Monit.* **2005**, *11*, Cr224–Cr229.
48. Gawaz, M.; Konrad, I.; Hauser, A.I.; Sauer, S.; Li, Z.; Wester, H.J.; Bengel, F.M.; Schwaiger, M.; Schömig, A.; Massberg, S.; et al. Non-invasive imaging of glycoprotein VI binding to injured arterial lesions. *Thromb. Haemost.* **2005**, *93*, 910–913. [CrossRef] [PubMed]
49. Goebel, S.; Li, Z.; Vogelmann, J.; Holthoff, H.P.; Degen, H.; Hermann, D.M.; Gawaz, M.; Ungerer, M.; Münch, G. The GPVI-Fc fusion protein Revacept improves cerebral infarct volume and functional outcome in stroke. *PLoS ONE* **2013**, *8*, e66960. [CrossRef] [PubMed]
50. Gruner, S.; Prostredna, M.; Koch, M.; Miura, Y.; Schulte, V.; Jung, S.M.; Moroi, M.; Nieswandt, B. Relative antithrombotic effect of soluble GPVI dimer compared with anti-GPVI antibodies in mice. *Blood* **2005**, *105*, 1492–1499. [CrossRef]
51. Jamasbi, J.; Megens, R.T.; Bianchini, M.; Münch, G.; Ungerer, M.; Faussner, A.; Sherman, S.; Walker, A.; Goyal, P.; Jung, S.; et al. Differential Inhibition of Human Atherosclerotic Plaque-Induced Platelet Activation by Dimeric GPVI-Fc and Anti-GPVI Antibodies: Functional and Imaging Studies. *J. Am. Coll. Cardiol.* **2015**, *65*, 2404–2415. [CrossRef]
52. Massberg, S.; Konrad, I.; Bültmann, A.; Schulz, C.; Münch, G.; Peluso, M.; Lorenz, M.; Schneider, S.; Besta, F.; Müller, I.; et al. Soluble glycoprotein VI dimer inhibits platelet adhesion and aggregation to the injured vessel wall in vivo. *FASEB J.* **2004**, *18*, 397–399. [CrossRef]
53. Schulz, C.; Penz, S.; Hoffmann, C.; Langer, H.; Gillitzer, A.; Schneider, S.; Brandl, R.; Seidl, S.; Massberg, S.; Pichler, B.; et al. Platelet GPVI binds to collagenous structures in the core region of human atheromatous plaque and is critical for atheroprogression in vivo. *Basic Res. Cardiol.* **2008**, *103*, 356–367. [CrossRef]
54. Ungerer, M.; Li, Z.; Baumgartner, C.; Goebel, S.; Vogelmann, J.; Holthoff, H.P.; Gawaz, M.; Münch, G. The GPVI-Fc fusion protein Revacept reduces thrombus formation and improves vascular dysfunction in atherosclerosis without any impact on bleeding times. *PLoS ONE* **2013**, *8*, e71193. [CrossRef]
55. Schönberger, T.; Ziegler, M.; Borst, O.; Konrad, I.; Nieswandt, B.; Massberg, S.; Ochmann, C.; Jürgens, T.; Seizer, P.; Langer, H.; et al. The dimeric platelet collagen receptor GPVI-Fc reduces platelet adhesion to activated endothelium and preserves myocardial function after transient ischemia in mice. *Am. J. Physiol. Cell Physiol.* **2012**, *303*, C757–C766. [CrossRef]
56. Ziegler, M.; Elvers, M.; Baumer, Y.; Leder, C.; Ochmann, C.; Schonberger, T.; Jurgens, T.; Geisler, T.; Schlosshauer, B.; Lunov, O.; et al. The bispecific SDF1-GPVI fusion protein preserves myocardial function after transient ischemia in mice. *Circulation* **2012**, *125*, 685–696. [CrossRef] [PubMed]
57. Krueger, I.; Gremer, L.; Mangels, L.; Klier, M.; Jurk, K.; Willbold, D.; Bock, H.H.; Elvers, M. Reelin amplifies GPVI activation and alphaiib beta3 integrin outside-in signaling via PLC gamma 2 and Rho GTPases. *Arterioscler. Thromb. Vasc. Biol.* **2020**. [CrossRef]
58. Severin, S.; Pollitt, A.Y.; Navarro-Nunez, L.; Nash, C.A.; Mourao-Sa, D.; Eble, J.A.; Senis, Y.A.; Watson, S.P. Syk-dependent phosphorylation of CLEC-2: A novel mechanism of hem-immunoreceptor tyrosine-based activation motif signaling. *J. Biol. Chem.* **2011**, *286*, 4107–4116. [CrossRef] [PubMed]
59. Manne, B.K.; Badolia, R.; Dangelmaier, C.; Eble, J.A.; Ellmeier, W.; Kahn, M.; Kunapuli, S.P. Distinct pathways regulate Syk protein activation downstream of immune tyrosine activation motif (ITAM) and hemITAM receptors in platelets. *J. Biol. Chem.* **2015**, *290*, 11557–11568. [CrossRef]
60. Hughes, C.E.; Navarro-Nunez, L.; Finney, B.A.; Mourao-Sa, D.; Pollitt, A.Y.; Watson, S.P. CLEC-2 is not required for platelet aggregation at arteriolar shear. *J. Thromb. Haemost. JTH* **2010**, *8*, 2328–2332. [CrossRef]
61. May, F.; Hagedorn, I.; Pleines, I.; Bender, M.; Vögtle, T.; Eble, J.; Elvers, M.; Nieswandt, B. CLEC-2 is an essential platelet-activating receptor in hemostasis and thrombosis. *Blood* **2009**, *114*, 3464–3472. [CrossRef] [PubMed]
62. Kleinschnitz, C.; Pozgajova, M.; Pham, M.; Bendszus, M.; Nieswandt, B.; Stoll, G. Targeting platelets in acute experimental stroke: Impact of glycoprotein Ib, VI, and IIb/IIIa blockade on infarct size, functional outcome, and intracranial bleeding. *Circulation* **2007**, *115*, 2323–2330. [CrossRef] [PubMed]

63. Pachel, C.; Mathes, D.; Arias-Loza, A.P.; Heitzmann, W.; Nordbeck, P.; Deppermann, C.; Lorenz, V.; Hofmann, U.; Nieswandt, B.; Frantz, S. Inhibition of Platelet GPVI Protects Against Myocardial Ischemia-Reperfusion Injury. *Arterioscler. Thromb. Vasc. Biol.* **2016**, *36*, 629–635. [CrossRef] [PubMed]
64. Donner, L.; Toska, L.M.; Kruger, I.; Groniger, S.; Barroso, R.; Burleigh, A.; Mezzano, D.; Pfeiler, S.; Kelm, M.; Gerdes, N.; et al. The collagen receptor glycoprotein VI promotes platelet-mediated aggregation of beta-amyloid. *Sci. Signal.* **2020**, *13*. [CrossRef] [PubMed]
65. Payne, H.; Ponomaryov, T.; Watson, S.P.; Brill, A. Mice with a deficiency in CLEC-2 are protected against deep vein thrombosis. *Blood* **2017**, *129*, 2013–2020. [CrossRef]
66. Nicolson, P.L.; Welsh, J.D.; Chauhan, A.; Thomas, M.R.; Kahn, M.L.; Watson, S.P. A rationale for blocking thromboinflammation in COVID-19 with Btk inhibitors. *Platelets* **2020**, *31*, 685–690. [CrossRef]
67. Yeung, J.; Li, W.; Holinstat, M. Platelet Signaling and Disease: Targeted Therapy for Thrombosis and Other Related Diseases. *Pharmacol. Rev.* **2018**, *70*, 526–548. [CrossRef]
68. Moers, A.; Nieswandt, B.; Massberg, S.; Wettschureck, N.; Gruner, S.; Konrad, I.; Schulte, V.; Aktas, B.; Gratacap, M.P.; Simon, M.I.; et al. G13 is an essential mediator of platelet activation in hemostasis and thrombosis. *Nat. Med.* **2003**, *9*, 1418–1422. [CrossRef]
69. Woulfe, D.S. Platelet G protein-coupled receptors in hemostasis and thrombosis. *J. Thromb. Haemost. JTH* **2005**, *3*, 2193–2200. [CrossRef]
70. Wong, S.K.; Parker, E.M.; Ross, E.M. Chimeric muscarinic cholinergic: Beta-adrenergic receptors that activate Gs in response to muscarinic agonists. *J. Biol. Chem.* **1990**, *265*, 6219–6224.
71. Kroeze, W.K.; Sheffler, D.J.; Roth, B.L. G-protein-coupled receptors at a glance. *J. Cell Sci.* **2003**, *116*, 4867–4869. [CrossRef] [PubMed]
72. Cheng, Y.; Austin, S.C.; Rocca, B.; Koller, B.H.; Coffman, T.M.; Grosser, T.; Lawson, J.A.; FitzGerald, G.A. Role of prostacyclin in the cardiovascular response to thromboxane A2. *Science* **2002**, *296*, 539–541. [CrossRef] [PubMed]
73. van der Meijden, P.E.J.; Heemskerk, J.W.M. Platelet biology and functions: New concepts and clinical perspectives. *Nat. Rev. Cardiol.* **2019**, *16*, 166–179. [CrossRef] [PubMed]
74. Fender, A.C.; Rauch, B.H.; Geisler, T.; Schror, K. Protease-Activated Receptor PAR-4: An Inducible Switch between Thrombosis and Vascular Inflammation? *Thromb. Haemost.* **2017**, *117*, 2013–2025. [CrossRef] [PubMed]
75. Quinton, T.M.; Murugappan, S.; Kim, S.; Jin, J.; Kunapuli, S.P. Different G protein-coupled signaling pathways are involved in alpha granule release from human platelets. *J. Thromb. Haemost. JTH* **2004**, *2*, 978–984. [CrossRef] [PubMed]
76. Quinton, T.M.; Kim, S.; Jin, J.; Kunapuli, S.P. Lipid rafts are required in Galpha(i) signaling downstream of the P2Y12 receptor during ADP-mediated platelet activation. *J. Thromb. Haemost. JTH* **2005**, *3*, 1036–1041. [CrossRef] [PubMed]
77. Nieman, M.T. Protease-activated receptors in hemostasis. *Blood* **2016**, *128*, 169–177. [CrossRef]
78. Holinstat, M.; Boutaud, O.; Apopa, P.L.; Vesci, J.; Bala, M.; Oates, J.A.; Hamm, H.E. Protease-activated receptor signaling in platelets activates cytosolic phospholipase A2alpha differently for cyclooxygenase-1 and 12-lipoxygenase catalysis. *Arterioscler. Thromb. Vasc. Biol.* **2011**, *31*, 435–442. [CrossRef]
79. Franchi, F.; Rollini, F.; Angiolillo, D.J. Antithrombotic therapy for patients with STEMI undergoing primary PCI. *Nat. Rev. Cardiol.* **2017**, *14*, 361–379. [CrossRef]
80. Kovacs, I.B.; Mayou, S.C.; Kirby, J.D. Infusion of a stable prostacyclin analogue, iloprost, to patients with peripheral vascular disease: Lack of antiplatelet effect but risk of thromboembolism. *Am. J. Med.* **1991**, *90*, 41–46. [CrossRef]
81. Gleissner, C.A.; von Hundelshausen, P.; Ley, K. Platelet chemokines in vascular disease. *Arterioscler. Thromb. Vasc. Biol.* **2008**, *28*, 1920–1927. [CrossRef] [PubMed]
82. Chatterjee, M.; Geisler, T. Inflammatory Contribution of Platelets Revisited: New Players in the Arena of Inflammation. *Semin. Thromb. Hemost.* **2016**, *42*, 205–214. [CrossRef] [PubMed]
83. Gear, A.R.; Camerini, D. Platelet chemokines and chemokine receptors: Linking hemostasis, inflammation, and host defense. *Microcirculation* **2003**, *10*, 335–350. [CrossRef] [PubMed]
84. Bakogiannis, C.; Sachse, M.; Stamatelopoulos, K.; Stellos, K. Platelet-derived chemokines in inflammation and atherosclerosis. *Cytokine* **2019**, *122*, 154157. [CrossRef] [PubMed]

85. Clemetson, K.J.; Clemetson, J.M.; Proudfoot, A.E.; Power, C.A.; Baggiolini, M.; Wells, T.N. Functional expression of CCR1, CCR3, CCR4, and CXCR4 chemokine receptors on human platelets. *Blood* **2000**, *96*, 4046–4054. [CrossRef] [PubMed]
86. Rath, D.; Chatterjee, M.; Borst, O.; Muller, K.; Langer, H.; Mack, A.F.; Schwab, M.; Winter, S.; Gawaz, M.; Geisler, T. Platelet surface expression of stromal cell-derived factor-1 receptors CXCR4 and CXCR7 is associated with clinical outcomes in patients with coronary artery disease. *J. Thromb. Haemost. JTH* **2015**, *13*, 719–728. [CrossRef]
87. Rath, D.; Chatterjee, M.; Borst, O.; Muller, K.; Stellos, K.; Mack, A.F.; Bongartz, A.; Bigalke, B.; Langer, H.; Schwab, M.; et al. Expression of stromal cell-derived factor-1 receptors CXCR4 and CXCR7 on circulating platelets of patients with acute coronary syndrome and association with left ventricular functional recovery. *Eur. Heart. J.* **2014**, *35*, 386–394. [CrossRef]
88. Borst, O.; Munzer, P.; Gatidis, S.; Schmidt, E.M.; Schonberger, T.; Schmid, E.; Towhid, S.T.; Stellos, K.; Seizer, P.; May, A.E.; et al. The inflammatory chemokine CXC motif ligand 16 triggers platelet activation and adhesion via CXC motif receptor 6-dependent phosphatidylinositide 3-kinase/Akt signaling. *Circ. Res.* **2012**, *111*, 1297–1307. [CrossRef]
89. Massberg, S.; Konrad, I.; Schurzinger, K.; Lorenz, M.; Schneider, S.; Zohlnhoefer, D.; Hoppe, K.; Schiemann, M.; Kennerknecht, E.; Sauer, S.; et al. Platelets secrete stromal cell-derived factor 1alpha and recruit bone marrow-derived progenitor cells to arterial thrombi in vivo. *J. Exp. Med.* **2006**, *203*, 1221–1233. [CrossRef]
90. Chatterjee, M.; Gawaz, M. Platelet-derived CXCL12 (SDF-1alpha): Basic mechanisms and clinical implications. *J. Thromb. Haemost. JTH* **2013**, *11*, 1954–1967. [CrossRef]
91. Wirtz, T.H.; Tillmann, S.; Strussmann, T.; Kraemer, S.; Heemskerk, J.W.; Grottke, O.; Gawaz, M.; von Hundelshausen, P.; Bernhagen, J. Platelet-derived MIF: A novel platelet chemokine with distinct recruitment properties. *Atherosclerosis* **2015**, *239*, 1–10. [CrossRef] [PubMed]
92. Strussmann, T.; Tillmann, S.; Wirtz, T.; Bucala, R.; von Hundelshausen, P.; Bernhagen, J. Platelets are a previously unrecognised source of MIF. *Thromb. Haemost.* **2013**, *110*, 1004–1013. [CrossRef] [PubMed]
93. Chatterjee, M.; von Ungern-Sternberg, S.N.; Seizer, P.; Schlegel, F.; Buttcher, M.; Sindhu, N.A.; Muller, S.; Mack, A.; Gawaz, M. Platelet-derived CXCL12 regulates monocyte function, survival, differentiation into macrophages and foam cells through differential involvement of CXCR4-CXCR7. *Cell Death. Dis.* **2015**, *6*, e1989. [CrossRef] [PubMed]
94. Sun, Y.; Chang, Z.; Zhang, S. Increased serum CXCL16 level is a marker for acute coronary syndromes. *Arch. Med. Res.* **2008**, *39*, 332–337. [CrossRef]
95. Lehrke, M.; Millington, S.C.; Lefterova, M.; Cumaranatunge, R.G.; Szapary, P.; Wilensky, R.; Rader, D.J.; Lazar, M.A.; Reilly, M.P. CXCL16 is a marker of inflammation, atherosclerosis, and acute coronary syndromes in humans. *J. Am. Coll. Cardiol.* **2007**, *49*, 442–449. [CrossRef]
96. Jansson, A.M.; Aukrust, P.; Ueland, T.; Smith, C.; Omland, T.; Hartford, M.; Caidahl, K. Soluble CXCL16 predicts long-term mortality in acute coronary syndromes. *Circulation* **2009**, *119*, 3181–3188. [CrossRef]
97. Tan, K.; Lu, S.; Chen, Y.; Song, X.; Wu, X.; Jin, Z.; Yuan, F.; Zhou, Y.; Li, H.; Yang, T.; et al. CXC chemokine ligand 16 as a prognostic marker in patients with intermediate coronary artery lesions: A 2-year follow-up study. *Tohoku J. Exp. Med.* **2011**, *223*, 277–283. [CrossRef]
98. Andersen, T.; Ueland, T.; Ghukasyan Lakic, T.; Åkerblom, A.; Bertilsson, M.; Aukrust, P.; Michelsen, A.E.; James, S.K.; Becker, R.C.; Storey, R.F.; et al. C-X-C Ligand 16 Is an Independent Predictor of Cardiovascular Death and Morbidity in Acute Coronary Syndromes. *Arterioscler. Thromb. Vasc. Biol.* **2019**, *39*, 2402–2410. [CrossRef]
99. Seizer, P.; Stellos, K.; Selhorst, G.; Kramer, B.F.; Lang, M.R.; Gawaz, M.; May, A.E. CXCL16 is a novel scavenger receptor on platelets and is associated with acute coronary syndrome. *Thromb. Haemost.* **2011**, *105*, 1112–1114. [CrossRef]
100. Meyer Dos Santos, S.; Blankenbach, K.; Scholich, K.; Dorr, A.; Monsefi, N.; Keese, M.; Linke, B.; Deckmyn, H.; Nelson, K.; Harder, S. Platelets from flowing blood attach to the inflammatory chemokine CXCL16 expressed in the endothelium of the human vessel wall. *Thromb. Haemost.* **2015**, *114*, 297–312. [CrossRef]
101. Chatterjee, M.; Borst, O.; Walker, B.; Fotinos, A.; Vogel, S.; Seizer, P.; Mack, A.; Alampour-Rajabi, S.; Rath, D.; Geisler, T.; et al. Macrophage migration inhibitory factor limits activation-induced apoptosis of platelets via CXCR7-dependent Akt signaling. *Circ. Res.* **2014**, *115*, 939–949. [CrossRef]

102. Chatterjee, M.; Seizer, P.; Borst, O.; Schonberger, T.; Mack, A.; Geisler, T.; Langer, H.F.; May, A.E.; Vogel, S.; Lang, F.; et al. SDF-1alpha induces differential trafficking of CXCR4-CXCR7 involving cyclophilin A, CXCR7 ubiquitination and promotes platelet survival. *FASEB J.* **2014**, *28*, 2864–2878. [CrossRef] [PubMed]
103. Rath, D.; Chatterjee, M.; Meyer, L.; Tekath, N.; Olma, C.; Krumm, P.; Adams, C.; Borst, O.; Muller, K.; Droppa, M.; et al. Relative survival potential of platelets is associated with platelet CXCR4/CXCR7 surface exposure and functional recovery following STEMI. *Atherosclerosis* **2018**, *278*, 269–277. [CrossRef]
104. Kraemer, B.F.; Schmidt, C.; Urban, B.; Bigalke, B.; Schwanitz, L.; Koch, M.; Seizer, P.; Schaller, M.; Gawaz, M.; Lindemann, S. High shear flow induces migration of adherent human platelets. *Platelets* **2011**, *22*, 415–421. [CrossRef]
105. Kraemer, B.F.; Borst, O.; Gehring, E.M.; Schoenberger, T.; Urban, B.; Ninci, E.; Seizer, P.; Schmidt, C.; Bigalke, B.; Koch, M.; et al. PI3 kinase-dependent stimulation of platelet migration by stromal cell-derived factor 1 (SDF-1). *J. Mol. Med. (Berl.)* **2010**, *88*, 1277–1288. [CrossRef]
106. Kowalska, M.A.; Ratajczak, M.Z.; Majka, M.; Jin, J.; Kunapuli, S.; Brass, L.; Poncz, M. Stromal cell-derived factor-1 and macrophage-derived chemokine: 2 chemokines that activate platelets. *Blood* **2000**, *96*, 50–57. [CrossRef]
107. Abi-Younes, S.; Sauty, A.; Mach, F.; Sukhova, G.K.; Libby, P.; Luster, A.D. The stromal cell-derived factor-1 chemokine is a potent platelet agonist highly expressed in atherosclerotic plaques. *Circ. Res.* **2000**, *86*, 131–138. [CrossRef]
108. Gear, A.R.; Suttitanamongkol, S.; Viisoreanu, D.; Polanowska-Grabowska, R.K.; Raha, S.; Camerini, D. Adenosine diphosphate strongly potentiates the ability of the chemokines MDC, TARC, and SDF-1 to stimulate platelet function. *Blood* **2001**, *97*, 937–945. [CrossRef]
109. Walsh, T.G.; Harper, M.T.; Poole, A.W. SDF-1alpha is a novel autocrine activator of platelets operating through its receptor CXCR4. *Cell. Signal.* **2015**, *27*, 37–46. [CrossRef]
110. Ohtsuka, H.; Iguchi, T.; Hayashi, M.; Kaneda, M.; Iida, K.; Shimonaka, M.; Hara, T.; Arai, M.; Koike, Y.; Yamamoto, N.; et al. SDF-1alpha/CXCR4 Signaling in Lipid Rafts Induces Platelet Aggregation via PI3 Kinase-Dependent Akt Phosphorylation. *PLoS ONE* **2017**, *12*, e0169609. [CrossRef]
111. Salim, J.P.; Goette, N.P.; Lev, P.R.; Chazarreta, C.D.; Heller, P.G.; Alvarez, C.; Molinas, F.C.; Marta, R.F. Dysregulation of stromal derived factor 1/CXCR4 axis in the megakaryocytic lineage in essential thrombocythemia. *Br. J. Haematol.* **2009**, *144*, 69–77. [CrossRef] [PubMed]
112. Chatterjee, M.; Rath, D.; Schlotterbeck, J.; Rheinlaender, J.; Walker-Allgaier, B.; Alnaggar, N.; Zdanyte, M.; Muller, I.; Borst, O.; Geisler, T.; et al. Regulation of oxidized platelet lipidome: Implications for coronary artery disease. *Eur. Heart. J.* **2017**, *38*, 1993–2005. [CrossRef] [PubMed]
113. Stellos, K.; Ruf, M.; Sopova, K.; Kilias, A.; Rahmann, A.; Stamatelopoulos, K.; Jorbenadze, R.; Geisler, T.; Gawaz, M.; Bigalke, B. Plasma levels of stromal cell-derived factor-1 in patients with coronary artery disease: Effect of clinical presentation and cardiovascular risk factors. *Atherosclerosis* **2011**, *219*, 913–916. [CrossRef] [PubMed]
114. Müller, I.I.; Müller, K.A.; Schönleber, H.; Karathanos, A.; Schneider, M.; Jorbenadze, R.; Bigalke, B.; Gawaz, M.; Geisler, T. Macrophage migration inhibitory factor is enhanced in acute coronary syndromes and is associated with the inflammatory response. *PLoS ONE* **2012**, *7*, e38376. [CrossRef] [PubMed]
115. Hynes, R.O. Integrins: Bidirectional, allosteric signaling machines. *Cell* **2002**, *110*, 673–687. [CrossRef]
116. Jennings, L.K.; Phillips, D.R. Purification of glycoproteins IIb and III from human platelet plasma membranes and characterization of a calcium-dependent glycoprotein IIb-III complex. *J. Biol. Chem.* **1982**, *257*, 10458–10466.
117. Hynes, R.O. Integrins: Versatility, modulation, and signaling in cell adhesion. *Cell* **1992**, *69*, 11–25. [CrossRef]
118. Lefkovits, J.; Plow, E.F.; Topol, E.J. Platelet glycoprotein IIb/IIIa receptors in cardiovascular medicine. *N. Engl. J. Med.* **1995**, *332*, 1553–1559. [CrossRef]
119. Hynes, R.O. Integrins: A family of cell surface receptors. *Cell* **1987**, *48*, 549–554. [CrossRef]
120. Wagner, C.L.; Mascelli, M.A.; Neblock, D.S.; Weisman, H.F.; Coller, B.S.; Jordan, R.E. Analysis of GPIIb/IIIa receptor number by quantification of 7E3 binding to human platelets. *Blood* **1996**, *88*, 907–914. [CrossRef]
121. Wencel-Drake, J.D.; Plow, E.F.; Kunicki, T.J.; Woods, V.L.; Keller, D.M.; Ginsberg, M.H. Localization of internal pools of membrane glycoproteins involved in platelet adhesive responses. *Am. J. Pathol.* **1986**, *124*, 324–334.
122. Favier, M.; Bordet, J.C.; Favier, R.; Gkalea, V.; Pillois, X.; Rameau, P.; Debili, N.; Alessi, M.C.; Nurden, P.; Raslova, H.; et al. Mutations of the integrin αIIb/β3 intracytoplasmic salt bridge cause macrothrombocytopenia and enlarged platelet α-granules. *Am. J. Hematol.* **2018**, *93*, 195–204. [CrossRef]

123. Nieswandt, B.; Moser, M.; Pleines, I.; Varga-Szabo, D.; Monkley, S.; Critchley, D.; Fassler, R. Loss of talin1 in platelets abrogates integrin activation, platelet aggregation, and thrombus formation in vitro and in vivo. *J. Exp. Med.* **2007**, *204*, 3113–3118. [CrossRef] [PubMed]
124. Moser, M.; Nieswandt, B.; Ussar, S.; Pozgajova, M.; Fassler, R. Kindlin-3 is essential for integrin activation and platelet aggregation. *Nat. Med.* **2008**, *14*, 325–330. [CrossRef] [PubMed]
125. Estevez, B.; Shen, B.; Du, X. Targeting integrin and integrin signaling in treating thrombosis. *Arterioscler. Thromb. Vasc. Biol.* **2015**, *35*, 24–29. [CrossRef] [PubMed]
126. Gao, J.; Huang, M.; Lai, J.; Mao, K.; Sun, P.; Cao, Z.; Hu, Y.; Zhang, Y.; Schulte, M.L.; Jin, C.; et al. Kindlin supports platelet integrin alphaIIbbeta3 activation by interacting with paxillin. *J. Cell Sci.* **2017**, *130*, 3764–3775. [CrossRef]
127. Huang, J.; Li, X.; Shi, X.; Zhu, M.; Wang, J.; Huang, S.; Huang, X.; Wang, H.; Li, L.; Deng, H.; et al. Platelet integrin αIIbβ3: Signal transduction, regulation, and its therapeutic targeting. *J. Hematol. Oncol.* **2019**, *12*, 26. [CrossRef]
128. Lahav, J.; Gofer-Dadosh, N.; Luboshitz, J.; Hess, O.; Shaklai, M. Protein disulfide isomerase mediates integrin-dependent adhesion. *FEBS Lett.* **2000**, *475*, 89–92. [CrossRef]
129. Lahav, J.; Jurk, K.; Hess, O.; Barnes, M.J.; Farndale, R.W.; Luboshitz, J.; Kehrel, B.E. Sustained integrin ligation involves extracellular free sulfhydryls and enzymatically catalyzed disulfide exchange. *Blood* **2002**, *100*, 2472–2478. [CrossRef]
130. Jasuja, R.; Passam, F.H.; Kennedy, D.R.; Kim, S.H.; van Hessem, L.; Lin, L.; Bowley, S.R.; Joshi, S.S.; Dilks, J.R.; Furie, B.; et al. Protein disulfide isomerase inhibitors constitute a new class of antithrombotic agents. *J. Clin. Investig.* **2012**, *122*, 2104–2113. [CrossRef]
131. Zwicker, J.I.; Schlechter, B.L.; Stopa, J.D.; Liebman, H.A.; Aggarwal, A.; Puligandla, M.; Caughey, T.; Bauer, K.A.; Kuemmerle, N.; Wong, E.; et al. Targeting protein disulfide isomerase with the flavonoid isoquercetin to improve hypercoagulability in advanced cancer. *JCI Insight* **2019**, *4*. [CrossRef] [PubMed]
132. Wu, Y.; Span, L.M.; Nygren, P.; Zhu, H.; Moore, D.T.; Cheng, H.; Roder, H.; DeGrado, W.F.; Bennett, J.S. The Tyrosine Kinase c-Src Specifically Binds to the Active Integrin αIIbβ3 to Initiate Outside-in Signaling in Platelets. *J. Biol. Chem.* **2015**, *290*, 15825–15834. [CrossRef]
133. Obergfell, A.; Eto, K.; Mocsai, A.; Buensuceso, C.; Moores, S.L.; Brugge, J.S.; Lowell, C.A.; Shattil, S.J. Coordinate interactions of Csk, Src, and Syk kinases with [alpha]IIb[beta]3 initiate integrin signaling to the cytoskeleton. *J. Cell Biol.* **2002**, *157*, 265–275. [CrossRef] [PubMed]
134. Suzuki-Inoue, K.; Hughes, C.E.; Inoue, O.; Kaneko, M.; Cuyun-Lira, O.; Takafuta, T.; Watson, S.P.; Ozaki, Y. Involvement of Src kinases and PLCgamma2 in clot retraction. *Thromb. Res.* **2007**, *120*, 251–258. [CrossRef] [PubMed]
135. Elvers, M.; Pozgaj, R.; Pleines, I.; May, F.; Kuijpers, M.J.; Heemskerk, J.M.; Yu, P.; Nieswandt, B. Platelet hyperreactivity and a prothrombotic phenotype in mice with a gain-of-function mutation in phospholipase Cgamma2. *J. Thromb. Haemost. JTH* **2010**, *8*, 1353–1363. [CrossRef]
136. Bye, A.P.; Unsworth, A.J.; Vaiyapuri, S.; Stainer, A.R.; Fry, M.J.; Gibbins, J.M. Ibrutinib Inhibits Platelet Integrin alphaIIbbeta3 Outside-In Signaling and Thrombus Stability but Not Adhesion to Collagen. *Arterioscler. Thromb. Vasc. Biol.* **2015**, *35*, 2326–2335. [CrossRef]
137. Dobie, G.; Kuriri, F.A.; Omar, M.M.A.; Alanazi, F.; Gazwani, A.M.; Tang, C.P.S.; Sze, D.M.; Handunnetti, S.M.; Tam, C.; Jackson, D.E. Ibrutinib, but not zanubrutinib, induces platelet receptor shedding of GPIb-IX-V complex and integrin alphaIIbbeta3 in mice and humans. *Blood Adv.* **2019**, *3*, 4298–4311. [CrossRef]
138. Jurk, K.; Kehrel, B.E. Platelets: Physiology and biochemistry. *Semin. Thromb. Hemost.* **2005**, *31*, 381–392. [CrossRef]
139. Huttner, K.M.; Bevins, C.L. Antimicrobial peptides as mediators of epithelial host defense. *Pediatric Res.* **1999**, *45*, 785–794. [CrossRef]
140. Horn, M.; Bertling, A.; Brodde, M.F.; Muller, A.; Roth, J.; Van Aken, H.; Jurk, K.; Heilmann, C.; Peters, G.; Kehrel, B.E. Human neutrophil alpha-defensins induce formation of fibrinogen and thrombospondin-1 amyloid-like structures and activate platelets via glycoprotein IIb/IIIa. *J. Thromb. Haemost. JTH* **2012**, *10*, 647–661. [CrossRef]
141. Abu-Fanne, R.; Stepanova, V.; Litvinov, R.I.; Abdeen, S.; Bdeir, K.; Higazi, M.; Maraga, E.; Nagaswami, C.; Mukhitov, A.R.; Weisel, J.W.; et al. Neutrophil alpha-defensins promote thrombosis in vivo by altering fibrin formation, structure, and stability. *Blood* **2019**, *133*, 481–493. [CrossRef] [PubMed]

142. Berridge, M.J.; Bootman, M.D.; Roderick, H.L. Calcium signalling: Dynamics, homeostasis and remodelling. *Nat. Rev. Mol. Cell Biol.* **2003**, *4*, 517–529. [CrossRef] [PubMed]
143. Rink, T.; Sage, S. Calcium signaling in human platelets. *Annu. Rev. Physiol.* **1990**, *52*, 431–449. [CrossRef] [PubMed]
144. Hathaway, D.R.; Adelstein, R.S. Human platelet myosin light chain kinase requires the calcium-binding protein calmodulin for activity. *Proc. Natl. Acad. Sci. USA* **1979**, *76*, 1653–1657. [CrossRef]
145. Gerrard, J.M.; White, J.G.; Peterson, D.A. The platelet dense tubular system: Its relationship to prostaglandin synthesis and calcium flux. *Thromb. Haemost.* **1978**, *40*, 224–231. [CrossRef]
146. Rosado, J. Acidic Ca2+ stores in platelets. *Cell Calcium* **2011**, *50*, 168–174. [CrossRef]
147. Tanabe, Y.; Masu, M.; Ishii, T.; Shigemoto, R.; Nakanishi, S. A family of metabotropic glutamate receptors. *Neuron* **1992**, *8*, 169–179. [CrossRef]
148. Chatterton, J.E.; Awobuluyi, M.; Premkumar, L.S.; Takahashi, H.; Talantova, M.; Shin, Y.; Cui, J.; Tu, S.; Sevarino, K.A.; Nakanishi, N. Excitatory glycine receptors containing the NR3 family of NMDA receptor subunits. *Nature* **2002**, *415*, 793–798. [CrossRef]
149. Chen, B.-S.; Roche, K.W. Regulation of NMDA receptors by phosphorylation. *Neuropharmacology* **2007**, *53*, 362–368. [CrossRef]
150. Morrell, C.N.; Sun, H.; Ikeda, M.; Beique, J.-C.; Swaim, A.M.; Mason, E.; Martin, T.V.; Thompson, L.E.; Gozen, O.; Ampagoomian, D.; et al. Glutamate mediates platelet activation through the AMPA receptor. *J. Exp. Med.* **2008**, *205*, 575–584. [CrossRef]
151. Hitchcock, I.S.; Skerry, T.M.; Howard, M.R.; Genever, P.G. NMDA receptor–mediated regulation of human megakaryocytopoiesis. *Blood* **2003**, *102*, 1254–1259. [CrossRef]
152. Kalev-Zylinska, M.L.; Green, T.N.; Morel-Kopp, M.-C.; Sun, P.P.; Park, Y.-E.; Lasham, A.; During, M.J.; Ward, C.M. N-methyl-D-aspartate receptors amplify activation and aggregation of human platelets. *Thromb. Res.* **2014**, *133*, 837–847. [CrossRef]
153. Green, T.N.; Hamilton, J.R.; Morel-Kopp, M.-C.; Zheng, Z.; Chen, T.-Y.T.; Hearn, J.I.; Sun, P.P.; Flanagan, J.U.; Young, D.; Barber, P.A.; et al. Inhibition of NMDA receptor function with an anti-GluN1-S2 antibody impairs human platelet function and thrombosis. *Platelets* **2017**, *28*, 799–811. [CrossRef] [PubMed]
154. Panchina, Y.; Kelmanson, I.; Matz, M.; Lukyanov, K.; Usman, N.; Lukyanov, S. A ubiquitous family of putative gap junction molecules. *Curr. Biol.* **2000**, *10*, R473–R474. [CrossRef]
155. Bruzzone, R.; Hormuzdi, S.G.; Barbe, M.T.; Herb, A.; Monyer, H. Pannexins, a family of gap junction proteins expressed in brain. *Proc. Natl. Acad. Sci. USA* **2003**, *100*, 13644–13649. [CrossRef] [PubMed]
156. Taylor, K.; Wright, J.; Vial, C.; Evans, R.; Mahaut-Smith, M. Amplification of human platelet activation by surface pannexin-1 channels. *J. Thromb. Haemost.* **2014**, *12*, 987–998. [CrossRef]
157. Molica, F.; Morel, S.; Meens, M.J.; Denis, J.-F.; Bradfield, P.F.; Penuela, S.; Zufferey, A.; Monyer, H.; Imhof, B.A.; Chanson, M. Functional role of a polymorphism in the Pannexin1 gene in collageninduced platelet aggregation. *Thromb. Haemost.* **2015**, *114*, 325–336. [CrossRef] [PubMed]
158. Molica, F.; Meens, M.J.; Pelli, G.; Hautefort, A.; Emre, Y.; Imhof, B.A.; Fontana, P.; Scemes, E.; Morel, S.; Kwak, B.R. Selective inhibition of Panx1 channels decreases hemostasis and thrombosis in vivo. *Thromb. Res.* **2019**, *183*, 56–62. [CrossRef] [PubMed]
159. Li, S.; Bjelobaba, I.; Stojilkovic, S.S. Interactions of Pannexin1 channels with purinergic and NMDA receptor channels. Biochimica et biophysica acta. *Biomembranes* **2018**, *1860*, 166–173. [CrossRef]
160. Weilinger, N.L.; Lohman, A.W.; Rakai, B.D.; Ma, E.M.; Bialecki, J.; Maslieieva, V.; Rilea, T.; Bandet, M.V.; Ikuta, N.T.; Scott, L. Metabotropic NMDA receptor signaling couples Src family kinases to pannexin-1 during excitotoxicity. *Nat. Neurosci.* **2016**, *19*, 432–442. [CrossRef]
161. Siller-Matula, J.M.; Trenk, D.; Schror, K.; Gawaz, M.; Kristensen, S.D.; Storey, R.F.; Huber, K.; EPA. Response variability to P2Y12 receptor inhibitors: Expectations and reality. *JACC Cardiovasc. Interv.* **2013**, *6*, 1111–1128. [CrossRef] [PubMed]
162. Armaganijan, L.V.; Alexander, K.P.; Huang, Z.; Tricoci, P.; Held, C.; Van de Werf, F.; Armstrong, P.W.; Aylward, P.E.; White, H.D.; Moliterno, D.J.; et al. Effect of age on efficacy and safety of vorapaxar in patients with non-ST-segment elevation acute coronary syndrome: Insights from the Thrombin Receptor Antagonist for Clinical Event Reduction in Acute Coronary Syndrome (TRACER) trial. *Am. Heart. J.* **2016**, *178*, 176–184. [CrossRef] [PubMed]

163. Magnani, G.; Bonaca, M.P.; Braunwald, E.; Dalby, A.J.; Fox, K.A.; Murphy, S.A.; Nicolau, J.C.; Oude Ophuis, T.; Scirica, B.M.; Spinar, J.; et al. Efficacy and safety of vorapaxar as approved for clinical use in the United States. *J. Am. Heart Assoc.* **2015**, *4*, e001505. [CrossRef] [PubMed]
164. Serebruany, V.L.; Fortmann, S.D. The FDA report on vorapaxar in the elderly: A convoluted dilemma. *Int. J. Cardiol.* **2015**, *201*, 601–603. [CrossRef]
165. Serebruany, V.L.; Tomek, A.; Pokov, A.N.; Kim, M.H. Clopidogrel, prasugrel, ticagrelor or vorapaxar in patients with renal impairment: Do we have a winner? *Expert Rev. Cardiovasc. Ther.* **2015**, *13*, 1333–1344. [CrossRef]
166. Bonaca, M.P.; Bhatt, D.L.; Steg, P.G.; Storey, R.F.; Cohen, M.; Im, K.; Oude Ophuis, T.; Budaj, A.; Goto, S.; Lopez-Sendon, J.; et al. Ischaemic risk and efficacy of ticagrelor in relation to time from P2Y12 inhibitor withdrawal in patients with prior myocardial infarction: Insights from PEGASUS-TIMI 54. *Eur. Heart J.* **2016**, *37*, 1133–1142. [CrossRef]
167. Parker, W.A.; Storey, R.F. Long-term antiplatelet therapy following myocardial infarction: Implications of PEGASUS-TIMI 54. *Heart* **2016**, *102*, 783–789. [CrossRef]
168. Kedhi, E.; Fabris, E.; van der Ent, M.; Buszman, P.; von Birgelen, C.; Roolvink, V.; Zurakowski, A.; Schotborgh, C.E.; Hoorntje, J.C.A.; Eek, C.H.; et al. Six months versus 12 months dual antiplatelet therapy after drug-eluting stent implantation in ST-elevation myocardial infarction (DAPT-STEMI): Randomised, multicentre, non-inferiority trial. *BMJ* **2018**, *363*, k3793. [CrossRef]
169. Watanabe, H.; Domei, T.; Morimoto, T.; Natsuaki, M.; Shiomi, H.; Toyota, T.; Ohya, M.; Suwa, S.; Takagi, K.; Nanasato, M.; et al. Effect of 1-Month Dual Antiplatelet Therapy Followed by Clopidogrel vs 12-Month Dual Antiplatelet Therapy on Cardiovascular and Bleeding Events in Patients Receiving PCI: The STOPDAPT-2 Randomized Clinical Trial. *JAMA* **2019**, *321*, 2414–2427. [CrossRef]
170. Costa, F.; van Klaveren, D.; James, S.; Heg, D.; Raber, L.; Feres, F.; Pilgrim, T.; Hong, M.K.; Kim, H.S.; Colombo, A.; et al. Derivation and validation of the predicting bleeding complications in patients undergoing stent implantation and subsequent dual antiplatelet therapy (PRECISE-DAPT) score: A pooled analysis of individual-patient datasets from clinical trials. *Lancet* **2017**, *389*, 1025–1034. [CrossRef]
171. Helft, G.; Steg, P.G.; Le Feuvre, C.; Georges, J.L.; Carrie, D.; Dreyfus, X.; Furber, A.; Leclercq, F.; Eltchaninoff, H.; Falquier, J.F.; et al. Stopping or continuing clopidogrel 12 months after drug-eluting stent placement: The OPTIDUAL randomized trial. *Eur. Heart J.* **2016**, *37*, 365–374. [CrossRef] [PubMed]
172. Wallentin, L. Commentary on the OPTIDUAL trial results: How to optimise prolonged dual antiplatelet treatment and independent randomised clinical trials. *Eur. Heart J.* **2016**, *37*, 375–377. [CrossRef] [PubMed]
173. Darmon, A.; Bhatt, D.L.; Elbez, Y.; Aboyans, V.; Anand, S.; Bosch, J.; Branch, K.R.; Connolly, S.J.; Dyal, L.; Eikelboom, J.W.; et al. External applicability of the COMPASS trial: An analysis of the reduction of atherothrombosis for continued health (REACH) registry. *Eur. Heart J.* **2018**, *39*, 750–757a. [CrossRef]
174. Bhatt, D.L.; Eikelboom, J.W.; Connolly, S.J.; Steg, P.G.; Anand, S.S.; Verma, S.; Branch, K.R.H.; Probstfield, J.; Bosch, J.; Shestakovska, O.; et al. Role of Combination Antiplatelet and Anticoagulation Therapy in Diabetes Mellitus and Cardiovascular Disease: Insights From the COMPASS Trial. *Circulation* **2020**, *141*, 1841–1854. [CrossRef]
175. Hernandez, J.L.; Lozano, F.S.; Riambau, V.; Almendro-Delia, M.; Cosin-Sales, J.; Bellmunt-Montoya, S.; Garcia-Alegria, J.; Garcia-Moll, X.; Gomez-Doblas, J.J.; Gonzalez-Juanatey, J.R.; et al. Reducing residual thrombotic risk in patients with peripheral artery disease: Impact of the COMPASS trial. *Drugs Context* **2020**, *9*. [CrossRef]
176. Ohman, E.M.; Roe, M.T.; Steg, P.G.; James, S.K.; Povsic, T.J.; White, J.; Rockhold, F.; Plotnikov, A.; Mundl, H.; Strony, J.; et al. Clinically significant bleeding with low-dose rivaroxaban versus aspirin, in addition to P2Y12 inhibition, in acute coronary syndromes (GEMINI-ACS-1): A double-blind, multicentre, randomised trial. *Lancet* **2017**, *389*, 1799–1808. [CrossRef]
177. Anand, S.S.; Bosch, J.; Eikelboom, J.W.; Connolly, S.J.; Diaz, R.; Widimsky, P.; Aboyans, V.; Alings, M.; Kakkar, A.K.; Keltai, K.; et al. Rivaroxaban with or without aspirin in patients with stable peripheral or carotid artery disease: An international, randomised, double-blind, placebo-controlled trial. *Lancet* **2018**, *391*, 219–229. [CrossRef]
178. Shyamkumar, K.; Hirsh, J.; Bhagirath, V.C.; Sinha, S.; Xu, K.; Ginsberg, J.S.; Diaz, R.; Eikelboom, J.W.; Chan, N.C. Mortality benefit in the COMPASS trial: Is it related to superior statistical power or better efficacy and safety? *Future Cardiol.* **2020**. [CrossRef]

179. Pirlet, C.; Legrand, V.; Nyssen, A.; Pierard, L.; Gach, O. Duration of dual anti-platelet therapy—State of the art after the DAPT and PEGASUS-TIMI 54 trials. *Acta Cardiol.* **2017**, *72*, 256–264. [CrossRef]
180. McFadyen, J.D.; Schaff, M.; Peter, K. Current and future antiplatelet therapies: Emphasis on preserving haemostasis. *Nat. Rev. Cardiol.* **2018**, *15*, 181–191. [CrossRef]
181. Jackson, S.P.; Schoenwaelder, S.M.; Goncalves, I.; Nesbitt, W.S.; Yap, C.L.; Wright, C.E.; Kenche, V.; Anderson, K.E.; Dopheide, S.M.; Yuan, Y.; et al. PI 3-kinase p110beta: A new target for antithrombotic therapy. *Nat. Med.* **2005**, *11*, 507–514. [CrossRef] [PubMed]
182. Gilio, K.; Harper, M.T.; Cosemans, J.M.; Konopatskaya, O.; Munnix, I.C.; Prinzen, L.; Leitges, M.; Liu, Q.; Molkentin, J.D.; Heemskerk, J.W.; et al. Functional divergence of platelet protein kinase C (PKC) isoforms in thrombus formation on collagen. *J. Biol. Chem.* **2010**, *285*, 23410–23419. [CrossRef] [PubMed]
183. Harper, M.T.; Poole, A.W. PKC inhibition markedly enhances Ca2+ signaling and phosphatidylserine exposure downstream of protease-activated receptor-1 but not protease-activated receptor-4 in human platelets. *J. Thromb. Haemost. JTH* **2011**, *9*, 1599–1607. [CrossRef]
184. Moore, S.F.; van den Bosch, M.T.; Hunter, R.W.; Sakamoto, K.; Poole, A.W.; Hers, I. Dual regulation of glycogen synthase kinase 3 (GSK3)alpha/beta by protein kinase C (PKC)alpha and Akt promotes thrombin-mediated integrin alphaIIbbeta3 activation and granule secretion in platelets. *J. Biol. Chem.* **2013**, *288*, 3918–3928. [CrossRef]
185. Tullemans, B.M.E.; Heemskerk, J.W.M.; Kuijpers, M.J.E. Acquired platelet antagonism: Off-target antiplatelet effects of malignancy treatment with tyrosine kinase inhibitors. *J. Thromb. Haemost. JTH* **2018**, *16*, 1686–1699. [CrossRef] [PubMed]
186. Busygina, K.; Jamasbi, J.; Seiler, T.; Deckmyn, H.; Weber, C.; Brandl, R.; Lorenz, R.; Siess, W. Oral Bruton tyrosine kinase inhibitors selectively block atherosclerotic plaque-triggered thrombus formation in humans. *Blood* **2018**, *131*, 2605–2616. [CrossRef] [PubMed]
187. Schwarz, M.; Katagiri, Y.; Kotani, M.; Bassler, N.; Loeffler, C.; Bode, C.; Peter, K. Reversibility versus persistence of GPIIb/IIIa blocker-induced conformational change of GPIIb/IIIa (alphaIIbbeta3, CD41/CD61). *J. Pharmacol. Exp. Ther.* **2004**, *308*, 1002–1011. [CrossRef] [PubMed]
188. Hagemeyer, C.E.; Peter, K. Targeting the platelet integrin GPIIb/IIIa. *Curr. Pharm. Des.* **2010**, *16*, 4119–4133. [CrossRef] [PubMed]
189. Shen, B.; Zhao, X.; O'Brien, K.A.; Stojanovic-Terpo, A.; Delaney, M.K.; Kim, K.; Cho, J.; Lam, S.C.; Du, X. A directional switch of integrin signalling and a new anti-thrombotic strategy. *Nature* **2013**, *503*, 131–135. [CrossRef]
190. Armstrong, P.C.; Peter, K. GPIIb/IIIa inhibitors: From bench to bedside and back to bench again. *Thromb. Haemost.* **2012**, *107*, 808–814. [CrossRef]
191. Ziegler, M.; Hohmann, J.D.; Searle, A.K.; Abraham, M.K.; Nandurkar, H.H.; Wang, X.; Peter, K. A single-chain antibody-CD39 fusion protein targeting activated platelets protects from cardiac ischaemia/reperfusion injury. *Eur. Heart J.* **2018**, *39*, 111–116. [CrossRef] [PubMed]
192. Hohmann, J.D.; Wang, X.; Krajewski, S.; Selan, C.; Haller, C.A.; Straub, A.; Chaikof, E.L.; Nandurkar, H.H.; Hagemeyer, C.E.; Peter, K. Delayed targeting of CD39 to activated platelet GPIIb/IIIa via a single-chain antibody: Breaking the link between antithrombotic potency and bleeding? *Blood* **2013**, *121*, 3067–3075. [CrossRef] [PubMed]
193. Granja, T.; Korner, A.; Gluck, C.; Hohmann, J.D.; Wang, X.; Kohler, D.; Streissenberger, A.; Nandurkar, H.H.; Mirakaj, V.; Rosenberger, P.; et al. Targeting CD39 Toward Activated Platelets Reduces Systemic Inflammation and Improves Survival in Sepsis: A Preclinical Pilot Study. *Crit. Care Med.* **2019**, *47*, e420–e427. [CrossRef] [PubMed]
194. Samudra, A.N.; Dwyer, K.M.; Selan, C.; Freddi, S.; Murray-Segal, L.; Nikpour, M.; Hickey, M.J.; Peter, K.; Robson, S.C.; Sashindranath, M.; et al. CD39 and CD73 activity are protective in a mouse model of antiphospholipid antibody-induced miscarriages. *J. Autoimmun.* **2018**, *88*, 131–138. [CrossRef] [PubMed]
195. Ungerer, M.; Rosport, K.; Bultmann, A.; Piechatzek, R.; Uhland, K.; Schlieper, P.; Gawaz, M.; Munch, G. Novel antiplatelet drug revacept (Dimeric Glycoprotein VI-Fc) specifically and efficiently inhibited collagen-induced platelet aggregation without affecting general hemostasis in humans. *Circulation* **2011**, *123*, 1891–1899. [CrossRef] [PubMed]

196. Nylander, S.; Kull, B.; Bjorkman, J.A.; Ulvinge, J.C.; Oakes, N.; Emanuelsson, B.M.; Andersson, M.; Skarby, T.; Inghardt, T.; Fjellstrom, O.; et al. Human target validation of phosphoinositide 3-kinase (PI3K)beta: Effects on platelets and insulin sensitivity, using AZD6482 a novel PI3Kbeta inhibitor. *J. Thromb. Haemost. JTH* **2012**, *10*, 2127–2136. [CrossRef]
197. Stopa, J.D.; Neuberg, D.; Puligandla, M.; Furie, B.; Flaumenhaft, R.; Zwicker, J.I. Protein disulfide isomerase inhibition blocks thrombin generation in humans by interfering with platelet factor V activation. *JCI Insight* **2017**, *2*, e89373. [CrossRef]
198. Jurk, K.; Lahav, J.; Van Aken, H.; Brodde, M.F.; Nofer, J.R.; Kehrel, B.E. Extracellular protein disulfide isomerase regulates feedback activation of platelet thrombin generation via modulation of coagulation factor binding. *J. Thromb. Haemost. JTH* **2011**, *9*, 2278–2290. [CrossRef]
199. Moeckel, D.; Jeong, S.S.; Sun, X.; Broekman, M.J.; Nguyen, A.; Drosopoulos, J.H.; Marcus, A.J.; Robson, S.C.; Chen, R.; Abendschein, D. Optimizing human apyrase to treat arterial thrombosis and limit reperfusion injury without increasing bleeding risk. *Sci. Transl. Med.* **2014**, *6*, 248ra105. [CrossRef]
200. Tardif, J.C.; Tanguay, J.F.; Wright, S.R.; Duchatelle, V.; Petroni, T.; Gregoire, J.C.; Ibrahim, R.; Heinonen, T.M.; Robb, S.; Bertrand, O.F.; et al. Effects of the P-selectin antagonist inclacumab on myocardial damage after percutaneous coronary intervention for non-ST-segment elevation myocardial infarction: Results of the SELECT-ACS trial. *J. Am. Coll. Cardiol.* **2013**, *61*, 2048–2055. [CrossRef]
201. Oikonomou, E.; Leopoulou, M.; Theofilis, P.; Antonopoulos, A.S.; Siasos, G.; Latsios, G.; Mystakidi, V.C.; Antoniades, C.; Tousoulis, D. A link between inflammation and thrombosis in atherosclerotic cardiovascular diseases: Clinical and therapeutic implications. *Atherosclerosis* **2020**, *309*, 16–26. [CrossRef] [PubMed]
202. d'Alessandro, E.; Becker, C.; Bergmeier, W.; Bode, C.; Bourne, J.H.; Brown, H.; Buller, H.R.; Ten Cate-Hoek, A.J.; Ten Cate, V.; van Cauteren, Y.J.M.; et al. Thrombo-Inflammation in Cardiovascular Disease: An Expert Consensus Document from the Third Maastricht Consensus Conference on Thrombosis. *Thromb. Haemost.* **2020**, *120*, 538–564. [CrossRef] [PubMed]
203. Rayes, J.; Bourne, J.H.; Brill, A.; Watson, S.P. The dual role of platelet-innate immune cell interactions in thrombo-inflammation. *Res. Pract. Thromb. Haemost.* **2020**, *4*, 23–35. [CrossRef]
204. Ridker, P.M.; Everett, B.M.; Thuren, T.; MacFadyen, J.G.; Chang, W.H.; Ballantyne, C.; Fonseca, F.; Nicolau, J.; Koenig, W.; Anker, S.D.; et al. Antiinflammatory Therapy with Canakinumab for Atherosclerotic Disease. *N. Engl. J. Med.* **2017**, *377*, 1119–1131. [CrossRef] [PubMed]
205. Muller, K.A.; Chatterjee, M.; Rath, D.; Geisler, T. Platelets, inflammation and anti-inflammatory effects of antiplatelet drugs in ACS and CAD. *Thromb. Haemost.* **2015**, *114*, 498–518. [CrossRef] [PubMed]
206. Tunjungputri, R.N.; Li, Y.; de Groot, P.G.; Dinarello, C.A.; Smeekens, S.P.; Jaeger, M.; Doppenberg-Oosting, M.; Cruijsen, M.; Lemmers, H.; Toenhake-Dijkstra, H.; et al. The Inter-Relationship of Platelets with Interleukin-1beta-Mediated Inflammation in Humans. *Thromb. Haemost.* **2018**, *118*, 2112–2125. [CrossRef]

3

Role of GRK6 in the Regulation of Platelet Activation through Selective G Protein-Coupled Receptor (GPCR) Desensitization

Preeti Kumari Chaudhary [1], Sanggu Kim [1], Youngheun Jee [2], Seung-Hun Lee [1], Kyung-Mee Park [1] and Soochong Kim [1,*]

1 College of Veterinary Medicine, Chungbuk National University, Cheongju 28644, Korea; chaudharypreety11@gmail.com (P.K.C.); tkdrnfld@naver.com (S.K.); dvmshlee@cbu.ac.kr (S.-H.L.); parkkm@chungbuk.ac.kr (K.-M.P.)

2 College of Veterinary Medicine and Veterinary Medical Research Institute, Jeju National University, Jeju 63243, Korea; yhjee@jejunu.ac.kr

* Correspondence: skim0026@cbu.ac.kr

Abstract: Platelet G protein-coupled receptors (GPCRs) regulate platelet function by mediating the response to various agonists, including adenosine diphosphate (ADP), thromboxane A_2, and thrombin. Although GPCR kinases (GRKs) are considered to have the crucial roles in most GPCR functions, little is known regarding the regulation of GPCR signaling and mechanisms of GPCR desensitization by GRKs in platelets. In this study, we investigated the functional role of GRK6 and the molecular basis for regulation of specific GPCR desensitization by GRK6 in platelets. We used GRK6 knockout mice to evaluate the functional role of GRK6 in platelet activation. Platelet aggregation, dense- and α-granule secretion, and fibrinogen receptor activation induced by 2-MeSADP, U46619, thrombin, and AYPGKF were significantly potentiated in $GRK6^{-/-}$ platelets compared to the wild-type (WT) platelets. However, collagen-related peptide (CRP)-induced platelet aggregation and secretion were not affected in $GRK6^{-/-}$ platelets. Interestingly, platelet aggregation induced by co-stimulation of serotonin and epinephrine which activate G_q-coupled $5HT_{2A}$ and G_z-coupled α_{2A} adrenergic receptors, respectively, was not affected in $GRK6^{-/-}$ platelets, suggesting that GRK6 was involved in specific GPCR regulation. In addition, platelet aggregation in response to the second challenge of ADP and AYPGKF was restored in $GRK6^{-/-}$ platelets whereas re-stimulation of the agonist failed to induce aggregation in WT platelets, indicating that GRK6 contributed to $P2Y_1$, $P2Y_{12}$, and PAR4 receptor desensitization. Furthermore, 2-MeSADP-induced Akt phosphorylation and AYPGKF-induced Akt, extracellular signal-related kinase (ERK), and protein kinase Cδ (PKCδ) phosphorylation were significantly potentiated in $GRK6^{-/-}$ platelets. Finally, $GRK6^{-/-}$ mice exhibited an enhanced and stable thrombus formation after $FeCl_3$ injury to the carotid artery and shorter tail bleeding times, indicating that $GRK6^{-/-}$ mice were more susceptible to thrombosis and hemostasis. We conclude that GRK6 plays an important role in regulating platelet functional responses and thrombus formation through selective GPCR desensitization.

Keywords: GRK6; GPCR; PAR; ADP; desensitization; platelets

---

1. Introduction

Platelet aggregation is important to maintain hemostasis and thrombosis. Various agonists including thrombin, adenosine diphosphate (ADP), and thromboxane, mediate their cellular effects through the G protein-coupled receptors (GPCRs) to induce platelet activation, and these GPCRs have been the most common target for anti-thrombotic drug development. ADP-induced platelet aggregation requires co-activation of both the $P2Y_1$ and $P2Y_{12}$ receptors that couple to G_q and G_i, respectively, and concomitant signaling from G_q and G_i is necessary for ADP-induced platelet aggregation [1]. Thromboxane A_2 (TxA_2) exerts its actions via the G protein-coupled thromboxane A_2 receptor (TP receptor) and requires co-activation of G_q and G_i pathways to cause platelet aggregation [2–4]. However, it has been shown to activate G_i pathways through the $P2Y_{12}$ receptor activation by the secreted ADP. Similarly, thrombin and thrombin receptor-activating peptides mediate their effect via the G protein-coupled protease-activated receptors (PARs), by direct coupling to G_q and $G_{12/13}$. PARs also cause G_i stimulation indirectly through the $P2Y_{12}$ receptor activation by the secreted ADP [5]. It has also been shown that co-stimulation of serotonin and epinephrine, which acts via the G_q-coupled $5HT_{2A}$ receptor and the G_z-coupled α_{2A} adrenergic receptor, respectively, induces platelet aggregation [6]. On the other hand, collagen and collagen-related peptide (CRP) activate platelets by stimulating tyrosine phosphorylation of the Fc receptor (FcR) γ-chain that contains an immuno-receptor tyrosine-based activation motif (ITAM) through the Ig receptor, glycoprotein (GP)VI [7].

The G-protein-coupled receptor kinases (GRKs) consist of seven different genes, of which two are exclusively expressed in rod and cone cells (GRK1 and 7), respectively. One has a very limited expression in the cerebellum, testis, and kidneys (GRK4), and the remaining four are expressed ubiquitously in mammalian tissues (GRK2, GRK3, GRK5, and GRK6) [8–10]. All GRKs share a similar basic structure with an N-terminal, a central catalytic domain, and a distinguishing C-terminal domain [11]. The N-terminus shares considerable homology, which has been considered to be responsible for receptor recognition, rendering it a critical structure for the selective targeting of GPCRs, while the C-terminal region consists of phosphorylation sites and key binding sites to PIP2 and $G\beta\gamma$ [12]. GRK6 is constitutively associated with the cell membrane through the N-terminal and C-terminal basic amino acid residue motifs. While there is some evidence of substrate specificity among the different members of the GRK family, most show activity towards a wide variety of agonist-occupied receptors in vitro.

The GRKs are known to have the ability to interact with the activated conformation of seven-transmembrane receptors besides heterotrimeric G proteins [13]. GRKs are recruited concurrently to agonist bound-activated GPCRs. Upon agonist stimulation, GRKs phosphorylate agonist-activated GPCRs on serine and/or threonine residues located in the C-terminal tail region and/or the third cytoplasm loop, leading to lower active coupling to the heterotrimeric G protein and the termination of the signaling, excluding further interaction between the GPCR and G protein [12]. This process is termed as receptor desensitization. It is used by the vast majority of GPCRs to turn off receptor-mediated signal transduction pathways [11,12,14]. GRK5 has been shown to play a crucial role in the mechanism of thrombin-induced desensitization of PAR1 in endothelial cells [15], and co-expression of GRK3 (but not GRK2) with PAR1 has been shown to inhibit thrombin-activated G_q signaling in *Xenopus* oocytes [16]. Desensitization of $P2Y_{12}$ receptor in 1321N1 cells has been shown to be mediated by GRK2 and GRK6 [17]. It has been shown that GRK6 plays a major role in oxytocin receptor desensitization in the uterine smooth muscles [18]. GRK5 and GRK6 have been shown to regulate phosphorylation and desensitization of the $TP\alpha$ receptor [19], while $TP\beta$ receptor internalization is regulated by GRK2 in human embryonic kindey 293 (HEK293) cells [20]. In addition, it has been demonstrated that GRK2 and GRK3 are primarily responsible for agonist-dependent receptor phosphorylation and functional uncoupling, whereas GRK5 and GRK6 make lesser contributions to this outcome [21].

Although the contribution of GRK isoforms in the regulation of specific GPCR desensitization have been reported in other cells, little is known about the role of GRKs in GPCR-mediated desensitization in platelets. It is important that the responsiveness of platelets to various agonists be tightly regulated to avoid inappropriate thrombosis or excessive bleeding. Given the crucial role of agonists such as ADP, TxA_2, and thrombin in platelet activation, characterization of the role of GRKs in GPCR-mediated platelet responses may reveal novel regulatory mechanisms regulating platelet function.

This study was therefore undertaken to evaluate the functional role of GRK6 and its molecular basis for regulation of GPCR desensitization in platelets using GRK6 knockout mice. We have shown that GRK6 selectively regulates specific GPCR-mediated platelet aggregation and secretion. We have further shown that GRK6 plays an important role in ADP and PAR4 receptor desensitization and regulates both G_q- and G_i-mediated signaling in platelets. Moreover, GRK6 contributes to thrombus formation *in vivo*. Therefore, we conclude that GRK6 is crucial for regulating platelet functional responses through selective GPCR desensitization.

2. Results

2.1. GRK6 Selectively Regulates GPCR-Mediated Platelet Functional Responses

To determine the contribution of GRK6 to platelet function, we measured the various agonists-induced platelet aggregations and dense granule secretions in the wild-type (WT) and the GRK6-deficient mouse platelets. As shown in Figure 1, platelet aggregation and dense granule secretion induced by GPCR agonists including 2-MeSADP, U46619, AYPGKF, and thrombin were significantly potentiated in the GRK6-deficient platelets compared to those in the WT platelets. We found that the dose response curves were left-shifted, and there was little difference between WT and GRK6$^{-/-}$ platelet in response to high dose of agonists. The extent of potentiation of the platelet function in response to U46619 was not as significant as it was for the other GPCR agonists. However, platelet aggregation and dense granule secretion in response to Glycoprotein VI (GPVI) agonist CRP were not affected in the GRK6-deficient platelets, indicating that GRK6 selectively regulated platelet aggregation and secretion in response to GPCR agonists.

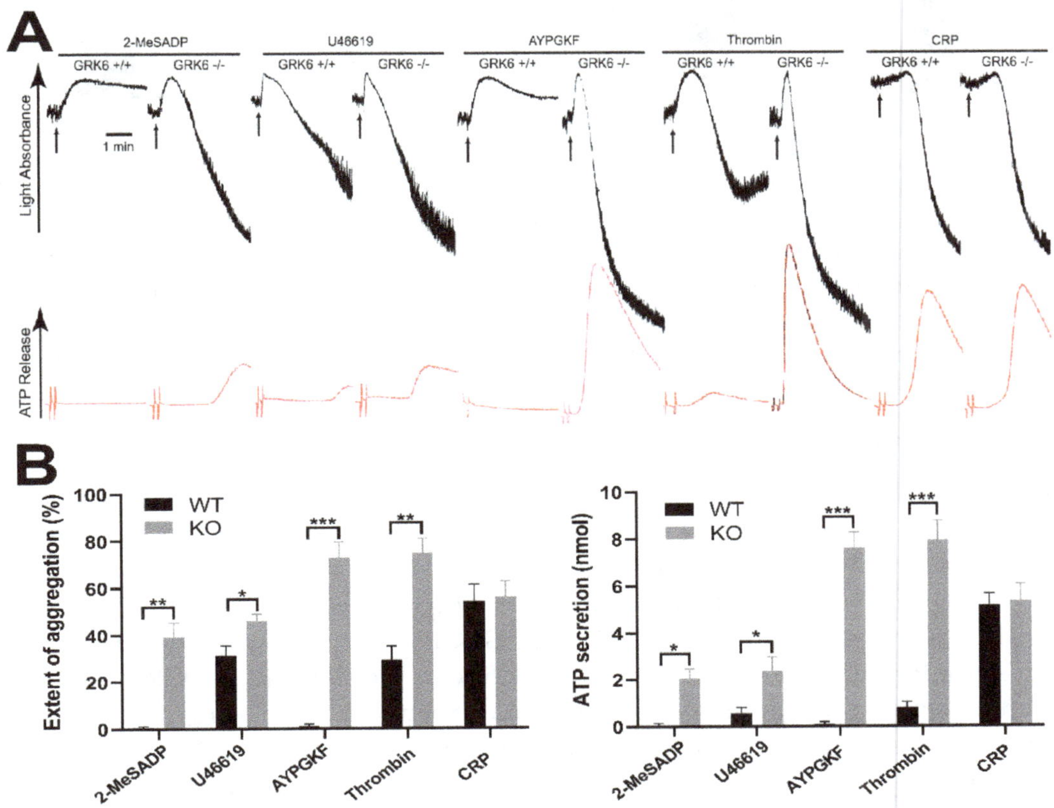

Figure 1. Agonist-induced platelet aggregation and dense granule secretion in the GPCR kinase (GRK)6-deficient platelets. Washed platelets from GRK6$^{-/-}$ mice and GRK6$^{+/+}$ littermates were stimulated with G-protein-coupled receptor (GPCR) agonists 30 nM 2MeSADP, 50 nM U46619, 60 μM AYPGKF, 0.1 unit/mL thrombin and GPVI agonist 2.5 μg/mL collagen-related peptide (CRP) for 3.5 min under stirring conditions. (**A**) Platelet aggregation (top) and ATP secretion (bottom) were measured in a lumi-aggregometer. All tracings shown are representative of at least three different experiments. (**B**) Quantification of extent of aggregation and dense granule secretion from panel A. Data are presented as mean ± SE *, $p < 0.05$; **, $p < 0.01$; ***, $p < 0.005$.

A similar approach was adopted to analyze the activation of integrin αIIbβ3 by measuring the binding of the conformation-dependent antibody JON/A, and α-granule secretion by measuring the binding of an fluorescein isothiocyanate (FITC)-labeled anti-P-selectin antibody that recognizes the α-granule protein: P-selectin, by flow cytometry. Consistent with the results shown in Figure 1, we found that both JON/A binding and P-selectin expression induced by AYPGKF were significantly increased in GRK6-deficient platelets compared to WT platelets (Figure 2A,B). However, CRP-induced JON/A binding and P-selectin binding to GRK6$^{-/-}$ platelets were not affected compared to WT platelets (Figure 2A,B). These results show that GRK6 plays an important role in the regulation of platelet functional responses downstream of GPCR pathways.

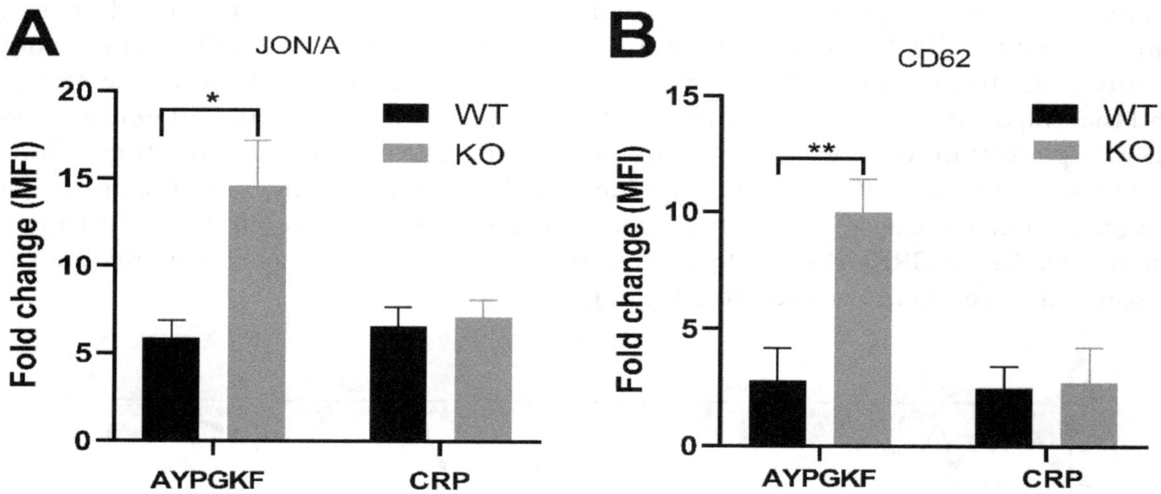

Figure 2. Integrin αIIbβ3 activation and α granule secretion in GRK6-deficient platelets. Washed platelets from GRK6$^{-/-}$ (knockout, KO) mice and GRK6$^{+/+}$ (wild-type, WT) littermates were stimulated with 100 μM AYPGKF and 2.5 μg/mL CRP in the presence of phycoerythrin (PE)-labeled JON/A antibody to determine integrin $\alpha_{IIb}\beta_3$ activation (**A**) or FITC-labeled anti-P-selectin (CD62) antibody to evaluate α-granule secretion (**B**) by flow cytometry. Mean fluorescence intensities (MFI) are shown as fold change from unstimulated control. Data are representative of three independent experiments and are presented as mean ± SE. *, $p < 0.05$; **, $p < 0.01$.

2.2. *GRK6 Regulates Specific GPCR-Mediated Platelet Aggregation*

It is known that ADP-induced platelet aggregation requires co-activation of the G_q-coupled P2Y$_1$ and the G_i-coupled P2Y$_{12}$ receptors while platelet aggregation induced by co-stimulation of serotonin (5HT) and epinephrine bypasses the P2Y$_1$ and P2Y$_{12}$ receptors to activate the G_q-coupled 5HT$_{2A}$ receptor and the G_z-coupled α_{2A} adrenergic receptor, respectively [6]. In order to evaluate the role of GRK6 in the differential regulation of GPCRs, we measured the platelet aggregation induced by co-stimulation of serotonin and epinephrine in the WT and the GRK6-deficient platelets. Interestingly, either serotonin or epinephrine alone failed to induce platelet aggregation in both WT and GRK6-deficient platelets (Figure 3A). In addition, compared to the effect of GRK6 in 2-MeSADP-induced aggregation, platelet aggregation induced by co-stimulation of serotonin and epinephrine was not affected in the GRK6-deficient platelets (Figure 3B), suggesting that GRK6 was not involved in the regulation of 5HT$_{2A}$- and α_{2A} adrenergic receptor-induced platelet aggregation.

We further evaluated the contribution of GRK6 on P2Y$_1$ and P2Y$_{12}$ receptors. Platelet aggregation induced by co-stimulation of G_q plus G_i with serotonin and 2-MeSADP in the presence of P2Y$_1$-receptor antagonist MRS2179 or G_q plus G_z by 2-MeSADP in the presence of P2Y$_{12}$-receptor antagonist AR-C69931MX and epinephrine was significantly potentiated in GRK6-deficient platelets compared to WT platelets (Figure 3C). However, the extent of the potentiation of platelet aggregation was smaller compared to the extent of platelet aggregation induced by 2-MeSADP in GRK6-deficient platelets in Figure 3B, confirming that GRK6 selectively regulates $P2Y_1$- and $P2Y_{12}$-mediated platelet functions.

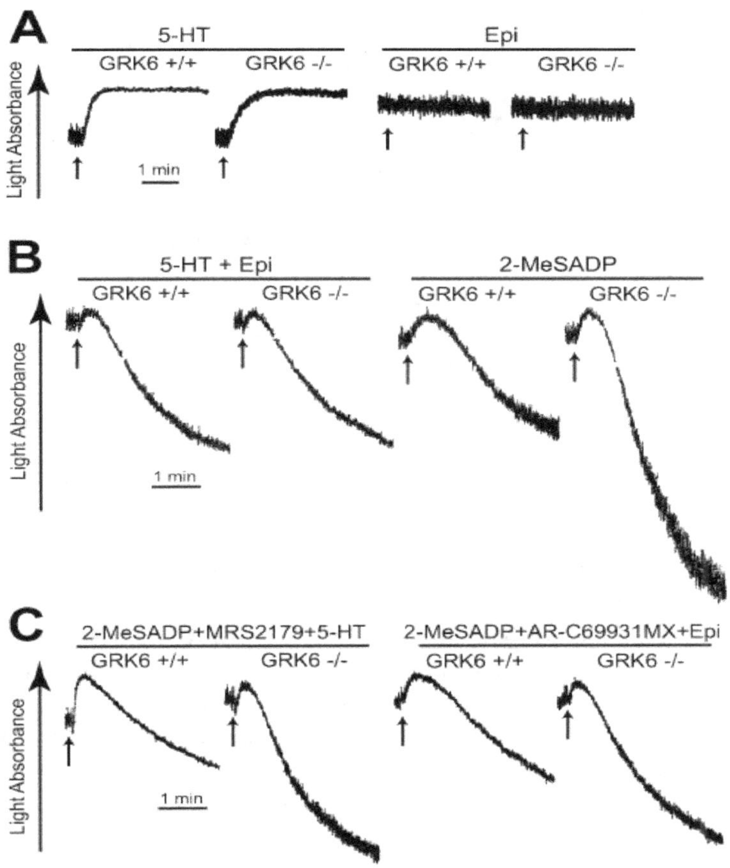

Figure 3. GRK6 selectively regulates $P2Y_1$ and $P2Y_{12}$ receptor-mediated platelet responses. Platelets from $GRK6^{+/+}$ and $GRK6^{-/-}$ mice were stimulated with (**A**) 20 μM 5-HT and 100 μM epinephrine, (**B**) 20 μM 5-HT + 100 μM epinephrine and 30 nM 2-MeSADP, and (**C**) 30 nM 2-MeSADP + 100 μM MRS2179 + 20 μM 5-HT and 30 nM 2-MeSADP + 100 nM AR-C69931MX + 100 μM epinephrine, and platelet aggregation was measured. Tracings are representative of three independent experiments.

2.3. GRK6 Plays an Important Role in ADP and PAR4 Receptor Desensitization in Platelets

In order to determine the role of GRK6 in GPCR desensitization in platelets, we measured platelet aggregation in response to re-stimulation in the WT and the GRK6-deficient platelets. As shown in Figure 4A, re-stimulation with ADP immediately after disaggregation, 3 min following the first addition of ADP, did not induce shape change or aggregation in the WT platelets. In contrast, platelet aggregation in response to the second challenge of ADP was restored in the GRK6-deficient platelets. Moreover, there was an increase in ADP-induced Ca^{2+} release in $GRK6^{-/-}$ platelets compared to the WT (Figure 4B). Slight increase in Ca^{2+} release was also observed in response to re-stimulation of ADP in $GRK6^{-/-}$ platelets, indicating that GRK6 regulated ADP-induced $P2Y_1$ receptor desensitization in platelets. Similarly, platelet aggregation was restored in response to ADP after pretreatment with ADP in the GRK6-deficient platelets compared to that in the WT platelets (Figure 4C), indicating that GRK6 played a role in the $P2Y_1$ and $P2Y_{12}$ receptor desensitization in platelets. In addition, platelet aggregation was restored in response to a second challenge of AYPGKF after pretreatment with AYPGKF in the GRK6-deficient platelets, whereas re-stimulation of AYPGKF failed to induce aggregation in the WT platelets (Figure 4D), suggesting that GRK6 contributed to PAR4 desensitization in platelets.

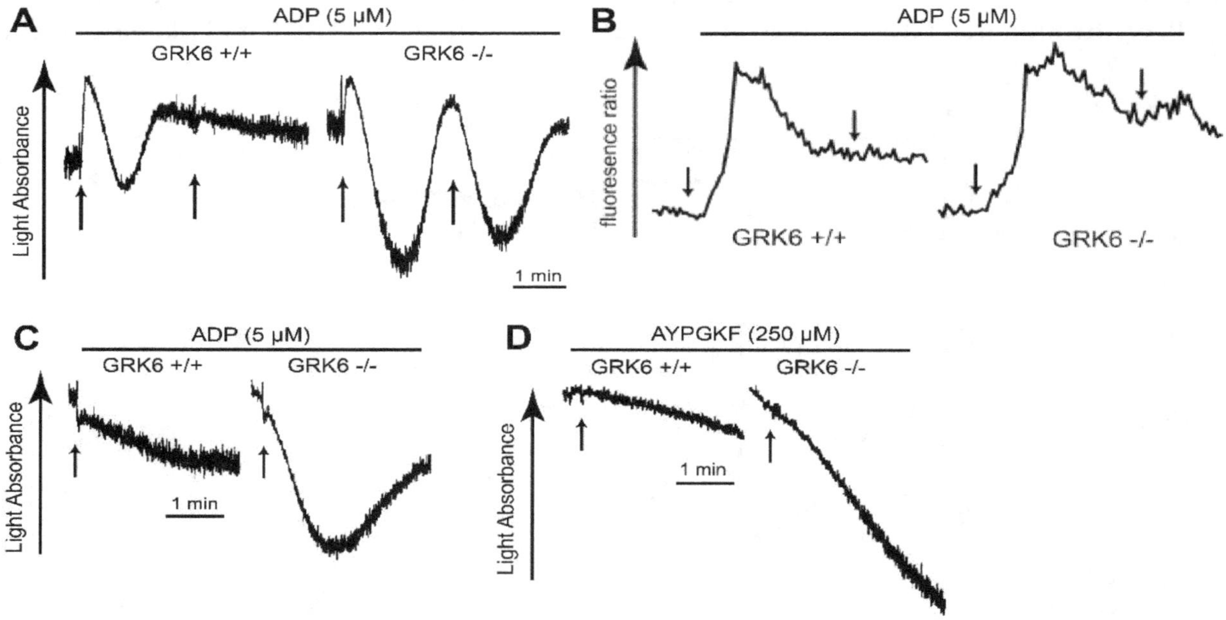

Figure 4. Desensitization of platelet aggregation in response to adenosine diphosphate (ADP) and AYPGKF in GRK6-deficient platelets. (**A**) Platelets from $GRK6^{+/+}$ and $GRK6^{-/-}$ mice were stimulated with 5 μM ADP for 3 min. Platelets were re-stimulated with 5 μM ADP for 3 min. (**B**) Fura-2-AM-loaded platelets from $GRK6^{+/+}$ and $GRK6^{-/-}$ mice were stimulated with 5 μM ADP for 3 min and re-stimulated with 5 μM ADP. Intracellular Ca^{2+} increase was measured. (**C**) Platelets from $GRK6^{+/+}$ and $GRK6^{-/-}$ mice were stimulated with 5 μM ADP after 3 min pre-incubation with 5 μM ADP. (**D**) Platelets from $GRK6^{+/+}$ and $GRK6^{-/-}$ mice were stimulated with 250 μM AYPGKF after 30 min pre-incubation with 250 μM AYPGKF. Results are representative of three independent experiments.

2.4. GPCR-Mediated Signaling Events are Potentiated in GRK6-Deficient Platelets

In order to evaluate the effect of GRK6 on platelet signaling, we stimulated platelets with 2-MeSADP and AYPGKF in the WT and the GRK6-deficient platelets and measured the phosphorylation of protein kinase Cδ (PKCδ), Akt, and extracellular signal-related kinase (ERK) as per a method described previously [22]. These proteins were selected for their known mechanism and effect on platelet activation [23–27]. Moreover, 2-MeSADP-induced Akt phosphorylation (Figure 5A) and AYPGKF-induced Akt, ERK, and PKCδ phosphorylations (Figure 5B) were potentiated in the GRK6-deficient platelets compared to WT platelets, suggesting a role of GRK6 in regulating both G_q- and G_i-mediated signaling events in platelets.

2.5. Contribution of GRK6 in the Platelet Function In Vivo

ADP, thromboxane, and thrombin are important agonists that contribute to thrombus growth and stability. To evaluate the contribution of GRK6 to the platelet function in vivo, we used in-vivo thrombosis models using $FeCl_3$ injury of the carotid artery [28,29] in WT and $GRK6^{-/-}$ mice. As shown in Figure 6A, WT mice formed a thrombus 12.5 min after the vascular injury whereas the $GRK6^{-/-}$ mice formed a thrombus 7 min after the injury, suggesting the enhanced thrombus formation in $GRK6^{-/-}$ mice after the vascular injury. In addition, 90% of the $GRK6^{-/-}$ mice showed stable thrombus formation, while 80% of the WT mice showed unstable thrombus formation (Figure 6B). We also measured tail-bleeding time in $GRK6^{+/+}$ and $GRK6^{-/-}$ mice, which reflects primary hemostatic function. We observed that average time taken for complete blockade of tail bleeding was 36 s in $GRK6^{-/-}$ compared to 58 s in $GRK6^{+/+}$ mice, suggesting that GRK6 negatively regulates hemostatic function in vivo (Figure 6C). Taken together, these results indicate the critical role of GRK6 in platelet function in vivo.

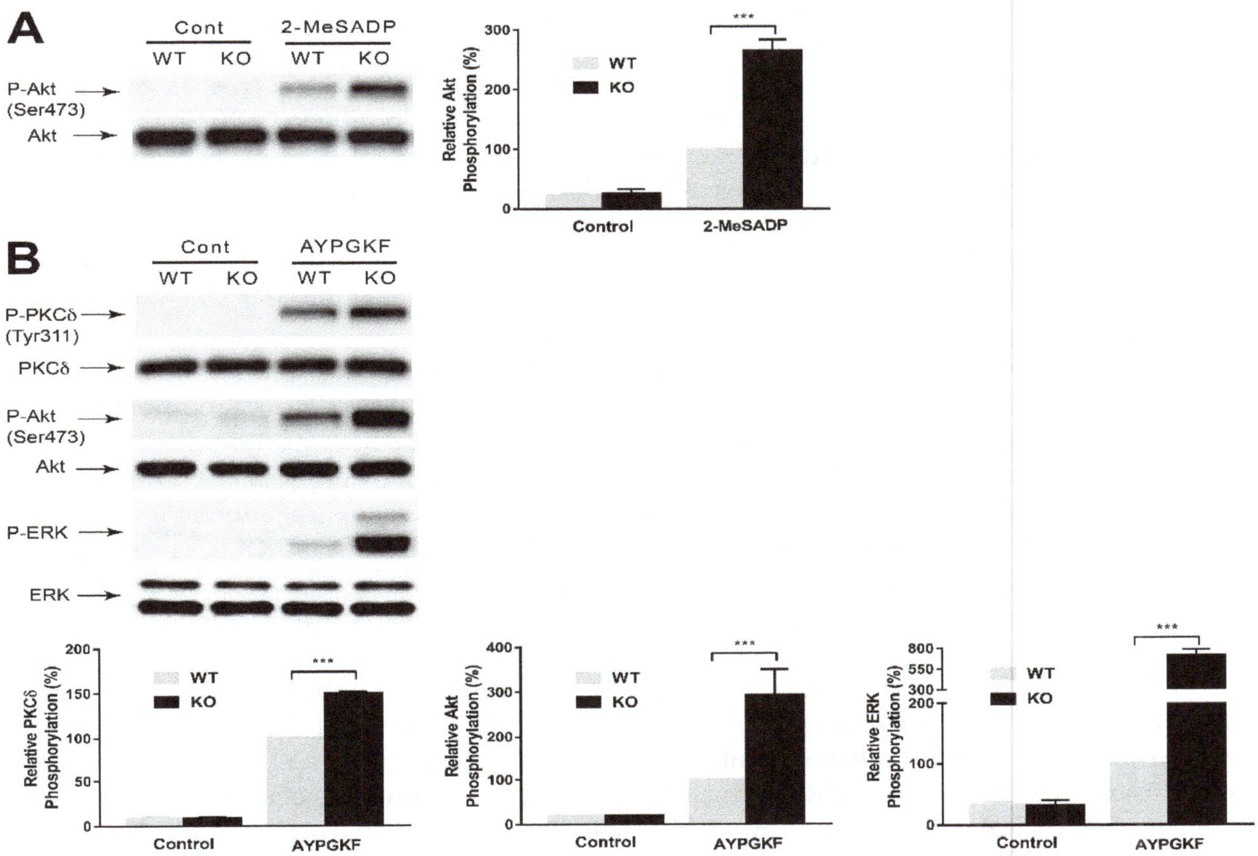

Figure 5. Potentiation of GPCR-mediated signaling events in GRK6-deficient platelets. Washed platelets from GRK6$^{-/-}$ (KO) mice and GRK6$^{+/+}$ (WT) littermates were stimulated with (**A**) 50 nM 2-MeSADP and (**B**) 100 μM AYPGKF for 2 min and probed with anti-phospho-PKCδ (Tyr^{311}), anti-phospho-Akt (Ser^{473}), anti-phospho-extracellular signal-related kinase (ERK), anti-Akt, anti-protein kinase Cδ (PKCδ), or anti-ERK antibodies by Western blotting. Blots are representative of three independent experiments and are presented as mean ± SE. ***, $p < 0.005$.

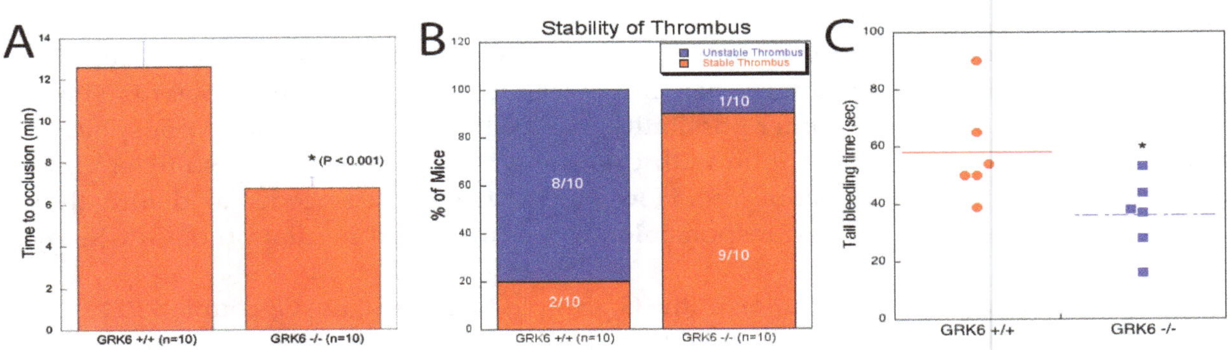

Figure 6. Enhanced in vivo platelet functional responses, thrombus growth, and stability in GRK6$^{-/-}$ mice in the $FeCl_3$ injury model. The carotid arteries of WT and GRK6$^{-/-}$ mice were injured by 5% $FeCl_3$ for 90 s. Flow rates through the carotid artery were measured for 30 min with a Doppler flow probe following vascular injury. (**A**) Time to occlusion of the carotid artery in GRK6$^{+/+}$ and GRK6$^{-/-}$ mice was measured. Data are presented as mean ± SE. Statistical analysis was performed using a Student *t* test (* $p < 0.001$). (**B**) Thrombus stability was assessed by determining the percentage of mice that maintain stable thrombi for at least 10 min. (**C**) Tail bleeding time was measured in GRK6$^{+/+}$ and GRK6$^{-/-}$ mice following excision of the distal 4 mm of the tail. Data are mean ± SE. *, $p < 0.05$.

3. Discussion

GRKs are ubiquitously expressed in tissues, and most of them show activity towards a wide variety of agonist-occupied receptors in vitro. Although GRKs were known to be involved in the desensitization of GPCRs in other cell types, the functional role of these signaling molecules and their mechanisms in regulating platelet physiological events have not been fully understood. It has been demonstrated that human platelets can become refractory to activation after a major surgery, possibly leading to an increased risk of post-surgical bleeding [30], suggesting that desensitization mechanisms play a critical role in regulating platelet responsiveness. The absence of a specific inhibitor of GRK6 has made it difficult to precisely determine its role in platelet function. Therefore, we used the GRK6-deficient platelets to evaluate the role of GRK6 in platelet function and characterized the novel mechanisms involved in GPCR desensitization in platelets.

It has been shown in other cells that different GRKs are involved in the regulation of specific GPCR desensitization. It has been demonstrated that the PAR1 and PAR4 receptors are internalized after desensitization following activation in platelets, but the basis of GRKs in regulation of platelet function has not been determined [31,32]. Thus, we first have investigated whether GRK6 had a role in the regulation of GPCR-mediated platelet function. We found that platelet aggregation and dense granule secretion induced by GPCR agonists including 2-MeSADP, U46619, AYPGKF, and thrombin were potentiated in the GRK6-deficient platelets compared to those in the WT platelets. We also observed that the GPCR-mediated platelet response was potentiated in the aspirin-treated GRK6-deficient platelets, indicating that the enhanced platelet aggregation in the GRK6-deficient platelet is not due to the effect of GRK6 on the positive feedback effect of generated TxA_2 (data not shown). Consistently, both JON/A binding and P-selectin expression that recognizes integrin $\alpha IIb\beta 3$ activation and α-granule secretion, respectively, were significantly enhanced upon stimulation with AYPGKF in GRK6-deficient platelets. However, GPVI agonist CRP-induced platelet aggregation and secretion were not affected in the GRK6-deficient platelets suggesting that GRK6 did not have any role in non-GPCR-mediated platelet function. Taken together, our data suggest that GRK6 plays an important role in the regulation of GPCR-mediated platelet function.

GRK6 had the least effect on the TxA_2-induced platelet aggregation and secretion. TxA_2 receptor exists as two alternatively spliced variants: $TP\alpha$ (platelets) and $TP\beta$ (endothelial cells), that differ in length and sequence of the C-terminal tail [33]. Different GRK isoforms have been shown to have different affinities toward these variants in other cells [34]. In contrast, it has been reported that the C-terminus of $TP\alpha$ is not capable of being phosphorylated by GRKs in other nucleated cell types [19,20]. Differences in tail sequence and number of the phosphorylation-amenable serine residues also allow for differential affinity for its agonists. It is known that $TP\alpha$ has a shorter C-terminus with lesser phosphorylatable serine residues, which might lead to decreased affinity towards $TP\alpha$ receptor than other GPCRs in platelets. Another possibility may be that receptors tend to phosphorylate and internalize rapidly following stimulation with U46619, which does not allow sufficient binding of GRK6 for phosphorylation of the residue [35]. However, further study is needed to identify the relation between serine residues and affinity, responsible for the functional regulation of GPCRs by GRK6 in platelets.

It has been shown that co-stimulation of the G_q-coupled $P2Y_1$ and the G_i-coupled $P2Y_{12}$ receptors is essential for ADP-induced platelet aggregation [1,36]. Moreover, co-stimulation of serotonin and epinephrine induces aggregation in platelets. The α_{2A} adrenergic receptor, when stimulated with epinephrine, couples primarily to G_z, which belongs to the G_i class of the G proteins [37,38] that can mimic the $P2Y_{12}$ receptor activation pathway. On the other hand, serotonin acts on the $5HT_{2A}$ receptor, which bypasses the $P2Y_1$ receptor to supplement G_q signaling [39]. Although concomitant signaling through the serotonin and epinephrine causes platelet aggregation, neither of them can cause platelet aggregation by itself [1,40,41]. Interestingly, serotonin only induced shape change while epinephrine induced neither shape change nor aggregation in both GRK6-deficient and WT platelets. Moreover, we found that in contrast to an ADP-induced aggregation, platelet aggregation induced by co-stimulation of serotonin and epinephrine was not affected in the GRK6-deficient platelets. These suggest that GRK6 is not involved in the regulation of α_{2A} and $5HT_{2A}$ receptors and plays a role in regulating specific GPCR function.

It is possible that the enhanced platelet response in the GRK6-deficient platelets is due to an altered expression of receptors in the GRK6-deficient platelets. To clarify this, we assessed the PAR4 receptor expression in both the WT and the GRK6-deficient platelets and found that the PAR4 receptor level in GRK6-deficient platelets was similar to that in the WT platelets. This confirms that the expression of PAR4 receptor is not affected by the absence of GRK6.

GRK6 has been implicated in various receptor desensitizations in vivo and in vitro [42,43]. It has been suggested that re-sensitization of the platelet aggregation response to ADP is due to desensitization of the $P2Y_1$ receptor, while the $P2Y_{12}$ receptor remains functional [44]. Another study suggested that both the $P2Y_1$ and $P2Y_{12}$ receptors undergo desensitization by different mechanisms involving PKC and GRK2 and GRK6, respectively [17]. We found that re-stimulation of platelets with ADP and AYPGKF restored platelet aggregation in GRK6-deficient platelets, suggesting that GRK6 played a role in the desensitization of ADP and PAR4 receptors. At this stage, it is not clear whether the effect of GRK6 on ADP receptors is due to the desensitization of both the $P2Y_1$ and $P2Y_{12}$ receptors or the selective desensitization of either the $P2Y_1$ receptor or the $P2Y_{12}$ receptor, because ADP requires co-activation of both the $P2Y_1$ and $P2Y_{12}$ receptors. We found that platelet aggregation was potentiated in response to serotonin or epinephrine in combination with selective activation of $P2Y_1$ or $P2Y_{12}$, respectively, in GRK6-deficient mice, while serotonin and epinephrine failed to restore aggregation. We also found the increased Ca^{2+} release in response to both initial and re-stimulation of platelets with ADP in GRK6-deficient platelets compared to the WT platelets, indicating the role of GRK6 in $P2Y_1$ receptor desensitization. These data confirm that the desensitization of both $P2Y_1$ and $P2Y_{12}$ receptors is responsible for platelet desensitization to ADP, and GRK6 is involved in regulation of both the $P2Y_1$ and $P2Y_{12}$ receptors in platelets.

GPCR-mediated activation of Akt, ERK, and PKCδ is important and necessary to enhance and promote platelet aggregation. Previous studies have shown that ADP-induced Akt activation is dependent on G_i-mediated $P2Y_{12}$ receptor stimulation [45] and AYPGKF-mediated Akt and ERK phosphorylation has been shown to be regulated by the G_i pathway by the secreted ADP in platelets [26]. It has been also shown that PKCδ, downstream of G_q pathway, plays a positive role in AYPGKF-induced platelet aggregation and dense granule secretion [25]. We found that the deletion of GRK6 potentiated ADP-induced Akt phosphorylation and AYPGKF-induced Akt, ERK, and PKCδ phosphorylation, suggesting that GRK6 regulates both the G_q- and G_i-mediated signaling events in platelets.

ADP, thromboxane, and thrombin are important agonists that contribute to thrombus growth and stability. Since we observed the potentiation of these agonist-induced platelet responses in GRK6-deficient platelets, we then evaluated the contribution of GRK6 to the function of platelets in vivo. We found that enhanced platelet function in $GRK6^{-/-}$ mice resulted in stable thrombus formation and rapid time to occlusion in in vivo thrombosis models using $FeCl_3$ injury of the carotid artery, indicating that $GRK6^{-/-}$ mice are more susceptible to thrombosis. The bleeding time was also significantly reduced in $GRK6^{-/-}$ mice. These results confirm the critical role of GRK6 in platelet function in vivo. Since GRK6 is ubiquitously expressed, we cannot rule out the contribution of GRK6 of other cells including leukocytes, endothelial cells, and smooth muscle cells on thrombus formation.

While we were preparing this manuscript, Chen et al. [46] has published a study suggesting that GRK6 regulates the hemostatic response to injury through its rate-limiting effects on GPCR signaling in platelets. Similar to our study, they have shown that GRK6 regulates $P2Y_{12}$ and PAR4 receptor signaling. Importantly, we have investigated the effect of GRK6 on $5HT_{2A}$ and α_{2A} adrenergic receptor in $GRK6^{-/-}$ platelets with serotonin and epinephrine and have found that these receptors are not regulated by GRK6 in platelets. This approach with the addition of either $P2Y_1$ and $P2Y_{12}$ receptor antagonists also helped us to confirm that both $P2Y_1$ and $P2Y_{12}$ receptors are regulated by GRK6 in platelets. Moreover, we have demonstrated the effect of GRK6 on desensitization of platelet aggregation in response to ADP and AYPGKF.

In conclusion, we demonstrate that GRK6 plays an important role in the regulation of GPCR-mediated platelet activation and in vivo thrombus formation by selectively regulating desensitization of specific GPCRs.

4. Materials and Methods

4.1. Materials

U46619, 2-MeSADP, ADP, thrombin, serotonin, epinephrine, apyrase (type V), prostaglandin E_1 (PGE_1), sodium citrate, and acetylsalicylic acid were from Sigma (St. Louis, MO, USA). CRP was obtained from Dr. Richard Farndale at the University of Cambridge. Phycoerythrin-conjugated antibody JON/A and FITC-conjugated anti-P-selectin antibody were from Emfret Analytics (Sterntalerweg, Wurzburg, Germany). Hexapeptide AYPGKF was custom synthesized by Invitrogen (Carlsbad, CA, USA). Fura-2-AM was from Millipore (Temecula, CA, USA). Anti-phospho-Akt (Ser^{473}), anti-phospho-ERK (Thr202/Tyr204), anti-phospho-PKCδ (Tyr311), anti-Akt, anti-PKCδ, and anti-ERK antibodies were purchased from Cell Signaling Technology (Beverly, MA, USA). Horseradish peroxidase-labeled secondary antibody was bought from Santa Cruz Biotechnology (Santa Cruz, CA, USA). All other reagents were of analytical grade, and deionized water was used throughout the study.

4.2. Animals

All animal experiments were performed in accordance with the approval obtained from the Chungbuk National University Animal Ethics Committee (CBNUA-873-15-02, approved on July 1, 2019). $GRK6^{-/-}$ mice were obtained from Dr. Walter Koch (Temple University, Philadelphia, PA, USA). WT littermates were used as controls. The platelet count was normal in $GRK6^{-/-}$ mice.

4.3. Preparation of Murine Platelets

Blood was collected from equal number of both male and female mice and platelet was prepared as described [47]. Briefly, the collected citrated blood was centrifuged at 100× g for 10 min at room temperature (RT) to obtain Platelet Rich Plasma (PRP) and the obtained PRP was centrifuged at 400× g for 10 min to pellet the platelets. Platelet pellets were re-suspended in the Tyrode's buffer (pH = 7.4) containing 0.05 units/mL of apyrase, and cells were adjusted to 2×10^8 cells/mL.

4.4. Platelet Aggregation and Secretion

Agonist-induced platelet aggregation and secretion were measured using a lumi-aggregometer (Chrono-Log, Havertown, PA, USA) at 37 °C under stirring conditions (900 rpm) as per a method described previously [47]. Washed platelets (0.5 mL) were stimulated with the different agonists, and tracings of light transmission were measured. Platelet secretion was determined by measuring the release of ATP after adding the luciferin/luciferase reagent.

4.5. Flow Cytometry

Washed murine platelets were used to measure integrin $\alpha_{II}b\beta_3$ activation with phycoerythrin-conjugated JON/A antibodies and the surface exposure of P-selectin with FITC-conjugated P-selectin antibodies as per a method described previously [48]. All measurements were determined using a FACSCalibur flow cytometer (BD Biosciences, San Jose, CA, USA). Appropriate isotype controls were used.

4.6. Calcium Mobilization

Intracellular Ca^{2+} concentration was measured in Fura-2-AM loaded platelets as previously described [49].

4.7. Western Blotting

Platelets were stimulated with different agonists for the appropriate time, and phosphorylation events were measured as described previously [22]. Briefly, platelets were stimulated with 2-MeSADP or AYPGKF and the reaction was stopped by the addition of 3× sodium dodecyl sulfate (SDS) buffer. Platelet lysates were separated by 10% SDS/PAGE and transferred onto polyvinylidene difluoride (PVDF) membranes. Membranes were incubated overnight at 4 °C with anti-phospho-Akt (Ser473), anti-phospho-ERK (Thr202/Tyr204), anti-phospho-PKCδ (Tyr311), anti-Akt, anti-PKCδ, or anti-ERK antibodies. Membranes were then probed with goat anti-rabbit antibody and immune-reactivity was detected using the Fuji-Film Luminescent Image Analyzer (LAS-3000 CH, Tokyo, Japan).

4.8. $FeCl_3$-Induced In Vivo Thrombosis Model

Adult mice (10–12 weeks old) were anesthetized and $FeCl_3$-induced thrombosis was assessed as per a method described previously [29]. Briefly, the carotid artery was injured by 5% $FeCl_3$ for 90 s, and blood flow through the carotid artery was measured for 30 min using the Doppler flow probe following vascular injury. The occlusion time due to thrombus formation was recorded and plotted as time to occlusion.

4.9. Tail Bleeding Time Assay

The tail bleeding time assay was performed as described previously [29]. Briefly, a longitudinal cut was made in anesthetized mouse's tail and tail was immediately immersed in 0.9% isotonic saline at 37 °C. The bleeding time was defined as the time required for the blockade of blood flow.

4.10. Statistical Analysis

All statistical analyses were performed using the Prism software (version 3.0). Data have been presented as mean ± standard error (SE). Statistical significance was determined by the Student's *t*-test.

Author Contributions: Conceptualization, P.K.C. and S.K. (Soochong Kim); methodology, P.K.C. and S.K. (Sanggu Kim); validation, Y.J. and S.K. (Soochong Kim); investigation, P.K.C., S.-H.L., K.-M.P., and S.K. (Soochong Kim); writing—original draft, P.K.C.; writing—review and editing, Y.J., S.-H.L., K.-M.P., and S.K. (Soochong Kim); supervision, S.K. (Soochong Kim). All authors have read and agreed to the published version of the manuscript.

Abbreviations

GPCR	G protein-coupled receptor
GRK	G protein-coupled receptor kinase
ADP	Adenosine diphosphate
CRP	Collagen-related peptide
PAR	Protease-activated receptor
ERK	Extracellular signal-related kinase
PKCδ	Protein Kinase Cδ
TP	Thromboxane receptor
PRP	Platelet-rich plasma

References

1. Jin, J.; Kunapuli, S.P. Coactivation of two different G protein-coupled receptors is essential for ADP-induced platelet aggregation. *Proc. Natl. Acad. Sci. USA* **1998**, *95*, 8070–8074. [CrossRef] [PubMed]
2. Paul, B.Z.; Jin, J.; Kunapuli, S.P. Molecular mechanism of thromboxane A(2)-induced platelet aggregation. Essential role for p2t(ac) and alpha(2a) receptors. *J. Biol. Chem.* **1999**, *274*, 29108–29114. [CrossRef] [PubMed]
3. Offermanns, S.; Laugwitz, K.-L.; Spicher, K.; Schultz, G. G proteins of the G12 family are activated via thromboxane A2 and thrombin receptors in human platelets. *Proc. Natl. Acad. Sci. USA* **1994**, *91*, 504–508. [CrossRef] [PubMed]
4. Shenker, A.; Goldsmith, P.; Unson, C.G.; Spiegel, A.M. The G protein coupled to the thromboxane A2 receptor in human platelets is a member of the novel Gq family. *J. Biol. Chem.* **1991**, *266*, 9309–9313.
5. Kim, S.; Foster, C.; Lecchi, A.; Quinton, T.M.; Prosser, D.M.; Jin, J.; Cattaneo, M.; Kunapuli, S.P. Protease-activated receptors 1 and 4 do not stimulate Gi signaling pathways in the absence of secreted ADP and cause human platelet aggregation independently of Gisignaling. *Blood* **2002**, *99*, 3629–3636. [CrossRef]
6. de Chaffoy de Courcelles, D.; Roevens, P.; van Belle, H.; de Clerck, F. The synergistic effect of serotonin and epinephrine on the human platelet at the level of signal transduction. *FEBS Lett.* **1987**, *219*, 283–288. [CrossRef]
7. Nieswandt, B.; Watson, S.P. Platelet-collagen interaction: Is GPVI the central receptor? *Blood* **2003**, *102*, 449–461. [CrossRef]
8. Benovic, J.L.; Gomez, J. Molecular cloning and expression of GRK6. A new member of the G protein-coupled receptor kinase family. *J. Biol. Chem.* **1993**, *268*, 19521–19527.
9. Hisatomi, O.; Matsuda, S.; Satoh, T.; Kotaka, S.; Imanishi, Y.; Tokunaga, F. A novel subtype of G-protein-coupled receptor kinase, GRK7, in teleost cone photoreceptors. *FEBS Lett.* **1998**, *424*, 159–164. [CrossRef]
10. Reiter, E.; Lefkowitz, R.J. GRKs and β-arrestins: Roles in receptor silencing, trafficking and signaling. *Trends Endocrinol. Metab.* **2006**, *17*, 159–165. [CrossRef]
11. Willets, J.M.; Challiss, R.A.; Nahorski, S.R. Non-visual GRKs: Are we seeing the whole picture? *Trends Pharm. Sci.* **2003**, *24*, 626–633. [CrossRef] [PubMed]
12. Pitcher, J.A.; Freedman, N.J.; Lefkowitz, R.J. G protein-coupled receptor kinases. *Annu. Rev. Biochem.* **1998**, *67*, 653–692. [CrossRef] [PubMed]
13. Santulli, G.; Trimarco, B.; Iaccarino, G. G-protein-coupled receptor kinase 2 and hypertension. *High Blood Press. Cardiovasc. Prev.* **2013**, *20*, 5–12. [CrossRef] [PubMed]
14. Freedman, N.J.; Lefkowitz, R.J. Desensitization of G protein-coupled receptors. *Recent Prog. Horm. Res.* **1996**, *51*, 319–351; discussion 352–313.
15. Tiruppathi, C.; Yan, W.; Sandoval, R.; Naqvi, T.; Pronin, A.N.; Benovic, J.L.; Malik, A.B. G protein-coupled receptor kinase-5 regulates thrombin-activated signaling in endothelial cells. *Proc. Natl. Acad. Sci. USA* **2000**, *97*, 7440–7445. [CrossRef]
16. Ishii, K.; Chen, J.; Ishii, M.; Koch, W.J.; Freedman, N.J.; Lefkowitz, R.J.; Coughlin, S.R. Inhibition of thrombin receptor signaling by a G-protein coupled receptor kinase. Functional specificity among G-protein coupled receptor kinases. *J. Biol. Chem.* **1994**, *269*, 1125–1130.
17. Hardy, A.R.; Conley, P.B.; Luo, J.; Benovic, J.L.; Poole, A.W.; Mundell, S.J. P2Y1 and P2Y12 receptors for ADP desensitize by distinct kinase-dependent mechanisms. *Blood* **2005**, *105*, 3552–3560. [CrossRef]
18. Willets, J.M.; Brighton, P.J.; Mistry, R.; Morris, G.E.; Konje, J.C.; Challiss, R.J. Regulation of oxytocin receptor responsiveness by G protein-coupled receptor kinase 6 in human myometrial smooth muscle. *Mol. Endocrinol.* **2009**, *23*, 1272–1280. [CrossRef]
19. Zhou, H.; Yan, F.; Tai, H.-H. Phosphorylation and desensitization of the human thromboxane receptor-α by G protein-coupled receptor kinases. *J. Pharmacol. Exp. Ther.* **2001**, *298*, 1243–1251.
20. Parent, J.-L.; Labrecque, P.; Orsini, M.J.; Benovic, J.L. Internalization of the TXA2 Receptor α and β Isoforms Role of the differentially spliced COOH terminus in agonist-promoted receptor internalization. *J. Biol. Chem.* **1999**, *274*, 8941–8948. [CrossRef]
21. Kim, J.; Ahn, S.; Ren, X.-R.; Whalen, E.J.; Reiter, E.; Wei, H.; Lefkowitz, R.J. Functional antagonism of different G protein-coupled receptor kinases for β-arrestin-mediated angiotensin II receptor signaling. *Proc. Natl. Acad. Sci. USA* **2005**, *102*, 1442–1447. [CrossRef] [PubMed]

22. Kim, S.; Jin, J.; Kunapuli, S.P. Relative contribution of G-protein-coupled pathways to protease-activated receptor-mediated Akt phosphorylation in platelets. *Blood* **2006**, *107*, 947–954. [CrossRef] [PubMed]
23. Chen, J.; De, S.; Damron, D.S.; Chen, W.S.; Hay, N.; Byzova, T.V. Impaired platelet responses to thrombin and collagen in AKT-1–deficient mice. *Blood* **2004**, *104*, 1703–1710. [CrossRef] [PubMed]
24. Kim, S.; Jin, J.; Kunapuli, S.P. Akt activation in platelets depends on Gi signaling pathways. *J. Biol. Chem.* **2004**, *279*, 4186–4195. [CrossRef] [PubMed]
25. Murugappan, S.; Tuluc, F.; Dorsam, R.T.; Shankar, H.; Kunapuli, S.P. Differential role of protein kinase Cδ isoform in agonist-induced dense granule secretion in human platelets. *J. Biol. Chem.* **2004**, *279*, 2360–2367. [CrossRef] [PubMed]
26. Shankar, H.; Garcia, A.; Prabhakar, J.; Kim, S.; Kunapuli, S. P2Y12 receptor-mediated potentiation of thrombin-induced thromboxane A2 generation in platelets occurs through regulation of Erk1/2 activation. *J. Thromb. Haemost.* **2006**, *4*, 638–647. [CrossRef] [PubMed]
27. Woulfe, D.; Jiang, H.; Morgans, A.; Monks, R.; Birnbaum, M.; Brass, L.F. Defects in secretion, aggregation, and thrombus formation in platelets from mice lacking Akt2. *J. Clin. Investig.* **2004**, *113*, 441–450. [CrossRef]
28. Nagy, B.; Bhavaraju, K.; Getz, T.; Bynagari, Y.S.; Kim, S.; Kunapuli, S.P. Impaired activation of platelets lacking protein kinase C-θ isoform. *Blood* **2009**, *113*, 2557–2567. [CrossRef]
29. Bynagari-Settipalli, Y.S.; Lakhani, P.; Jin, J.; Bhavaraju, K.; Rico, M.C.; Kim, S.; Woulfe, D.; Kunapuli, S.P. Protein kinase C isoform epsilon negatively regulates ADP-induced calcium mobilization and thromboxane generation in platelets. *Arterioscler. Thromb. Vasc. Biol.* **2012**, *32*, 1211–1219. [CrossRef]
30. O'Brien, J.R.; Etherington, M.; Jamieson, S. Refractory state of platelet aggregation with major operations. *Lancet* **1971**, *2*, 741–743. [CrossRef]
31. Shapiro, M.J.; Weiss, E.J.; Faruqi, T.R.; Coughlin, S.R. Protease-activated receptors 1 and 4 are shut off with distinct kinetics after activation by thrombin. *J. Biol. Chem.* **2000**, *275*, 25216–25221. [CrossRef] [PubMed]
32. Molino, M.; Bainton, D.F.; Hoxie, J.A.; Coughlin, S.R.; Brass, L.F. Thrombin receptors on human platelets initial localization and subsequent redistribution during platelet activation. *J. Biol. Chem.* **1997**, *272*, 6011–6017. [CrossRef] [PubMed]
33. Raychowdhury, M.K.; Yukawa, M.; Collins, L.J.; McGrail, S.H.; Kent, K.C.; Ware, J.A. Alternative splicing produces a divergent cytoplasmic tail in the human endothelial thromboxane A2 receptor. *J. Biol. Chem.* **1994**, *269*, 19256–19261. [PubMed]
34. Kelley-Hickie, L.P.; O'Keeffe, M.B.; Reid, H.M.; Kinsella, B.T. Homologous desensitization of signalling by the alpha (α) isoform of the human thromboxane A2 receptor: A specific role for nitric oxide signalling. *Biochim. Biophys. Acta (BBA)-Mol. Cell Res.* **2007**, *1773*, 970–989. [CrossRef]
35. Habib, A.; Vezza, R.; Créminon, C.; Maclouf, J.; FitzGerald, G.A. Rapid, Agonist-dependent Phosphorylation in Vivo of Human Thromboxane Receptor Isoforms Minimal Involvement of Protein Kinase C. *J. Biol. Chem.* **1997**, *272*, 7191–7200. [CrossRef]
36. Hechler, B.; Leon, C.; Vial, C.; Vigne, P.; Frelin, C.; Cazenave, J.P.; Gachet, C. The P2Y1 receptor is necessary for adenosine 5′-diphosphate-induced platelet aggregation. *Blood* **1998**, *92*, 152–159. [CrossRef]
37. Smith, S.K.; Limbird, L.E. Solubilization of human platelet alpha-adrenergic receptors: Evidence that agonist occupancy of the receptor stabilizes receptor—Effector interactions. *Proc. Natl. Acad. Sci. USA* **1981**, *78*, 4026–4030. [CrossRef]
38. Yang, J.; Wu, J.; Kowalska, M.A.; Dalvi, A.; Prevost, N.; O'Brien, P.J.; Manning, D.; Poncz, M.; Lucki, I.; Blendy, J.A. Loss of signaling through the G protein, Gz, results in abnormal platelet activation and altered responses to psychoactive drugs. *Proc. Natl. Acad. Sci. USA* **2000**, *97*, 9984–9989. [CrossRef]
39. Cerrito, F.; Lazzaro, M.P.; Gaudio, E.; Arminio, P.; Aloisi, G. 5HT2-receptors and serotonin release: Their role in human platelet aggregation. *Life Sci.* **1993**, *53*, 209–215. [CrossRef]
40. Daniel, J.L.; Dangelmaier, C.; Jin, J.; Kim, Y.B.; Kunapuli, S.P. Role of intracellular signaling events in ADP-induced platelet aggregation. *Thromb. Haemost.* **1999**, *82*, 1322–1326. [CrossRef]
41. Kim, S.; Kunapuli, S.P. P2Y12 receptor in platelet activation. *Platelets* **2011**, *22*, 54–58. [CrossRef] [PubMed]
42. Aiyar, N.; Disa, J.; Dang, K.; Pronin, A.N.; Benovic, J.L.; Nambi, P. Involvement of G protein-coupled receptor kinase-6 in desensitization of CGRP receptors. *Eur. J. Pharmacol.* **2000**, *403*, 1–7. [CrossRef]
43. Gainetdinov, R.R.; Bohn, L.M.; Sotnikova, T.D.; Cyr, M.; Laakso, A.; Macrae, A.D.; Torres, G.E.; Kim, K.-M.; Lefkowitz, R.J.; Caron, M.G. Dopaminergic supersensitivity in G protein-coupled receptor kinase 6-deficient mice. *Neuron* **2003**, *38*, 291–303. [CrossRef]

44. Baurand, A.; Eckly, A.; Bari, N.; Léon, C.; Hechler, B.; Cazenave, J.-P.; Gachet, C. Desensitization of the platelet aggregation response to ADP: Differential down-regulation of the P2Y1 and P2cyc receptors. *Thromb. Haemost.* **2000**, *84*, 484–491. [CrossRef] [PubMed]
45. Kahner, B.N.; Shankar, H.; Murugappan, S.; Prasad, G.L.; Kunapuli, S.P. Nucleotide receptor signaling in platelets. *J. Thromb. Haemost.* **2006**, *4*, 2317–2326. [CrossRef] [PubMed]
46. Chen, X.; Gupta, S.; Cooper, M.; de Helian, D.; Zhao, X.; Naik, M.U.; Wurtzel, J.G.; Stalker, T.J.; Goldfinger, L.E.; Benovic, J. GRK6 regulates the hemostatic response to injury through its rate-limiting effects on GPCR signaling in platelets. *Blood Adv.* **2020**, *4*, 76–86. [CrossRef] [PubMed]
47. Kim, S.; Cipolla, L.; Guidetti, G.; Okigaki, M.; Jin, J.; Torti, M.; Kunapuli, S.P. Distinct role of Pyk2 in mediating thromboxane generation downstream of both G12/13 and integrin αIIbβ3 in platelets. *J. Biol. Chem.* **2013**, *288*, 18194–18203. [CrossRef]
48. Daniel, J.L.; Dangelmaier, C.A.; Mada, S.; Buitrago, L.; Jin, J.; Langdon, W.Y.; Tsygankov, A.Y.; Kunapuli, S.P.; Sanjay, A. Cbl-b is a novel physiologic regulator of glycoprotein VI-dependent platelet activation. *J. Biol. Chem.* **2010**, *285*, 17282–17291. [CrossRef]
49. Kim, S.; Kunapuli, S.P. Negative regulation of G_q-mediated pathways in platelets by $G_{12/13}$ pathways through *Fyn kinase*. *J. Biol. Chem.* **2011**, *286*, 24170–24179. [CrossRef]

4

Platelet Concentrates in Musculoskeletal Medicine

Erminia Mariani [1,2,*] and Lia Pulsatelli [1]

1 Laboratorio di Immunoreumatologia e rigenerazione tissutale, IRCCS Istituto Ortopedico Rizzoli, Via di Barbiano 1/10, 40136 Bologna, Italy; lia.pulsatelli@ior.it

2 Dipartimento di Scienze Mediche e Chirurgiche, Alma Mater Studiorum-University of Bologna, Via Massarenti 9, 40138 Bologna, Italy

* Correspondence: erminia.mariani@unibo.it

Abstract: Platelet concentrates (PCs), mostly represented by platelet-rich plasma (PRP) and platelet-rich fibrin (PRF) are autologous biological blood-derived products that may combine plasma/platelet-derived bioactive components, together with fibrin-forming protein able to create a natural three-dimensional scaffold. These types of products are safely used in clinical applications due to the autologous-derived source and the minimally invasive application procedure. In this narrative review, we focus on three main topics concerning the use of platelet concentrate for treating musculoskeletal conditions: (a) the different procedures to prepare PCs, (b) the composition of PCs that is related to the type of methodological procedure adopted and (c) the clinical application in musculoskeletal medicine, efficacy and main limits of the different studies.

Keywords: platelet-rich plasma; platelet-rich fibrin; preparation; composition; musculoskeletal diseases

---

1. Introduction

In the last 10 years, autologous biological blood-derived products have been largely investigated as useful therapeutic tools for treating musculoskeletal conditions (such as osteoarthritis, muscle injuries, tendinopathies and intervertebral disc degeneration) [1–3]. Platelet concentrates (PCs), mostly represented by platelet-rich plasma (PRP) and platelet-rich fibrin (PRF), are included in this type of biology-oriented autologous therapeutic strategy that may combine plasma/platelet-derived bioactive components (cytokines, chemokines, growth-factors and enzymes) with fibrin-forming protein able to create a natural three-dimensional scaffold [4].

This approach allows us to deliver biomolecules released by a concentrated pool of activated platelets to the target tissue site of injury, thus effectively contributing to the modulation of inflammatory process, angiogenesis and immune response, as well as promoting the healing and repair of injured tissues [5,6]. Moreover, biological blood-derived products have been recognized to have antimicrobial effects, such as being able to inhibit and/or to inactivate different bacterial strains [6–8].

The potential clinical application of these biologic products in musculoskeletal medicine relies on their capability of modulating the joint environment and their beneficial role in reducing the local inflammation and promoting cartilage and synovium anabolism [5,9–12].

These types of therapeutic strategies provide advantages in clinical applications due to the autologous-derived source, safety profile, easiness to obtain and the minimally invasive application procedure. On the other hand, clinical efficacy is still controversial, and solid evidence and consensus supporting the therapeutic application are still to be achieved.

Indeed, there are already some issues to be addressed concerning the high variability of platelet concentrate products, which mainly depends on patients' characteristics (age, sex, circadian rhythms and drug regimen) [13–16], as well as on the lack of standardized methods for platelet isolation/collection/activation and on heterogeneity among therapeutic protocols applied in clinical practice.

In this narrative review, we focus on three main challenging topics concerning the use of platelet concentrate for treating musculoskeletal conditions: (a) the different procedures to prepare platelet concentrate, (b) the composition of these products that is mainly related to the type of methodological procedure adopted and (c) the clinical application in musculoskeletal conditions and level of efficacy.

Short History of Platelet Concentrates

The concept of PRP originally was developed in transfusion medicine. In this field, the PRP term was used in 1954 by Kingsley [17] to identify thrombocyte concentrate for treating patients with severe thrombopenia.

The history of the techniques to obtain blood-derived products for improving tissue healing started in 1970 with the studies of Matras [18] on fibrin glue use in a rat model.

Subsequently, an autologous product termed "platelet–fibrinogen–thrombin mixture" was developed, including, in fibrin glue, a significant concentration of platelets, in order to reinforce the fibrin polymerization [19].

In the following years, the role of platelets in supporting tissue healing was confirmed and clinically demonstrated by using a blood-derived product called "platelet-derived wound healing factors or formula-PDWHF" [20] for treating skin ulcers.

About ten years later, Whitman et al. [21] published a clinical study on the results obtained in oral and maxillofacial surgery by using a "platelet gel" obtained by a gradient density cell separator.

However, the term of PRP in regenerative medicine associated to the notion of platelet growth factors to promote tissue healing was truly introduced by Marx et al. in 1998 [22], in a study that reported the effect of platelet-rich product on bone healing in maxillofacial surgery.

After these publications, the term "PRP" was generically associated with all the multiple formulations of platelet concentrates. Afterward, an end-product characterized by a fibrin matrix denser and more stable than in other PRP formulations was produced and called platelet-rich fibrin matrix (PRFM) or pure platelet-rich fibrin (P-PRF).

In 2001, a different form of platelet concentrates was proposed and identified as leukocyte- and platelet-rich fibrin (L-PRF) [23]. These preparations are organized as a high-density fibrin and were considered as a "second generation" platelet concentrates. This family of platelet concentrates appears to be particularly suitable for oral clinical application.

2. Preparation Procedure

2.1. Platelet-Rich Plasma

PRP is obtained from autologous blood by using commercial kits or "in-house techniques", aiming to provide a product characterized by a supra-physiological platelet concentration that can be used as liquid or activated gel form [14,24–26].

Despite the broad spectrum of protocols for PRP preparation, a common sequence of key steps [27,28] can be identified involving peripheral blood drawing from the patients by venipuncture, blood centrifugation to retrieve platelet-enriched fraction and platelet stimulation to release bioactive molecules.

In each of these phases, potential sources of variability may be identified, mainly ascribed to volume of blood samples drawn, type of anticoagulant, centrifugation protocols, material of collection tubes and type of platelet-activating agents [14,24].

The great variability in the different procedures results in a wide heterogeneity among PRP preparations in terms of platelet concentration, presence/absence of leukocytes and erythrocytes, and ultimately in terms of biological potential [14,24].

2.1.1. Anticoagulants

There are multiple choices of anticoagulants (ethylene diamine tetra-acetic acid-EDTA, citrate dextrose-A, tri-sodium citrate and heparin) that are used for blood collection and that can differently affect PRP quality [29].

Lei and colleagues [30] investigated the effect of heparin, citrate, acid citrate dextrose (ACD) and citrate-theophylline-adenosine-dipyridamole (CTAD) on platelet-rich plasma quality, to determine the appropriate anticoagulants for PRP production.

ACD and CTAD appear to be more effective compared to heparin and citrate in maintaining the integrity of platelet structures and in preventing their spontaneous activation. ACD-PRP and CTAD-PRP released more TGF-beta1 and significantly increased the proliferation rate of human marrow stromal cells compared to heparin- and citrate-PRP, thus showing ACD and CTAD appropriate anticoagulants for PRP production [30].

An animal model study, aiming to investigate the influence of sodium citrate and ACD- solution A anticoagulants on cell count and growth factor concentration in pure platelet-rich gel supernatants, reported an increased number of platelets and leukocytes in sodium citrate PRP compared to homologous acid–citrate–dextrose solution A PRP fraction, but no difference concerning growth factor concentration [31].

Another "in vitro" study explored the effects of sodium citrate (SC), EDTA, or anticoagulant ACD- solution A, on PRP characteristics and on mesenchymal stromal cell (MSC) culture [29]. A higher platelet count was observed in blood collected with EDTA, even if an increase of mean platelet volume has been reported after the two centrifugation steps. Conversely, following the centrifugation procedure, platelet yield was higher in SC product. SC and ACD showed similar efficacy in inducing MSC proliferation [29].

These findings support the most frequent use of citrate-based anticoagulants for PRP preparations [29].

A very recent comparative study [32] evaluated the effects of EDTA, heparin sodium (HS) and SC on PRP quality and on bone marrow stem cells' functionality.

Compared to HS and SC, EDTA has been shown to preserve platelet structure, minimize their spontaneous activation and sustain growth factor release for a more extended time.

Overall, these findings underline that also the choice of the best anticoagulant represents an open issue to address for optimizing PRP formulation.

To overcome this criticism, a study published in 2018 described a novel approach of PRP preparation without any additive, named temperature controlled PRP (t-PRP), by which the coagulation was previously inhibited in hypothermic environment. In this study, t-PRP was compared to PRP obtained by ACD-A blood.

Overall, t-PRP showed a more physiologic pH, higher platelet yield, slower release and degradation of growth factors. Furthermore, animal model experiments demonstrated that t-PRP was able to promote wound healing [33].

2.1.2. Isolation Protocols

PRP can be obtained according to two basic protocols designed as plasma-based and buffy-coat-based procedures [14,34]. Plasma-based methods retrieve platelets, while minimizing leukocyte and erythrocyte fractions. For this purpose, a slower and shorter spin regimen is applied in plasma-based protocols. Platelets concentration is usually twofold to threefold increased above baseline whole blood levels (300,000 to 500,000 platelets/μL) [14,34].

Alternatively, the main goal of protocols for buffy-coat systems is to maximize platelet isolation during the centrifugation procedure, by high spin rates and long spin regimens. PRP obtained by this method is characterized by a high platelet recovery, increasing about threefold to eightfold compared to baseline levels (500,000 to 1,500,000 platelets/μL) and by the presence of variable concentrations of leukocytes and erythrocytes [14,34]. This type of PRP preparation is generally called leucocyte-rich PRP (L-PRP).

Specific protocols developed to obtain PRP by using either a commercial device/kit or manual/homemade procedures derive from multiple modifications of these two basic protocols (plasma-based and buffy-coat-based).

Most of commercially available systems produce PRP by buffy-coat-based method [35], and several comparative studies were reported, aiming to analyze different common commercial separation systems, essentially evaluating final PRP products in terms of platelets concentrations and growth factors release [24,35] (Table 1).

Table 1. Centrifugation protocol and composition of platelet-rich plasma (PRP) produced by common commercial PRP systems.

Device	Centrifugation Force (g)	Centrifugation Time (min)	Platelet Concentration $\times 10^3/\mu L$	Leukocyte Concentration $\times 10^3/\mu L$	PDGF-AB pg/mL	TGF-β1 pg/mL	VEGF pg/mL
ACP	350	5	500	<1	3133–22,180	456–73,867	59–246,78
GPSIII	1100	15	273.6–1560	15–52	5900–65,000	2647–153,863	1304–1991
Cascade	1100/1450	6/15	600–2900	<1	6100–13,300	20–180	0–600
SmartPrep	1250/1050	14/7–10	800–2600	<1–20	123,100–293,500	22,400–132,000	/
Magellan	610/1240	4/6	600–1500	8–35	23,700–45,100	100–300	400–2000
JP2000	1000/800	6/8	850	26.1	93,500	1563	42,000
GLO	1800/1800	3/6	891	10	67,300	1329	39,000
KIOCERA	600/2000	7/5	1312	14	76,200	1508.2	44,000
Selphyl	525	15	88	0.3	12,200	384	28,200
MyCells	2054	7	800	4.9	72,200	1328	39,200
Dr Shin's System	1720	8	650	14.9	37,000	938	31,000

Data (cumulative range or average) were obtained from the following references: [24,35,36].

As expected, overall findings underlined that commercially buffy-coat-based systems (such as SmartPrep, GPS III and Magellan systems) yield higher concentrations of platelets and leukocytes compared to plasma-based systems (such as ACP and Cascade). Among buffy-coat systems, generally, GPS III preparations demonstrated the highest concentration of platelets and leukocytes [35].

Wide variations of centrifugal force and total centrifugation time among the different common commercial systems were described, respectively, ranging from about 350 to 2000 g, and from 5 to 20 min [35]. The majority of the systems use a dual-spin method; the first centrifugation usually has a lower speed compared to the second one [24,35].

Conflicting results were reported concerning optimal centrifugation rate to maximize platelet concentration, avoiding their activation or damage. Indeed, there is evidence underlining that increasing centrifugation force results in higher platelet concentration [37]. Conversely, other studies reported an inverse relationship between platelet yields and gravitational force [38,39]; furthermore, an elevated centrifugal speed could induce platelet activation [40].

Very recently, Croisè et al. [40] performed a literature review, aiming to check multiple studies focused on PRP protocol optimization. Fourteen included studies were commented upon, and each of them suggested different centrifugation procedures in terms of speed and duration time, number of centrifugations and, consequently, variable platelet concentration enrichments (from no enrichment to about 8.5 times more than peripheral blood). Overall these results underline that, to date, there is no consensus on the optimal centrifugation regimen to obtain a good-quality PRP, in terms of best platelet yields, avoiding structural and/or functional alterations and optimal relative concentration of blood components.

Recently, in order to obtain a standardized PRP formulation, Gato-Calvo et al. [41] developed a novel methodology, defining the optimal content of PRP, based on absolute platelet concentration. This approach allows us to obtain an end-product not influenced by the variability of the donor basal platelet counts, thus improving the reproducibility of PRP effects.

Another source of variability may derive from the material of blood-collection tubes. Some studies have demonstrated that PCs obtained by blood collected in glass or silica-coated tubes presented different buffy-coat morphology, fibrin architecture and platelet/leukocyte distribution in the PC matrix [42]. Furthermore, silica micro-particles may be released by tube walls during centrifugation procedures, entrapped in PC matrix, thus modifying platelet distribution in the end-product [43].

2.1.3. Activation Process

Activation triggers two responses during PRP preparation: the release of the bioactive molecules stored in platelet alpha-granules, and the matrix formation by fibrinogen cleavage [44]. Clot formation entraps released growth factors (GFs), thus enabling bioactive molecules to be delivered and confined at the injured target site.

Activation process may be induced by endogenous and exogenous factors. Among exogenous factors, the most common activators are thrombin, calcium chloride and a mixture of calcium chloride plus thrombin [14,34,45]. Endogenous activation relies on the exposure of native collagen or other coagulation factor (such as adenosine diphosphate-ADP, thrombospondin and platelet-activator factor), spontaneously inducing clot formation at injured site [45].

In general, thrombin triggers a rapid platelet aggregation and stimulates a fast release of GFs [14,34,46]. Calcium chloride and collagen sustain a slower long-term release [34,46,47]. Furthermore, some findings reported that collagen activation results in a lower amount of released GFs compared to thrombin and calcium chloride [47].

A very recent study compared the effects of three different activation factors, thrombin, collagen I and ADP, on PRP quality and on bone marrow stem cells' (BMSCs) functionality. Collagen I-PRP has been shown to induce the most rapid increasing of BMSC number compared to the rate observed with ADP- or thrombin-activated PRP. In addition, BMSC seeded in Collagen-I-activated PRP induced a significantly higher gene expression of osteogenic differentiation markers, osteocalcin and RUNX2, compared to thrombin and ADP. Thrombin induced a rapid and direct GF release, while collagen-I-activated PRP showed a sustained and slow GF release. The lowest total release was observed for ADP-activated PRP [32].

The different kinetic release is a crucial issue that might influence the availability of bioactive molecules, so affecting treatment outcome. Indeed, given GFs' short half-life (from minutes to hours), if they are not promptly used upon platelet release, their degradation may occur before additional receptors, that are involved in the repair process, become available on cell surfaces [34].

Photo-activation has been suggested as an alternative method to trigger platelet activation: a very recent paper [48] described in vitro characterization of platelet photo-activation (polychromatic light source, in the range near-infrared region), in comparison with resting platelets and calcium chloride mediated PRP activation. That study showed that photo-activation of PRP induced a significantly more prolonged release and higher amount of platelet-derived growth factor (PDGF), basic fibroblast growth factor (FGF), and transforming growth factor (TGF)-beta than PRP activated with calcium chloride. Future clinical studies should be performed to verify the potential of using the photo-activation approach in PRP formulation.

2.2. Platelet-Rich Fibrin

This type of PC essentially includes two categories of different preparations organized as a high-density fibrin solid form: leukocyte-poor or pure platelet-rich fibrin (P-PRF) and leukocyte- and platelet-rich fibrin (L-PRF) [25,49]

Concerning P-PRF preparation, there is only one formulation, commercially known as Fibrinet (Platelet Rich Fibrin Matrix-PRFM, Cascade medical, Wayne, NJ, USA,) [25,49]. P-PRF is obtained by a double-centrifugation method analogous to other PRP protocol, but it differs since the clotting phase is a dynamic process occurring during the second centrifugation, after adding $CaCl_2$ [25,49].

L-PRF is a leukocyte-rich product, and compared to PRP, L-PRF preparation is easier and lacks biochemical modifications (no exogenous activation or anticoagulant are required), and unlike PRP, PRF end-products are characteristically organized in tridimensional architecture [25,49].

L-PRF protocol was developed by Choukroun et al. [23] as an open-access technique, based on one-step centrifugation without anticoagulant and blood activators. L-PRF is considered to be a second- generation platelet concentrate [25,50]. Briefly, venous blood collected in glass tube without anticoagulants is centrifuged at low speed, and clot formation is immediately triggered. Three layers become evident after centrifugation: the red blood cells (RBCs) bottom layer, a PRF clot in the middle and the acellular plasma top layer [50].

This procedure allows to harvest almost all the platelets and more than 50% of the leukocytes from the peripheral blood [50]. L-PRF clot appears to be organized in a strong fibrin architecture and presents a specific tridimensional distribution of the platelets and leukocytes [50].

The original open-access experimental method has evolved into a regulated medical device system and is marketed with CE/FDA clearance (Intra-Lock, Boca-Raton, FL, USA). This system is the only certificated L-PRF system available on the market, and it uses the original protocol and devices [51]. This method shows a high efficiency in platelet and leukocyte collection and in leukocyte preservation [25].

Many variations of the original method were proposed, using different centrifuges and/or different protocols. These modifications result in modified-PRF product compared to the original L-PRF.

P-PRF procedure is more expensive and complex compared to L-PRF protocol. Furthermore, this latter procedure allows to simultaneously obtain a large number of end-products [25].

To the best of our knowledge only one paper [52] compared PRFM and PRF products, in terms of growth factor release. In this study, PRFM and PRF were obtained by "home-made" protocols and appeared to have a different kinetic release. PRFM presented an early robust boost of growth factors, while PRF release was more gradual and constant up to 23 days. On the contrary, Lucarelli et al. [53] has shown that Fibrinet PRFM releases elevated levels of growth factors (such as PDGF, TGFβ and

VEGF) in the first 24 h, whereas other growth factors, such as bone morphogenic protein (BMP)-2 and -7 were undetectable.

Conversely, L-PRF products sustained a large growth factor release for up to seven days [50]. Interestingly, BMP-2 was detected in L-PRF releasate strengthening the regenerative potential of this PC [51]. It is hypothesized that the presence of leukocytes may have a relevant impact on the amount and the pattern of the released growth factors, and a potential synergistic effect between leukocytes and platelets has been suggested [25,50,51].

Centrifuge characteristics and centrifugation protocols have been shown to impact fibrin architecture, cellular distribution and growth factor release. Therefore, various PRF preparations could be associated to different biological profile and clinical potential [51]. Up to now, the different PRF preparations are not clearly characterized, and further investigations on the effects of protocol modifications need to be provided.

3. Classification Systems

The heterogeneity of PC preparation methods can impact on the functional characteristics and on the potential therapeutic efficacy of the final products, giving each PC formulation unique properties. The majority of the studies do not provide a full characterization of the various PC composition, so a reliable comparison among studies still remains a challenging issue [54].

Several classification systems (Table 2) have been developed over the years in attempt to help comparison among studies and to foster standardization of PC preparation process. However currently, no consensus on classification systems has yet been achieved [54].

Table 2. Summary of classification systems for platelet concentrates (PCs).

Study	Classification	Parameters
Dohan Ehrenfest et al. (2009) [25] (2012, 2014) [49,55]	Pure PRP, Leukocyte-rich PRP; Pure PRF, Leukocyte-rich PRF	• Leukocyte content • Presence/absence of fibrin
DeLong et al. (2012) [34]	PAW (Platelet Activation, White blood cells)	• Platelet absolute number (from baseline to above 1250 × $10^3/\mu L$ • Activation method • White Blood Cells and neutrophil content (above/below baseline)
Mishra et al. (2012) [56]	Sports medicine classification of platelet rich plasma.	• Platelet concentration (< or ≥5 times baseline) • White Blood Cell presence/absence • Activation or no activation prior to application
Mautner et al. (2015) [57]	PLRA (Platelet Leukocyte Red blood cells and activation)	• Platelet count (absolute number/μL) • Leukocyte content (as positive/negative) • Percentage of neutrophils • Red Blood Cells contents (as positive/negative) • Activation (yes or no for exogenous activation)

Table 2. *Cont.*

Study	Classification	Parameters
Magalon et al. (2012) [58]	DEPA (Dose of platelet Efficiency, Purity and activation)	• Dose (platelet number × PRP volume) • Efficiency (proportion of platelet recovery) • Purity (proportion of platelet compared with Red Blood Cells and leukocytes) • Exogenous activation (yes/no)
Lana et al. (2017) [59]	MARSPILL (Method, Activation, Red blood cells, Spin, Platelets, Image guidance, Leukocytes and Light activation)	• Method (automated manner or manually) • Number of spins • Platelet concentration (Fold basal) • Leukocyte content (< or ≥15 times baseline) • Red Blood Cell content (< or >baseline) • Photo-activation (yes/no) • Image guidance (yes/no)
Harrison P (2018) [60]	ISTH (International Society on Thrombosis and Hemostasis) classification	• Activation • Platelet count ($<900 \times 10^3$ μL; $900–1700 \times 10^3$ μL; $>1700 \times 10^3$ μL) • Preparation method • Leukocyte contents (as positive/negative) • Red Blood Cells contents (as positive/negative)

4. Composition

4.1. Platelets

The human blood platelet normal concentration ranges from 150,000 to 400,000/μL [61]. There is no consensus on the optimal concentration of platelets in PCs.

Platelet concentration was compared for its healing effect, and different optimal levels were identified for different applications [14,34].

PRP platelet concentration greatly differs in PRP obtained by the various commercial systems.

Plasma-based PRP systems usually contain a platelet concentration between baseline and 3x baseline (less or equal to 750×10^3 platelets/μL), and they are defined as low-yielding devices (such as ACP, Cascade, Endoret and RegenPrep) [35]. On the other hand, buffy-coat-based systems yield platelet concentration above 3x, ranging from 4x to 6x (greater than 750×10^3platelets/μL to 1800×10^3 platelets/μL). These systems are classified as high-yielded devices that produce PRP (GPS III, SmartPrep and Magellan) [35].

In vitro, in vivo and clinical studies have demonstrated successful results for PRP formulations with both a moderate (2× and 3×) and high platelet concentrations (from 4× to 6×) [14]. In particular, an in vitro study evidenced that the best angiogenic effect of PRP was obtained with 1500×10^3 platelets/μL, thus underlining the role of platelet concentrations on the clinical application when the increased angiogenesis contributes to the healing process [14,62].

Platelet concentration greater than 6x ($>1800 \times 10^3$ platelets/μL) may be detrimental or have side effects [63]. In fact, an excessive platelet amount may lead to cellular apoptosis, downregulation and desensitization of growth factor receptors, resulting in a paradoxical inhibitory effect [34].

Another source of variation is the platelet-counting mode. Indeed, it has been reported that, to achieve accurate platelet count, proper sample preparation is required and manual mode in the hematology analyzer is recommended, because automatic mode, allowing the sample to settle, may underestimate the absolute platelet count [34,64].

4.2. Leukocytes

As previously stated, leukocyte content in PCs depends on PRP preparation procedures.

Plasma-based process reduced leukocyte count up to 22 times the baseline, almost eliminating this cellular fraction. Buffy-coat-based procedures actively concentrate leucocytes from threefold to fivefold the baseline [65]. Furthermore, different buffy-coat methods produce a PRP formulation with different proportions of neutrophils, lymphocytes and monocytes [65]. Indeed, it has recently been reported that different centrifugation regimens, in terms of spin numbers and speed, modified lymphocyte/granulocyte ratio in the final products [66].

The inclusion of leukocytes in PC preparations remains a widely debated concern, as both beneficial and detrimental effects have been suggested.

Deleterious effects are mainly ascribed to leukocyte capacity to release inflammatory cytokines and metallo-proteinases, which can promote pro-inflammatory and catabolic effects on targeted tissue [67–70]. Furthermore, the massive release of reactive oxygen species by neutrophils causes tissue damage, by inhibiting healing process [71,72].

On the other hand, potential beneficial effects rely on leukocyte's role in tissue healing, in regulating inflammatory process [73–75] and in antibacterial activity [76,77] that may switch the inflammatory process toward a regenerative phase.

These potential effects are suggested and corroborated by the following main evidence:

- The presence of leukocytes contributes to potentiate total amount of released GFs [35]. Indeed, several studies have reported a positive correlation between leukocyte count and GF concentration [35,78–80].
- Leukocytes have anti-nociceptive action by releasing anti-inflammatory cytokines (IL-4, IL-10 and IL-13) and opioid peptides (beta-endorphin, Met-enkephalin and dynorphin-A) [25,81].
- Circulating monocytes differentiate into macrophage once they migrate into connective tissue and may switch from M1 (pro-inflammatory) to M2 (anti-inflammatory) phenotype [82,83] in response to micro-environmental signals and stimuli (such as neutrophil-derived micro-vesicles [4,84]).
- M2 macrophages have several functions in tissue remodeling, promoting angiogenesis, cell proliferation and extracellular matrix deposition [83,85], and they may contribute to resolution of inflammation [4].
- Proteinases secreted by leukocytes are able to modulate the activity of secreted growth factors, converting inactive form to active one and contributing to matrix remodeling in tissue healing [71,75].
- Neutrophils are essential for killing bacteria and other microorganisms [86]. Since platelets also contribute to the antibacterial response [6–8], leukocytes may synergize with platelets and potentiate PRP antimicrobial effects.

Furthermore, growing evidence on the relevance of leucocyte–platelet interaction and of their relative proportions in PRP preparation has been reported [4,44,50,66,87]. Indeed, leucocyte–platelet interaction may promote biosynthesis of other factors that facilitate the resolution of inflammation, such as lipoxins that are potent anti-inflammatory proteins able to limit neutrophil activation, so promoting the resolution phase of the healing process [44,88,89].

In addition, the interrelationship between platelets, blood cellular components and fibrin may have a key role in proper platelet function and growth factor release [4,50,87], and the relative platelet/leukocyte and lymphocyte/granulocyte ratios might drive the balance between catabolic and anabolic factors [66].

Therefore, future research efforts should not focalize on the concentrations of single PC component but on the optimal relative combination of platelets, leukocytes, growth factors and fibrin within the final preparation for the different clinical application fields.

4.3. Red Blood Cells

Red Blood Cells (RBCs) can be damaged as a result of high shear force during blood collection or during inadequate centrifugation process, so causing hemolysis with the release of hemoglobin and its degradation products, hemin and iron. The presence of these hemolytic-related products lead to several deleterious effects, such as radical oxygen reactions, endothelial disfunction, vascular endothelium damage, pro-inflammation response and tissue injury [90].

RBC damage also causes the release of migration inhibitory factor (MIF), which has been recognized as a very strong inflammatory cytokine [90]. MIF concentration in whole blood is 1000-fold increased than in plasma. Since leukocytes and platelets have been shown to minimally contribute to MIF concentration, RBCs represent the major reservoir of this factor [91], which is also functionally active [91].

MIF plays a pathophysiological role in promoting and maintaining OA pain [92]. Furthermore, MIF levels in plasma and synovial fluid have been found to be positively correlated to disease severity in knee OA [93]. Blood-induced joint damage has been highlighted by various in vitro studies. In fact, blood exposure results in increased synoviocyte cell death and pro-inflammatory mediator production [94], induction of chondrocyte apoptosis and cartilage degradation [95–97].

On the other hand, effects of free heme may be inhibited by its degradation or by specific binding proteins. The heme–heme oxygenase (HO) system is formed after HO-mediated heme degradation. Growing evidence support the protective HO system activity and its effector molecules against oxidative and inflammatory responses and cell damage and suggest that the heme-HO system may represents a novel and important target in the control of wound healing [98–100].

Even if RBC content is reduced or absent in PC preparations, the detrimental effect of RBCs should be addressed for optimization of PC performance.

4.4. Growth Factors

GFs and protein are stored in the platelet alpha-granules and are released by activation of the platelets. Over 300 proteins were identified in the platelet releasate [101].

Multiple pieces of evidence have suggested that platelet-derived growth factor (PDGF), transforming growth factor beta (TGF-beta), vascular endothelial growth factor (VEGF), insulin-like growth factor (IGF) and epidermal growth factor (EGF) are the most crucial factors implicated in tissue repair [102]. PDGF, TGF-beta and VEGF appear to be the most investigated, and the concentration of these GFs is often considered as a marker of PC preparation quality [24,35,102].

In PRP preparations, approximately 70% of platelet growth factors are secreted within the first 10 min following activation, and almost 95% within the first hour [103,104]. Platelets may continue to produce small amounts of growth factors during the residual life span (8–10 days) [103,104]. Conversely, PRF presents a more intense, slow and constant long-term release, up to 5–7 days [50,105].

Together with platelets, leukocytes also contribute to the release of some growth factors, as highlighted by several studies that reported a positive correlation between the amounts of released GFs and the number of leukocytes [35,78,79].

Multiple comparative studies have investigated GF released by PRP obtained by various commercial separation systems. A large heterogeneity in the GF concentrations and kinetic release have been shown when comparing multiple PRP preparations obtained by different commercial separation systems [106].

A recently published review underlined that growth-factor concentrations reported by the different studies appeared to be hardly comparable, due to wide variations of these results, not only among the different systems but also when comparing the same separation systems among the different studies [35]. This variability may be essentially ascribed to two criticisms: the different commercial kits used for growth-factor dosage [35], and the incomplete removal of platelets and erythrocytes that may impact the results [107]. Due to these limitations, a comparison between studies appears to be barely reliable, not allowing consistent evidence-based results concerning growth factor content profile of different PRP preparations.

Furthermore, the great inter-individual variability of GF concentration needs to be taken into consideration [107,108]. A study performed on a large number of OA patients (n = 105) showed a wide individual variation of PRP growth factors, with a coefficient of variation ranging from 5.30 to 78.45. In particular, basic FGF and TGF-beta1 showed, respectively, the highest and the lowest variation [109].

Concerning PRF, different GF releases by different formulations have been shown. Comparing original L-PRF to modified-PRF formulations, conflicting results were reported. Kobayashi et al. [105] demonstrated that significantly higher GF levels were released by advanced-PRF (A-PRF) compared to original L-PRF. On the other hand, Dohan Ehrenfest et al. [51] reported a much stronger release of GFs from original L-PRF than from A-PRF membrane.

Nowadays, the literature data highlights that biological profiles in terms of content, amount and release kinetics associated to different PCs need to be further investigated, in order to better understand GF potentiality of the various PCs in clinical applications.

5. Clinical Efficacy

5.1. Osteoarthritis

Osteoarthritis (OA) is a debilitating osteo-articular disease, triggered by a trauma to the joint, and it is associated with a progressive erosion of articular cartilage, subchondral bone sclerosis, excessive stiffness and pain.

Numerous clinical trials and case series, carried out using PRP administration in patients with OA, supported PRP for the symptomatic effect, reduction of pain, improvement in the degenerative injuries and safety of administration, but they have not reached an univocal consensus.

5.1.1. Knee Osteoarthritis

Knee OA is a chronic disease of joints that is characterized by pain and progressive disabilities, usually developing as the sufferer ages [110]. The most common treatments, both non-pharmacological and pharmacological, show positive outcomes, but their effectiveness is not long-lasting. Thus, surgical knee replacement is often the last chance for the relief of symptoms [111,112]

One of the first PRP studies establishing the safety of intra-articular use of this autologous preparation dates back to 2008 [113] (Table 3).

Table 3. Evidence of PRP treatment in knee OA (reported by year of study and grouped by treatment).

Treatment/ Control	Reference	Patient Number	Main Results
PRGF/HA	Sanchez et al. [113] Sanchez et al. [114] Vaquerizo et al. [115]	60 176 96	Significantly higher rate of response to PRGF than HA treatment as concerning knee pain, stiffness and physical function scores, up to 24 and 48 weeks
	Raeissadat et al. [116]	69	No difference between PRGF and HA treatments in alleviating pain and improving function
PRGF	Wang-Saegusa et al. [117]	261	Improvement in function and QoL were described
PRGF (1 cycle)/PRGF (2 cycles)	Vaquerizo et al. [118]	48	PRGF 2 cycles showed improved stiffness and QoL, but not pain decrease
PRP	Kon et al. [119] Filardo et al. [120] Gobbi et al. [121] Halpern et al. [122] Gobbi et al. [123] Hassan et al. [124] Bottegoni et al. [125] Chen et al. [126] Huang et al. [127] Fawzy et al. [128] Taniguchi et al. [129] Burchard et al. [130] Socuoğlu et al. [131]	100 90 50 22 93 20 60 24 127 60 10 59 42	Significant improvement of knee pain, function and QoL during therapy; improvement has been described to last for a short (2 months) or a medium/long period (6–12 months) follow-up and subsequently worsen; however, the improvement remained higher than the basal condition; further improvement at 18 months can be obtained by yearly repetition of PRP injection; better results were obtained in younger patients, lower degree of cartilage degeneration and short disease duration; worse results were observed in over-80-years-old patients; PRP injection was associated with inflammation decrease and anti-ageing physiological function increase; improved symptoms and pain were not dependent on the cartilage damage degree, as determined by MRI
PRP/HA	Cerza et al. [132] Filardo et al. [133] Spakova et al. [134] Say et al. [135] Guler et al. [136] Raeissadat et al. [137] Montanez-Heredia et al. [138] Ahmad et al. [139] Louis et al. [140]	120 109 120 90 132 160 55 89 56	PRP compared with HA showed better clinical outcomes and QoL; clinical improvement was evident at 3–6 months and up to 12 months of follow-up; PRP treatment was effective in initial stages/low grade of knee OA but not in patients with grade III arthrosis; in middle-aged subjects with moderate OA, PRP and HA induced similar improvements
	Filardo et al. [141] Di Martino et al. [142]	192 192	Both treatments were effective in improving knee clinical scores. PRP did not demonstrate a clinical superiority compared with HA at any follow-up (up to 6 years, at least)
PRP/High MWHA /Low MW HA	Kon et al. [143]	150	PRP displayed greater and longer efficacy than HA, as concerning pain, symptom and function improvement; better outcomes were obtained in young and active subjects and lower degree of cartilage damage; worse results were achieved in older patients and more damaged cartilage; in older patients, effects similar to viscosupplementation were obtained
PRP/ PRP+HA/HA	Lana et al. [144]	105	PRP was effective in mild/moderate knee OA; PRP+HA displayed a greater pain and functional limitation decrease than HA alone at 1 year post-injection; increased function compared to PRP alone at 1 and 3 months
PRP/ PRP+HA/HA /normal saline	Yu et al. [145]	360	Combined PRP + HA treatment improved pain, stiffness and physical function compared with PRP or HA alone
PRP/HA /normal saline	Lin et al. [146]	87	Leukocyte-poor PRP provided functional improvement for at least 1 year in mild/moderate OA
PRP/HA/ ozone	Duymus et al. [147]	102	PRP was more effective than HA and ozone
PRP/ PRP+ozone	Dernek et al. [148]	80	Similar efficacy was demonstrated by PRP alone or PRP+ozone; PRP+ozone-treated patients experienced less post-injection pain and a faster recovery

Table 3. *Cont.*

Treatment/ Control	Reference	Patient Number	Main Results
PRP/HA /CS	Huang et al. [149]	120	Pain decrease was significant in all groups compared to baseline; PRP showed a better recovery in physical function and decreasing pain at 6, 9 and 12 months
MP+PRP/ PRP/MP	Camurcu et al. [150]	115	MP+PRP injection determined better clinical improvement compared to PRP and MP alone
PRP double spinning/ PRGF single spinning	Filardo et al. [151]	144	Both treatments displayed similar clinical improvement compared to the baseline and at the follow-up; more pain and swelling reaction were present in PRP patients; younger patients with a low degree of cartilage degeneration showed better results
PRP (6x) + maintenance dose (3x)	Hart et al. [152]	50	PRP decreased pain and improved QoL in low-degree cartilage degeneration. MRI did not confirm cartilage improvement
PRP (1x) /PRP (2x)/ normal saline	Patel et al. [153]	78	Improvement in clinical parameters in both PRP groups; no difference between 1 or 2 injections; results deteriorated after 6 months
PRP (1x) /PRP(3×)/ HA/normal saline	Gormeli et al. [154]	162	PRP and HA treatments are proposed for all OA stages; multiple PRP injections achieved better clinical results in early OA, but did not influence results in advanced OA
PRP large volume	Guillibert et al. [155]	57	Large PRP volume was associated with functional and pain improvement. No MRI difference was reported
PRP+exercise/ exercise	Rayegani et al. [156]	62	Short-term improvement of pain, stiffness and QoL in PRP-treated patients compared to the control group was shown
LP-PRP	Duif et al. [157]	58	Improvement of pain and knee function was reported
LP-PRP /saline	Smith et al. [158]	30	Scores in the LP-PRP group were better than in the saline group, starting at 2 weeks throughout
LP-PRP/ acetaminophen	Simental-Mendia et al. [159]	65	Better clinical outcomes following LP-PRP treatment were reported
LP-PRP /HA	Cole et al. [160]	99	Similar primary outcomes between HA and PRP were observed at any time point; patient-reported outcome favored PRP; mild OA and low BMI displayed better outcome.
LP-PRP/ HA/NSAID	Buendia-Lopez et al. [161]	106	PRP decreased pain and improved physical function; PRP displayed better results; no modification in cartilage MRI was observed
PRP/ normal saline	Huang et al. [162] Elik et al. [163]	366 60	PRP improved clinical symptoms, improved QoL, decreased joint inflammation and did not increase thickness of cartilage
PRP/SH	Li et al. [164] (Chinese)	30	Significant differences pre- and post-injection in both groups; PRP was better than SH at 6 months
PRP/CS	Forogh et al. [165]	41	Pain, ADL and QoL improvement in the PRP-treated group was greater than in the CS group
PRP/PRL	Rahimzadeh et al. [166]	42	Decreased pain and improved physical function and QoL were observed after both treatments; PRP was more effective
Photo-activated PRP/HA	Paterson et al. [167]	23	Feasibility and safety of PA-PRP treatment were demonstrated; PA-PRP improved pain, symptoms and function; no differences between PA-PRP and HA were observed
PRP+SVF from adipose tissue	Bansal et al. [168]	10	PRP+SVF decreased pain, particularly after 3 months
PRP+intra osseous	Sanchez et al. [169]	14	Knee-joint function improvement and pain decrease were observed in patients with severe OA
PRP+intra-osseous/PRP	Sanchez et al. [170]	60	Intraosseous +intra-articular PRP injections induced better clinical outcome
PRP+intra-osseous/PRP/HA	Su et al. [171]	86	Intra-articular +intraosseous PRP infiltrations were not superior at 2 months, but they were superior at 6 and 12 months

Afterward, different studies demonstrated the positive effects of PRGF/PRP injection, either when used alone or when compared to hyaluronic acid (HA) one, in the knee OA patients [114,115,117,119–140]. These PCs were reported not only to have an effect on clinical symptoms (by decreasing pain and improving function), but also on synovial fluid and protein amounts, as well as on cartilaginous degeneration.

However, a recent study reporting results of a follow-up up to six years does not confirm superiority of PRP [142].

The superiority of PRP was also established by comparison with normal saline (physiological control), as indicated by early improving WOMAC (the Western Ontario and McMaster Universities Osteoarthritis) scores, and maintained up to six months [153,158,163], but slightly decreased afterward, in agreement with the anti-inflammatory action supposed for PRP [172].

Similarly, in a trial including 366 young patients (18–30 years old), positive outcomes were reported after intra-lesional PRP administration [162]. In general, better results were obtained in young patients, with low body mass index [117,122].

PRP was reported as better in terms of clinical improvement compared to oral NSAID administration [161], as synergistic and protective, when added to methylprednisolone [150] and comparable to HA and corticosteroids after three months, superior to both the other treatments in the long-term [149].

Both PRP and HA have a biological origin and may be critical for tissue healing at the beginning of OA development. In in vitro studies, the combination of PRP with HA may display synergistic effects on fibroblast migration [173,174], thus suggesting a better effect of PRP–HA combination than PRP alone [175].

In agreement, a recent randomized clinical trial in mild/moderate knee OA reported better outcomes of the patients treated with PRP–HA combination when compared to PRP (up to three months) or to HA (up to 12 months) groups [144].

Furthermore, the synergy between combined PRP and HA treatment was further investigated and compared with each of them alone and with a placebo, via intra-articular injections in a total of 360 patients with knee osteoarthritis [145], demonstrating significantly reduced pain and decreased immune response, as well as PRP treatment compared with low and high molecular weight HA [143].

Even if clinical studies on PRP–HA combined therapy are limited and there are several peculiar aspects of HA alone (such as molecular weight), of the PRP–HA mix (such as ideal combination and dosage schedule), the preliminary data are worth of being deepened.

The PRP administration schedule in OA knee, widely reported with different numbers of injections, different time intervals and duration, represents a further aspect to be defined.

Patel [153], first compared the effect of one with two PRP injections and showed similarly improved WOMAC scores. A following double-blind placebo-controlled randomized trial demonstrated that the patient group that had undergone three PRP injections presented a better score than groups treated with a single dose of PRP or HA [154].

A clinical efficacy of PRP was also described when PRP was alternatively used at annual intervals or at the request of the patient when the effect ended [123]. Moreover, the administration in two phases foreseeing six doses at weekly intervals, and then a three month suspension and a maintenance dose (three injections at three-month intervals), presented interesting functional improvements [152].

A single administration of very pure PRP offered a significant clinical benefit as one injection of HA [140], and a similar improvement was obtained by a single administration of about 9 mL of PRP [155].

An enlarged delivery approach was also described, firstly for the treatment of severe OA [169] and more recently for the treatment of mild to moderate forms [170,171]. In these studies, the intra-articular injection of PRP was associated with concomitant intraosseous PRP injections into the subchondral bone, obtaining significant results.

A significant improvement of pain and functional scores, as well as decreases of the inflammatory response, were also obtained by the concomitant injection of PRP both intra-articular and in peri-meniscal soft tissue structures, thus widening the PRP effect on pes anserine tendons, bursa, medial collateral ligament and medial meniscus [176].

A systematic review on PRGF [177] reported the efficacy of PRGF in pain improvement, but also pointed out the limits of the included studies that prevented to perform a meta-analysis. The heterogeneity of the primary outcomes, PRGF and HA administration schedules, HA molecular weight, the small number of studies fulfilling the eligibility criteria and the lack of placebo treated group were the main drawbacks.

PRP was described as effective, alternative and superior to HA treatment for long-term improvement of joint function and pain in patient with knee osteoarthritis, mainly in early-moderate disease compared to advanced disease. The limits reported in a narrative review [178], in a recent meta-analyses [179,180] and in a systematic review [181] evidenced the variability of OA severity (K-L I-IV), as well as age, sex and BMI in patients treated in the different studies. In addition, main criticisms concerned the number of injections, optimal dosage of PRP, administration schedule, heterogeneous PRP preparations and formulation discrepancies, absence of published studies supporting specific protocols of injection and lack of indications on the appropriate regimen for different OA severity degrees. The limited size of pooled patients that can under-power the statistical analysis to reach a significant threshold of difference in outcome measures, and the lack of a placebo group shades the evidence of PRP effects.

5.1.2. Hip Osteoarthritis

Although various trials have faced up to the use of on PRP use for knee OA, few studies have focused on the treatment of hip OA with PRP. These studies are summarized in Table 4.

Table 4. Evidence of PRP treatment in hip osteoarthritis (reported by year of study and grouped by treatment).

Treatment/ Control	Reference	Patient Number	Main Results
PRP	Sanchez et al. [182]	40	Study supported safety, tolerability and efficacy of PRP treatment; PRR improved pain and function in mild/moderate OA, up to six months
	Singh et al. [183]	36	
PRP/HA	Battaglia et al. [184]	100	PRP showed immediate short-term improvement of pain and function; at 12 months, HA effect was more evident
	Di Sante et al. [185]	43	
	Doria et al. [186]	80	PRP did not display better results than HA in patients with moderate OA
PRP/PRP+ HA/HA	Dallari et al. [187]	111	PRP induced a significant stable pain relief, functional recovery and QoL improvement, up to 12 months; side effects were not observed; improvement was better than PRP+HA or HA alone
PRP+intra-osseous/ PRP	Fitz et al. [188]	Not reported	Intra-articular + intraosseous PRP infiltrations induced improvements at 6 months, but not in the long-term

A recent study [188] described the intraosseous infiltration of PRP for the treatment of hip osteoarthritis, in agreement with knee reported ones. Future studies are required to confirm the potential advantage of this new application of PRP.

Meta-analysis results of a randomized clinical trial that compared the effectiveness of PRP versus hyaluronic acid (HA) in hip OA underlined that PRP treatment was related to a significant reduction of VAS at two months. Both PRP and hyaluronic acid appeared to be comparable in terms of functional recovery [189].

The systematic review on the use of ultrasound-guided PRP injections in the treatment of hip osteoarthritis concluded that this route of administration appears to be well tolerated. Furthermore, though the level of evidence is relatively low, PRP treatment may lead to efficacious long-term and clinically significant reduction of pain and functional improvement [190].

Overall, intra-articular injection of PRP in hip OA patients has been demonstrated to be safe and have some efficacy in pain reduction and in functional improvement. When compared with HA, PRP showed to induce a better early pain relief; however, over 12 months, PRP and HA had comparable effects.

Future large-size trials that include a placebo group are needed. These studies should increase the level of evidence for the actual potential efficacy of PRP as an alternative conservative treatment to delay surgery in hip OA patients.

5.1.3. Ankle Osteoarthritis

Osteoarthritis of the ankle is less common than the previously described localization of OA. Data concerning the use of PRP in ankle OA are obtained by case series. Four injections of PRP at weekly intervals induced improvement of function, pain and patient satisfaction [191], and similar improvements in pain and function up to 24 weeks after treatment were obtained after the administration of three injections every two weeks [192].

The limited data show some benefit in short–medium time, demonstrate the safety of the therapy and can be considered to be an alternative to postpone the need for surgery, but the comparisons with other injectable controls are lacking; therefore, no definitive conclusion can be made about the benefit of PRP in ankle OA.

5.2. Tendinopathies

Tendon tissue is poorly vascularized, and this characteristic is responsible for the limited healing capacity and the lesion irreversibility resulting in tendinopathies, which frequently occur in athletes [193].

5.2.1. Achilles Tendinopathy

Achilles tendinopathy is a painful condition. Physical stress leads to tendon micro-trauma, and the inflammatory and degenerative responses that follow are responsible for local pain, swelling and stiffness [194]. Its treatment is difficult, and sufferers easily relapse due to the poor curative effects of the conservative treatment approach. The reason for PRP application lies in the tendency of the tendinopathy to became chronic after the use of nonsurgical approaches.

The outcomes after PRP administration are variable, and the main results are reported in Table 5.

Table 5. Evidence of PRP treatment in Achilles tendinopathy (reported by year of study and grouped by treatment).

Treatment/ Control	Reference	Patient number	Main results
PRP	Gaweda et al. [195]	14	Lasting improvement of the clinical symptoms and imaging results were obtained; improvement was maintained at least for two years from treatment; low complication rate was reported; US-guided tenotomy, followed by PRP treatment, was safe, effective and associated with US improvement; PRP led to tendon matrix healing; effective also in patients who failed to respond to traditional non operative techniques; retrospective study demonstrated that 78% of PRP-injected patients presented clinical improvement and averted surgical intervention at 6-month follow-up; response was less evident in old subjects
	Volpi et al. [196]	15	
	Finoff et al. [197]	41	
	Deans et al. [198]	26	
	Ferrero et al. [199]	30	
	Monto et al. [200]	30	
	Murawski et al. [201]	32	
	Guelfi et al. [202]	73	
	Salini et al. [203]	44	
	Owens et al. [204]	10	Moderate improvement in functional outcome was reported; MRI remained largely unchanged
PRP repeated	Filardo et al. [205]	27	Repeated PRP injections produced overall good outcomes, with stable results up to a midterm follow-up; prolonged symptomatology indicated a difficult return to sport
PRP/normal saline	de Jonge et al. [206]	54	PRP injection in addition to eccentric exercises did not result in clinical and/or ultra-sonographic improvement; tendon diameter increased
	Krogh et al. [207]	24	
LP-PRP/ LR-PRP	Hanisch et al. [208]	84	No significant differences were observed between patients treated with LR-PRP and LP-PRP
PRP+HA	Gentile et al. [209]	10	Treatment was efficacious for tissue healing and regeneration in post-surgical complications of Achilles tendon
PRP+ (ESWT)	Erroi et al. [210]	45	Both PRP and ESWT treatments were similarly efficacious and safe in physically active people
PRP/surgery+ PRP	Oloff et al. [211]	26	Both PRP alone or PRP+ surgical debridement improved clinical outcomes and MRI
PRP/eccentric loading	Kearney et al. [212]	20	No differences between PRP and eccentric loading program as concerning clinical effectiveness
PRP+ eccentric exercise/normal saline+ eccentric exercise	de Vos et al. [213]	54	Patients treated with PRP+ eccentric exercises did not present greater improvement in pain and activity; PRP did not increment tendon structure or modified neovascularization degree
	de Vos et al. [214]	54	
PRP/HVI steroid/ normal saline	Boesen et al. [215]	60	Both HVI steroid or PRP seemed efficacious in improving pain and activity and in decreasing tendon thickness and intra-tendinous vascularity

Case series for chronic Achilles tendinopathy [195–200,211], retrospective studies [201,204] and prospective studies [208–210,215] have described promising efficacy of PRP treatment with lasting improvements [205].

Other studies did not show a superiority of PRP injection over saline solution [206,207,213] and no differences between patients treated with leukocyte-rich or -poor PRP [208]

Evidence for the efficacy of PRP in Achilles tendinopathy is not in agreement, and despite the important clinical significance, a strong basis for the use of PRP for Achilles tendinopathy was not demonstrated by meta-analyses and a systematic review [216–219].

5.2.2. Lateral Epicondyle Tendinopathy

Lateral epicondyle tendinopathy, also known as "tennis elbow" is a common cause of pain and disability. Symptoms have been attributed to micro-trauma to extensor carpi radialis brevis tendon and the resulting angiofibroblastic tendinosis [220].

Different therapeutic approaches have been used, and steroid injections are considered to be the gold standard. Recently, PRP also became popular in treating this disease, with effects opposite to those of steroids, by stimulating the healing process and down-modulating inflammatory response.

The majority of the studies compared PRP efficacy with steroid one; however, other treatment comparisons have been reported (Table 6).

Table 6. Evidence of PRP treatment in lateral epicondyle tendinopathy (reported by year of study and grouped by treatment).

Treatment/Control	Reference	Patient Number	Main Results
PRP	Mishra et al. [221] Hachtman et al. [222] Brkljac et al. [223] Brkljac et al. [224]	140 31 34 31	PRP was successful in refractory forms and preventing the need for surgery; it was safe and improved function, with effects lasting five years after the initial injection
PRP/Autologous blood	Creaney et al. [225] Thanasas et al. [226]	150 28	PRP seemed to be an effective treatment, superior to autologous blood in short-term, but not in long-term, follow-up; PRP appeared useful in patients resistant to first-line physical therapy
	Raeissadat et al. [227]	75	Both PRP and autologous blood were effective methods; PRP effect was similar to autologous blood
PRP/normal saline	Montalvan et al. [228] Schöffl et al. [229]	50 50	PRP treatment was not more effective than saline until 6- and 12-month follow-up
LR-PRP/LP-PRP	Yerlikaya et al. [230]	90	Neither LR-PRP nor LP-PRP did not seem to affect pain and function in the short-term; leukocyte number was not associated with local inflammation post-injection
PRP/active control	Mishra et al. [231]	230	No differences between treatments were observed at 12 weeks; clinical improvements in PRP-treated patients were observed at 24 weeks
PRP/CS	Peerbooms et al. [232] Gosens et al. [233] Gautam et al [234] Khaliq et al. [235] Varshney et al. [236] Gupta et al. [237]	100 100 30 102 83 80	PRP reduced pain and significantly increased function, exceeding the effect of corticosteroid injection, up to 2 years of follow-up; PRP enabled lesion heling; CS induced a short-term relief at 6 weeks, but favored tendon degeneration
PRP/beta-methasone	Lebiedzinski et al. [238]	120	PRP allowed better results at 12 months; PRP therapeutic effect was long-lasting; betamethasone gave more rapid improvement
PRP/dexa-methasone	Palacio et al. [239]	60	Both treatments were similarly effective
PRP/methyl-prednisolone	Yadav et al. [240]	65	Both PRP and methyl-prednisolone were effective; PRP showed a more prolonged efficacy
PRP/bupivacaine	Behera et al. [241]	25	Leukocyte-poor PRP injection enabled good improvement in pain and function
PRP/laser therapy	Tonk et al. [242]	81	Better results were obtained following PRP injection on the long-term period; low-level laser therapy was better in the short-term
PRP/ESWT	Alessio-Mazzola et al. [243]	63	PRP injection showed a more rapid efficacy than ESWT
PRP/triamcinolone/normal saline	Seetharamaiah et al. [244]	80	Better pain relief were obtained following PRP injection over a short-term period
PRP/glucocorticoids/normal saline	Krogh et al. [245]	60	No treatment was superior to saline in regard to pain reduction; glucocorticoids had a short-term pain-relief effect and reduced both color Doppler activity and tendon thickness, compared with PRP and saline
PRP+dry needling/dry needling	Stenhouse et al. [246]	28	Additional PRP showed a trend to greater clinical improvement in the short-term; no difference between the two treatments was demonstrated at each follow-up
PRP+arthroscopic debridement	Merolla et al. [247]	101	Both PRP injections and arthroscopic debridement were efficacious in short-/medium-term; pain intensified at 2 years in PRP patients; arthroscopic administration favored pain and grip-strength improvement
PRP/US-guided percutaneous tenotomy	Boden et al. [248]	62	PRP and US-guided percutaneous tenotomy were both successful in improving pain, function and QoL

Initial results have been promising [221,222]. The first randomized controlled trials displayed PRP treatment improvements in function and pain, exceeding the effect of steroid injections up to one [232] and two [233] years

Following trials, comparing PRP treatment with saline [228,229,245], steroid [232–240,245] autologous whole blood [225–227] and bupivacaine [241] showed variable effectiveness in reducing pain and improving function.

Studies showing similar therapeutic effects between PRP and whole blood [225–227] suggest that circulating platelet concentrations are enough for obtaining recovery. However, the limited patient number and the absence of placebo arm make questionable these results.

As far as we know, the results of a multicenter randomized controlled IMPROVE trial are not yet available. The four-arms of lateral epicondylitis treatment will compare PRP, whole blood injection and tendon fenestration, each associated with physical therapy and sham superficial subcutaneous soft tissue injection, plus physical therapy. Expected results should significantly impact clinical practice [249].

Despite the heterogeneity of data, a seven-year retrospective study [250] and several meta-analyses, differing for inclusion criteria are available for evaluation the effectiveness of PRP in the treatment of lateral epicondylitis [251–255].

These reviews demonstrated short-term benefits for corticosteroids, but a long-term effectiveness for PRP in regard to improving functional capacity and alleviating pain. The critical factors identified mostly mirror those evidenced in other anatomical sites. Volume and number of administrations, various treatment combination, lack of standardization for PRP preparation and for exercise protocol, different measures for outcome evaluation and different follow-up times need deeper assessments.

5.2.3. Plantar Fasciopathy

Plantar fasciopathy (PF), also known as "plantar fasciitis", affects the proximal insertion of the plantar fascia in the os calcis, causing pain. Tissue thickening and degenerative structural changes are more common than inflammatory findings, so the "plantar fasciopathy" definition better identifies this disorder [256].

The fascia plays a role of primary importance in the transmission of body weight to the foot while walking and running. Plantar fasciitis is very common in athletes, but can also occur in overweight or obese subjects.

Corticosteroids, autologous blood injection and extracorporeal shock wave therapy (ESWT) represent treatment options that have been used with varying results.

At present, a uniform therapy for the management of Plantar fasciopathy is missing; therefore, many studies have considered PRP to be an intriguing alternative option to favor healing in the plantar fascia without significant risk [257] (Table 7).

Table 7. Evidence of PRP treatment in plantar fasciitis (reported by year of study and treatment type).

Treatment/ Control	Reference	Patient Number	Main Results
PRP	Ragab et al. [258] Kumar et al. [259] Martinelli et al. [260] O'Malley et al. [261] Wilson et al. [262]	25 44 14 23 24	PRP injection may have a reparative effect, leading to resolution of symptoms; findings indicated a role in the management of chronic intractable plantar fasciitis; QoL improved; PRP injection was safe; it cannot impair the biomechanical function of the foot; no side effects were reported
PRP/PPP	Malahias et al. [263]	36	PRP and PPP gave similar results; both treatments provided improvement at 3- and 6-month follow-up
PRP/normal saline	Johnson-Lynn et al. [264]	28	PRP and placebo gave similar improvement in symptoms
PRP/CS	Aksahin et al. [265] Tiwari et al. [266] Omar et al. [267] Shetty et al. [268] Jain et al. [269] Sherpy et al. [270] Vahdatpour et al. [271] Acosta-Olivo et al. [272] Jain et al. [273]	60 60 30 60 60 50 32 28 80	Both treatments were safe and effective in improving pain and function at 3 and 6 months; at 12 months, PRP was significantly more effective, making it better and more durable than CS injection; taking into consideration the potential complication of corticosteroid treatment, PRP injection seemed to be safer and had, at least, the same effectivity in the treatment
	Monto et al. [274] Say et al. [275] Peerbooms et al. [276]	40 50 115	PRP appeared more effective and durable than CS injection in improving pain and function for the treatment of chronic recalcitrant cases
PRP/ methyl-prednisolone	Jiménez-Pérez et al. [277]	40	PRP injection showed better, long-lasting clinical and imaging effects than methylprednisolone
PRP/CS/ normal saline	Mahindra et al. [278]	75	PRP was as effective as, or more effective than, corticosteroid injection at 3-months follow-up
	Shetty et al. [279]	90	PRP and corticosteroids showed superior results to placebo; long-term results and low reinjection and/or surgery rate make PRP more attractive than CS
PRP+ct/ESWT+ct	Chew et al. [280]	54	Either PRP or ESWT treatment resulted in modestly and similarly improved pain and functional scores, compared with conventional treatments alone, over a 6-month follow-up; PRP demonstrated greater improvements in plantar fascia thickness reduction

Table 7. *Cont.*

Treatment/ Control	Reference	Patient Number	Main Results
PRP/DP	Kim et al. [281]	21	Each treatment was effective in chronic recalcitrant cases; PRP also may lead to a better initial improvement compared with DP
PRP/KT	Gonnade et al. [282]	64	PRP injection of high platelet counts was more effective and long-lasting than phonophoresis with kinesiotaping; no adverse effects were reported
PRP/LDR	Gogna et al. [283]	40	PRP and LDR showed similar improvement in pain, functional activity and fascia thickness

Early cohort studies have described the positive effect of PRP injection on relieving pain [260] and improving function [259], as well as on tissue structure [258] for chronic plantar fasciopathy.

The most recent randomized controlled trials comparing PRP, corticosteroids and normal saline administration describe a similar or a superior effect of PRP compared to corticosteroid injection and normal saline in reducing pain and increasing functional scores for chronic plantar fasciopathy [278,279].

Numerous other studies obtained variable results by the comparison of PRP and corticosteroid treatments: PRP was described as being either able to favor early pain relief and functional improvement [267,275] with prolonged effects [266,269,271,274,278] or to be likewise effective up to six months [265,268,270,272,273,276].

Trials comparing PRP with other treatment options for plantar fasciopathy showed a better initial PRP response but similar effects at six months; when PRP was compared with prolotherapy [281], no significant differences compared to extracorporeal shockwave [280] or plasma injection [263], superior and long-lasting effects compared to KT [282].

The latest systematic reviews and meta-analyses comparing PRP to other therapeutic approaches supported the use of PRP for the lack of complications or side effects [284], but, above all, for its superiority to corticoids, especially in long-term pain relief [285,286]; however, small sample number, study heterogeneities, adverse events and the lack of recording PF recurrence following treatment may decrease reliability of outcome measures.

5.2.4. Patellar Tendinopathy

Inferior pole patellar tendinopathy, generally known as jumper's knee, is mostly common among athletes who engage in sports involving frequent jumping, such as volleyball and basketball, but it is also observed in people who do not carry out sporting activities [287]. The main evidence on PRP treatment in patellar tendinopathy is reported in Table 8.

Table 8. Evidence of PRP treatment in patellar tendinopathy (reported by year of study and treatment type).

Treatment/ Control	Reference	Patient Number	Main Results
PRP	Volpi et al. [196] Ferrero et al. [199] Mautner et al. [288] Crescibene et al. [289] Kaux et al.[290]	15 28 27 7 20	Significant pain and clinical improvement after 3 months, lasting results up to 2 years; MRI improvement in patellar tendon structure was observed
	Bowman et al. [291] Manfreda et al. [292]	3 17	Symptoms worsening were described following PRP treatment; poor benefit at 4 months
PRP (multiple)	Filardo et al. [293]	43	Multiple injections provided good clinical outcomes and stable results, up to medium-term follow-up; patients with bilateral disease and a long history of pain obtained poorer results
PRP(3x)	Charousset et al. [294]	28	Satisfactory results in athletes with chronic tendinopathy and faster return to previous sport practice were reported
PRP(2x)/PRP(1x)	Zayni et al. [295]	40	PRP (2x) determined better results than a single one injection
	Kaux et al. [296]	20	No differences between PRP (2x) and one injection were observed
PRP/Physiotherapy	Filardo et al. [297]	31	PRP treatment significantly improved knee function and quality of life
PRP/PRP+ previous treatment	Gosens et al. [298]	36	PRP provided a significant improvement; no differences were observed between groups
PRP/ESWT	Vetrano et al. [299]	46	PRP led to better midterm clinical results
PRP/HVI image guided saline	Abate et al. [300]	54	Association of both resulted in greater improvement and tendon repair
PRP+dry needling/dry needling	Dragoo et al. [301]	23	PRP provided faster recovery at 12 weeks; no clinical difference at the final 26-week follow-up was observed
LR-PRP/ LP-PRP/ normal saline	Scott et al. [302]	38	LR-PRP or LP-PRP were no more effective than saline for the improvement of symptoms

PRP has been administered in several studies as a biological therapy for patellar tendinopathy, improving pain and MRI tendon structure, and significantly increasing functional outcomes, with long-lasting stable results up to two years, thus improving quality of life [196,199,288–290,303],

Multiple injections were found to be better than a single one for patellar tendinopathy, either in case series [293,294] or in a randomized prospective study [295], but the effect of two repeated injections or one single injection was also reported to be similar [296].

PRP treatment displayed better results than ESWT [299] and physiotherapy [297]. Dry-needling used for PRP administration made recovery faster than dry-needling alone; however, beneficial effects on pain and function only lasted three months, without improvement in QoL [301]. Furthermore, no clinical differences were observed when PRP was administered following other inefficacious treatments [298], or among leukocyte-rich or -poor PRP and saline [302].

Not long ago, no randomized controlled quality studies supported the use of PRP over conservative therapies, except in therapy-resistant cases [293,304]. However, recently, a systematic review [305] and meta-analyses of randomized trials have recommended the use of PRP for the management of patellar tendinopathy, due to its superiority to other nonsurgical therapies [306], in long-term pain relief and improvement in knee function [307]. Even if eccentric exercises seem to be the strategic choice in the short-term, in complexes cases, multiple PRP injections can be considered to be an option [308]. Variability on follow-up length, or its absence, and number of interventions are the main limitations of these studies.

5.3. Muscle Injuries

The use of PRP for the treatment of muscle injuries raised significant interest in the last years.

Similar to tendon healing, the initial muscle healing begins with an inflammatory response, followed by proliferation and differentiation of cells and tissue remodeling.

Acute hamstring injury is one of the most common muscle injuries affecting athletic patients, causing a decline in competition performance [309,310].

Some studies described positive results after injection of PRP in patients with injured skeletal muscles, and no negative side effects were reported [311,312] (Table 9).

Table 9. Evidence of PRP treatment in muscle injuries (reported by year of study and treatment type).

Treatment/ Control	Reference	Patient Number	Main Results
PRP	Bernuzzi et al. [311] Zanon et al. [312]	53 25	PRP injection under US guide induced a complete muscle-function recovery; pain disappeared; PRP did not accelerate healing but showed excellent muscle repair and small scar
PRP/normal saline	Reurink et al. [313] Reurink et al. [314]	80 80	PRP injection did not demonstrate superiority to normal saline on short-term; no benefits were found up to 12 months in subjective, clinical, MRI measures, return to play and rate of re-injury
	Punduk et al. [315]	12	PRP administration improved inflammatory response induced by high-intensity muscle exercise
PRP/control	Martinez-Zapata et al. [316]	71	PRP did not significantly shorten the time of healing compared to the control group
PRP+conservative treatment/ conservative treatment	Bubnov et al. [317] Wetzel et al. [318]	30 15	PRP induced a better physical recovery, decreased pain and promoted faster regeneration than conventional conservative treatment
PRP/CS	Park et al. [319]	56	PRP injection induced more favorable response than CS one week after injection
PRP+rehabilitation/ rehabilitation	A Hamid et al. [320] Rossi et al. [321] Borrione et al. [322]	28 75 61	PRP injection+ rehabilitation program induced an earlier full recovery than rehabilitation alone; lower score of pain severity was observed in PRP group; PRP reduced time and costs to reach a complete functional recovery
	Guillodo et al. [323]	34	PRP injection+rehabilitation did not reduce the time to return to play
PRP+rehabilitation/PPP+ rehabilitation/ rehabilitation	Hamilton et al. [324]	90	PRP injection+rehabilitation did not show benefit on intensive standardized rehabilitation program alone; PRP induced a more rapid return to sport than PPP

Contrasting results were obtained when PRP was compared to saline [313–315].

In general, an earlier comeback to sports activity, together with lower scores of pain severity and no significant increase of the re-injury risk, has been observed in patients/athletes who have undergone PRP administration, combined with a rehabilitation program, compared to patients treated with a rehabilitation program alone [320–322].

In particular, as a randomized clinical trial, this study showed positive outcomes in the PRP group as concerning convalescence time and returning to play [321].

Despite some favorable results, these studies do not have enough statistical power to support evidence-based adoption of PRP administration for skeletal muscle injury in clinical practice, as recently widely debated [325,326]. In general, current clinical evidence are conflicting, and univocal findings on the efficacy of PRP injections in the treatment of muscle injuries have not been achieved. Therefore, further human studies are strongly required to assess and validate the effectiveness of PRP for skeletal muscle regenerative purposes.

Platelet growth factors, specifically myostatin and TGF-β1, have been shown to have harmful effects to muscle regeneration. Indeed, TGF-beta1 is involved in the regulation of the level of fibrosis during muscle-injury repair, which is an important link in the complete restoration of muscle function [327]. An vitro study [328] demonstrated that platelet-poor plasma (PPP) or PRP with a second spin to remove the platelets induced differentiation of myoblasts into muscle cells.

However, since experimental evidence has not received a large consensus [329,330], further studies are needed to define the exact PPP-growth-factor content, its effect on myogenic precursors and its role on skeletal muscle regeneration. In addition, human clinical trials will be required to further explore the potential beneficial effects of muscle injuries treated with PPP.

These overall findings underline that none of the therapeutic options so far adopted have led to reliable results [325,326]. Even if skeletal muscle tissue exhibits an intrinsic remarkable regenerative potentiality in response to injury, in the case of extended damage, a dysregulated activity of different muscle interstitial cells occurs, resulting in aberration of tissue repair and maladaptive fibrotic scar or adipose tissue infiltration [331]. In this context, the morpho-functional recovery of injured skeletal muscle still remains a scientific challenge, and the identification of strategies that efficaciously improve the endogenous skeletal muscle regenerative mechanisms represents an unmet need.

6. Conclusions and Future Perspectives

PC use has gained popularity for the treatment of musculoskeletal diseases, even if conflicting results have been reported concerning clinical efficacy. Inconsistencies of clinical results rely on the huge heterogeneity of PC preparations, mainly ascribed to individual characteristics, different preparation protocols and variability in composition, as well as on different methodological limits of the protocols adopted in the clinical studies that have been previously underlined.

In addition to the different critical aspects already considered, the indistinct employment of words to refer to fresh, frozen/thawed or activated preparations increases confusion. Therefore, also a simple aspect such as a classification nomenclature comprehensive of all PCs, with the same characteristics allowing an overall clinical outcome comparison, could contribute to define the clinical use and improve our knowledge of PRP.

Besides being a paramount component of PRP, platelets have been proposed as carriers of pharmacological or biological molecules [332]; therefore, "future" PRP could be implemented with suitable molecules favoring specific biological functions.

The possibility of encapsulating PRP with a combination of HA, gelatin and biodegradable scaffolds displayed interesting results in in vitro studies of bone regeneration [333], and a new delivery system linking fibrinogen with high molecular weight HA (RegenoGel™) (merging the respective regenerative/wound healing properties and viscoelastic characteristics) showed positive outcomes in mild/severe osteoarthritis. In addition, this system can be used as a carrier for microRNA or inhibitory molecules (ADAMTs), allowing the preparation of specifically targeted custom-made devices [334,335].

Encouraging in vitro and in animal model studies has demonstrated that PRP combined with different biomaterials prolonged and improved growth factor release [336]; however, the possibility to translate these engineered biomaterials in the clinical practice to develop novel therapeutic strategies remains a future perspective.

Author Contributions: The authors similarly contributed to preparation of the paper. All authors have read and agreed to the published version of the manuscript.

Abbreviations

ACD	acid citrate dextrose
ADL	activities of daily living
ADP	adenosine diphosphate
A-PRF	advanced-platelet-rich fibrin
ACP	autologous conditioned plasma
BMSCs	bone marrow stem cells
BMP	bone morphogenic protein
CTAD	citrate-theophylline-adenosine-dipyridamole
ct	conventional treatment
CS	Corticosteroid

DP	dextrose prolotherapy
DEPA	dose of platelet efficiency, purity and activation
EGF	epidermal growth factor
EDTA	ethylene diamine tetra-acetic acid
ESWT	extracorporeal shock wave therapy
FG	fibroblast growth factor
GFs	growth factors
HO	heme oxygenase
HS	heparin sodium
HVI	high volume injection
HA	hyaluronic acid
IGF	insulin-like growth factor
ISTH	International Society on Thrombosis and Haemostasis
KT	kinesio therapy
KOOS	knee injury and osteoarthritis outcome score
LE	lateral epicondyle
L-PRF	leukocyte- and platelet-rich fibrin
LP-PRP	leukocyte-poor PRP
L-PRP	leukocyte-rich PRP
LR-PRP	leukocyte-rich PRP
MRI	magnetic resonance imaging
MSC	mesenchimal stem cells
MARSPILL	method, activation, red blood cells, spin, platelets, image guidance, leukocytes and light activation
MIF	migration inhibitory factor
OA	osteoarthritis
PAW	photoactivated
PDWHF	platelet-derived wound healing factors or formula-
PF	plantar fasciopathy
PRFM	platelet-rich fibrin matrix
PRGF	plasma rich in growth factors
PAW	platelet activation, white blood cells
PCs	platelet concentrates
PLRA	platelet leukocyte red blood cells and activation
PRP	platelet rich-plasma
PDGF	platelet-derived growth factor
PPP	platelet-poor plasma
PRF	platelet-rich fibrin
P-PRF	pure platelet-rich fibrin
QoL	quality of life
RBCs	red blood cells
SC	sodium citrate
t-PRP	temperature controlled PRP
TGF	transforming growth factor
US	ultra-sound
VEGF	vascular endothelial growth factor
VAS	visual analogue scale
WOMAC	the Western Ontario and McMaster Universities Osteoarthritis Index

References

1. Le, A.D.K.; Enweze, L.; DeBaun, M.R.; Dragoo, J.L. Current Clinical Recommendations for Use of Platelet-Rich Plasma. *Curr. Rev. Musculoskelet. Med.* **2018**, *11*, 624–634. [CrossRef] [PubMed]
2. Dhillon, M.S.; Behera, P.; Patel, S.; Shetty, V. Orthobiologics and platelet rich plasma. *Indian J. Orthop.* **2014**, *48*, 1–9. [CrossRef] [PubMed]

3. O'Connell, B.; Wragg, N.M.; Wilson, S.L. The use of PRP injections in the management of knee osteoarthritis. *Cell Tissue Res.* **2019**, *376*, 143–152. [CrossRef] [PubMed]
4. Anitua, E.; Nurden, P.; Prado, R.; Nurden, A.T.; Padilla, S. Autologous fibrin scaffolds: When platelet- and plasma-derived biomolecules meet fibrin. *Biomaterials* **2019**, *192*, 440–460. [CrossRef] [PubMed]
5. Boswell, S.G.; Cole, B.J.; Sundman, E.A.; Karas, V.; Fortier, L.A. Platelet-rich plasma: A milieu of bioactive factors. *Arthroscopy* **2012**, *28*, 429–439. [CrossRef] [PubMed]
6. Deppermann, C.; Kubes, P. Start a fire, kill the bug: The role of platelets in inflammation and infection. *Innate Immun.* **2018**, *24*, 335–348. [CrossRef]
7. Pham, T.A.V.; Tran, T.T.P.; Luong, N.T.M. Antimicrobial Effect of Platelet-Rich Plasma against Porphyromonas gingivalis. *Int. J. Dent.* **2019**, *2019*, 7329103. [CrossRef]
8. Mariani, E.; Filardo, G.; Canella, V.; Berlingeri, A.; Bielli, A.; Cattini, L.; Landini, M.P.; Kon, E.; Marcacci, M.; Facchini, A. Platelet-rich plasma affects bacterial growth in vitro. *Cytotherapy* **2014**, *16*, 1294–1304. [CrossRef]
9. Rashid, H.; Kwoh, C.K. Should Platelet-Rich Plasma or Stem Cell Therapy Be Used to Treat Osteoarthritis? *Rheum. Dis. Clin. N. Am.* **2019**, *45*, 417–438. [CrossRef]
10. Qian, Y.; Han, Q.; Chen, W.; Song, J.; Zhao, X.; Ouyang, Y.; Yuan, W.; Fan, C. Platelet-Rich Plasma Derived Growth Factors Contribute to Stem Cell Differentiation in Musculoskeletal Regeneration. *Front. Chem.* **2017**, *5*, 89. [CrossRef]
11. Mazzocca, A.D.; McCarthy, M.B.; Chowaniec, D.M.; Dugdale, E.M.; Hansen, D.; Cote, M.P.; Bradley, J.P.; Romeo, A.A.; Arciero, R.A.; Beitzel, K. The positive effects of different platelet-rich plasma methods on human muscle, bone, and tendon cells. *Am. J. Sports Med.* **2012**, *40*, 1742–1749. [CrossRef] [PubMed]
12. Lim, W.; Park, S.H.; Kim, B.; Kang, S.W.; Lee, J.W.; Moon, Y.L. Relationship of cytokine levels and clinical effect on platelet-rich plasma-treated lateral epicondylitis. *J. Orthop. Res.* **2018**, *36*, 913–920. [CrossRef] [PubMed]
13. Weibrich, G.; Kleis, W.K.; Hafner, G.; Hitzler, W.E. Growth factor levels in platelet-rich plasma and correlations with donor age, sex, and platelet count. *J. Craniomaxillofac. Surg.* **2002**, *30*, 97–102. [CrossRef] [PubMed]
14. Lansdown, D.A.; Fortier, L.A. Platelet-Rich Plasma: Formulations, Preparations, Constituents, and Their Effects. *Oper. Tech. Sports Med.* **2017**, *25*, 7–12. [CrossRef]
15. Montagnana, M.; Salvagno, G.L.; Lippi, G. Circadian variation within hemostasis: An underrecognized link between biology and disease? *Semin. Thromb. Hemost.* **2009**, *35*, 23–33. [CrossRef]
16. Mannava, S.; Whitney, K.E.; Kennedy, M.I.; King, J.; Dornan, G.J.; Klett, K.; Chahla, J.; Evans, T.A.; Huard, J.; LaPrade, R.F. The Influence of Naproxen on Biological Factors in Leukocyte-Rich Platelet-Rich Plasma: A Prospective Comparative Study. *Arthroscopy* **2019**, *35*, 201–210. [CrossRef]
17. Kingsley, C.S. Blood coagulation; evidence of an antagonist to factor VI in platelet-rich human plasma. *Nature* **1954**, *173*, 723–724. [CrossRef]
18. Matras, H. Effect of various fibrin preparations on reimplantations in the rat skin. *Osterreichische Z. Stomatol.* **1970**, *67*, 338–359.
19. Rosenthal, A.R.; Egbert, P.R.; Harbury, C.; Hopkins, J.L.; Rubenstein, E. Use of platelet-fibrinogen-thrombin mixture to seal experimental penetrating corneal wounds. *Albrecht Graefes Archiv Klin. Exp. Ophthalmol.* **1978**, *207*, 111–115. [CrossRef]
20. Knighton, D.R.; Ciresi, K.F.; Fiegel, V.D.; Austin, L.L.; Butler, E.L. Classification and treatment of chronic nonhealing wounds. Successful treatment with autologous platelet-derived wound healing factors (PDWHF). *Ann. Surg.* **1986**, *204*, 322–330. [CrossRef]
21. Whitman, D.H.; Berry, R.L.; Green, D.M. Platelet gel: An autologous alternative to fibrin glue with applications in oral and maxillofacial surgery. *J. Oral Maxillofac. Surg.* **1997**, *55*, 1294–1299. [CrossRef]
22. Marx, R.E.; Carlson, E.R.; Eichstaedt, R.M.; Schimmele, S.R.; Strauss, J.E.; Georgeff, K.R. Platelet-rich plasma: Growth factor enhancement for bone grafts. *Oral Surg. Oral Med. Oral Pathol. Oral Radiol. Endodontol.* **1998**, *85*, 638–646. [CrossRef]
23. Choukroun, J.; Adda, F.; Schoeffler, C.; Vervelle, A. Une opportunite' en paro-implantologie: Le PRF. *Implantodontie* **2001**, *42*, 55–62.
24. Amin, I.; Gellhorn, A.C. Platelet-Rich Plasma Use in Musculoskeletal Disorders: Are the Factors Important in Standardization Well Understood? *Phys. Med. Rehabil. Clin. N. Am.* **2019**, *30*, 439–449. [CrossRef]

25. Dohan Ehrenfest, D.M.; Rasmusson, L.; Albrektsson, T. Classification of platelet concentrates: From pure platelet-rich plasma (P-PRP) to leucocyte- and platelet-rich fibrin (L-PRF). *Trends Biotechnol.* **2009**, *27*, 158–167. [CrossRef]
26. Yung, Y.L.; Fu, S.C.; Cheuk, Y.C.; Qin, L.; Ong, M.T.; Chan, K.M.; Yung, P.S. Optimisation of platelet concentrates therapy: Composition, localisation, and duration of action. *Asia Pac. J. Sports Med. Arthrosc. Rehabil. Technol.* **2017**, *7*, 27–36. [CrossRef]
27. Anitua, E.; Troya, M.; Zalduendo, M.; Orive, G. Personalized plasma-based medicine to treat age-related diseases. *Mater. Sci. Eng. C Mater. Biol. Appl.* **2017**, *74*, 459–464. [CrossRef]
28. Dhillon, R.S.; Schwarz, E.M.; Maloney, M.D. Platelet-rich plasma therapy-future or trend? *Arthritis Res. Ther.* **2012**, *14*, 219. [CrossRef]
29. Do Amaral, R.J.; da Silva, N.P.; Haddad, N.F.; Lopes, L.S.; Ferreira, F.D.; Filho, R.B.; Cappelletti, P.A.; de Mello, W.; Cordeiro-Spinetti, E.; Balduino, A. Platelet-Rich Plasma Obtained with Different Anticoagulants and Their Effect on Platelet Numbers and Mesenchymal Stromal Cells Behavior In Vitro. *Stem Cells Int.* **2016**, *2016*, 7414036. [CrossRef]
30. Lei, H.; Gui, L.; Xiao, R. The effect of anticoagulants on the quality and biological efficacy of platelet-rich plasma. *Clin. Biochem.* **2009**, *42*, 1452–1460. [CrossRef]
31. Gonzalez, J.C.; Lopez, C.; Carmona, J.U. Implications of anticoagulants and gender on cell counts and growth factor concentration in platelet-rich plasma and platelet-rich gel supernatants from rabbits. *Vet. Comp. Orthop. Traumatol.* **2016**, *29*, 115–124. [CrossRef] [PubMed]
32. Zhang, N.; Wang, K.; Li, Z.; Luo, T. Comparative study of different anticoagulants and coagulants in the evaluation of clinical application of platelet-rich plasma (PRP) standardization. *Cell Tissue Bank.* **2019**, *20*, 61–75. [CrossRef] [PubMed]
33. Du, L.; Miao, Y.; Li, X.; Shi, P.; Hu, Z. A Novel and Convenient Method for the Preparation and Activation of PRP without Any Additives: Temperature Controlled PRP. *Biomed. Res. Int.* **2018**, *2018*, 1761865. [CrossRef] [PubMed]
34. DeLong, J.M.; Russell, R.P.; Mazzocca, A.D. Platelet-rich plasma: The PAW classification system. *Arthroscopy* **2012**, *28*, 998–1009. [CrossRef]
35. Oudelaar, B.W.; Peerbooms, J.C.; Huis In 't Veld, R.; Vochteloo, A.J.H. Concentrations of Blood Components in Commercial Platelet-Rich Plasma Separation Systems: A Review of the Literature. *Am. J. Sports Med.* **2019**, *47*, 479–487. [CrossRef] [PubMed]
36. Weibrich, G.; Kleis, W.K.; Buch, R.; Hitzler, W.E.; Hafner, G. The Harvest Smart PRePTM system versus the Friadent-Schutze platelet-rich plasma kit. *Clin. Oral Implant. Res.* **2003**, *14*, 233–239. [CrossRef]
37. Fukaya, M.; Ito, A. A New Economic Method for Preparing Platelet-rich Plasma. *Plast. Reconstr. Surg. Glob. Open* **2014**, *2*, e162. [CrossRef]
38. Amable, P.R.; Carias, R.B.; Teixeira, M.V.; da Cruz Pacheco, I.; Correa do Amaral, R.J.; Granjeiro, J.M.; Borojevic, R. Platelet-rich plasma preparation for regenerative medicine: Optimization and quantification of cytokines and growth factors. *Stem Cell Res. Ther.* **2013**, *4*, 67. [CrossRef]
39. Yin, W.; Xu, H.; Sheng, J.; Zhu, Z.; Jin, D.; Hsu, P.; Xie, X.; Zhang, C. Optimization of pure platelet-rich plasma preparation: A comparative study of pure platelet-rich plasma obtained using different centrifugal conditions in a single-donor model. *Exp. Ther. Med.* **2017**, *14*, 2060–2070. [CrossRef]
40. Croise, B.; Pare, A.; Joly, A.; Louisy, A.; Laure, B.; Goga, D. Optimized centrifugation preparation of the platelet rich plasma: Literature review. *J. Stomatol. Oral Maxillofac. Surg.* **2019**. [CrossRef]
41. Gato-Calvo, L.; Hermida-Gomez, T.; Romero, C.R.; Burguera, E.F.; Blanco, F.J. Anti-Inflammatory Effects of Novel Standardized Platelet Rich Plasma Releasates on Knee Osteoarthritic Chondrocytes and Cartilage in vitro. *Curr. Pharm. Biotechnol.* **2019**, *20*, 920–933. [CrossRef] [PubMed]
42. Bonazza, V.; Borsani, E.; Buffoli, B.; Castrezzati, S.; Rezzani, R.; Rodella, L.F. How the different material and shape of the blood collection tube influences the Concentrated Growth Factors production. *Microsc. Res. Tech.* **2016**, *79*, 1173–1178. [CrossRef] [PubMed]
43. Tsujino, T.; Masuki, H.; Nakamura, M.; Isobe, K.; Kawabata, H.; Aizawa, H.; Watanabe, T.; Kitamura, Y.; Okudera, H.; Okuda, K.; et al. Striking Differences in Platelet Distribution between Advanced-Platelet-Rich Fibrin and Concentrated Growth Factors: Effects of Silica-Containing Plastic Tubes. *J. Funct. Biomater.* **2019**, *10*. [CrossRef] [PubMed]

44. Parrish, W.R.; Roides, B. Physiology of Blood Components in Wound Healing: An Appreciation of Cellular Co-Operativity in Platelet Rich Plasma Action. *J. Exerc. Sports Orthop.* **2017**, *4*, 1–14. [CrossRef]
45. Davis, V.L.; Abukabda, A.B.; Radio, N.M.; Witt-Enderby, P.A.; Clafshenkel, W.P.; Cairone, J.V.; Rutkowski, J.L. Platelet-rich preparations to improve healing. Part II: Platelet activation and enrichment, leukocyte inclusion, and other selection criteria. *J. Oral Implantol.* **2014**, *40*, 511–521. [CrossRef]
46. Harrison, S.; Vavken, P.; Kevy, S.; Jacobson, M.; Zurakowski, D.; Murray, M.M. Platelet activation by collagen provides sustained release of anabolic cytokines. *Am. J. Sports Med.* **2011**, *39*, 729–734. [CrossRef]
47. Cavallo, C.; Roffi, A.; Grigolo, B.; Mariani, E.; Pratelli, L.; Merli, G.; Kon, E.; Marcacci, M.; Filardo, G. Platelet-Rich Plasma: The Choice of Activation Method Affects the Release of Bioactive Molecules. *Biomed. Res. Int.* **2016**, *2016*, 6591717. [CrossRef]
48. Irmak, G.; Demirtas, T.T.; Gumusderelioglu, M. Sustained Release of Growth Factors from Photoactivated Platelet Rich Plasma (PRP). *Eur. J. Pharm. Biopharm.* **2019**. [CrossRef]
49. Dohan Ehrenfest, D.M.; Andia, I.; Zumstein, M.A.; Zhang, C.Q.; Pinto, N.R.; Bielecki, T. Classification of platelet concentrates (Platelet-Rich Plasma-PRP, Platelet-Rich Fibrin-PRF) for topical and infiltrative use in orthopedic and sports medicine: Current consensus, clinical implications and perspectives. *Muscles Ligaments Tendons J.* **2014**, *4*, 3–9. [CrossRef]
50. Dohan Ehrenfest, D.M.; Bielecki, T.; Jimbo, R.; Barbe, G.; Del Corso, M.; Inchingolo, F.; Sammartino, G. Do the fibrin architecture and leukocyte content influence the growth factor release of platelet concentrates? An evidence-based answer comparing a pure platelet-rich plasma (P-PRP) gel and a leukocyte- and platelet-rich fibrin (L-PRF). *Curr. Pharm. Biotechnol.* **2012**, *13*, 1145–1152. [CrossRef]
51. Dohan Ehrenfest, D.M.; Pinto, N.R.; Pereda, A.; Jimenez, P.; Corso, M.D.; Kang, B.S.; Nally, M.; Lanata, N.; Wang, H.L.; Quirynen, M. The impact of the centrifuge characteristics and centrifugation protocols on the cells, growth factors, and fibrin architecture of a leukocyte- and platelet-rich fibrin (L-PRF) clot and membrane. *Platelets* **2018**, *29*, 171–184. [CrossRef] [PubMed]
52. Chatterjee, A.; Debnath, K. Comparative evaluation of growth factors from platelet concentrates: An in vitro study. *J. Indian Soc. Periodontol.* **2019**, *23*, 322–328. [CrossRef] [PubMed]
53. Lucarelli, E.; Beretta, R.; Dozza, B.; Tazzari, P.L.; O'Connel, S.M.; Ricci, F.; Pierini, M.; Squarzoni, S.; Pagliaro, P.P.; Oprita, E.I.; et al. A recently developed bifacial platelet-rich fibrin matrix. *Eur. Cell Mater.* **2010**, *20*, 13–23. [CrossRef] [PubMed]
54. Rossi, L.A.; Murray, I.R.; Chu, C.R.; Muschler, G.F.; Rodeo, S.A.; Piuzzi, N.S. Classification systems for platelet-rich plasma. *Bone Jt. J.* **2019**, *101*, 891–896. [CrossRef] [PubMed]
55. Dohan Ehrenfest, D.M.; Bielecki, T.; Mishra, A.; Borzini, P.; Inchingolo, F.; Sammartino, G.; Rasmusson, L.; Everts, P.A. In search of a consensus terminology in the field of platelet concentrates for surgical use: Platelet-rich plasma (PRP), platelet-rich fibrin (PRF), fibrin gel polymerization and leukocytes. *Curr. Pharm. Biotechnol.* **2012**, *13*, 1131–1137. [CrossRef] [PubMed]
56. Mishra, A.; Harmon, K.; Woodall, J.; Vieira, A. Sports medicine applications of platelet rich plasma. *Curr. Pharm. Biotechnol.* **2012**, *13*, 1185–1195. [CrossRef]
57. Mautner, K.; Malanga, G.A.; Smith, J.; Shiple, B.; Ibrahim, V.; Sampson, S.; Bowen, J.E. A call for a standard classification system for future biologic research: The rationale for new PRP nomenclature. *PM R* **2015**, *7*, S53–S59. [CrossRef]
58. Magalon, J.; Chateau, A.L.; Bertrand, B.; Louis, M.L.; Silvestre, A.; Giraudo, L.; Veran, J.; Sabatier, F. DEPA classification: A proposal for standardising PRP use and a retrospective application of available devices. *BMJ Open Sport Exerc. Med.* **2016**, *2*, e000060. [CrossRef]
59. Lana, J.; Purita, J.; Paulus, C.; Huber, S.C.; Rodrigues, B.L.; Rodrigues, A.A.; Santana, M.H.; Madureira, J.L., Jr.; Malheiros Luzo, A.C.; Belangero, W.D.; et al. Contributions for classification of platelet rich plasma-proposal of a new classification: MARSPILL. *Regen. Med.* **2017**, *12*, 565–574. [CrossRef]
60. Harrison, P.; Subcommittee on Platelet, P. The use of platelets in regenerative medicine and proposal for a new classification system: Guidance from the SSC of the ISTH. *J. Thromb. Haemost.* **2018**, *16*, 1895–1900. [CrossRef]
61. Grozovsky, R.; Giannini, S.; Falet, H.; Hoffmeister, K.M. Regulating billions of blood platelets: Glycans and beyond. *Blood* **2015**, *126*, 1877–1884. [CrossRef] [PubMed]

62. Giusti, I.; Rughetti, A.; D'Ascenzo, S.; Millimaggi, D.; Pavan, A.; Dell'Orso, L.; Dolo, V. Identification of an optimal concentration of platelet gel for promoting angiogenesis in human endothelial cells. *Transfusion* **2009**, *49*, 771–778. [CrossRef] [PubMed]
63. Weibrich, G.; Hansen, T.; Kleis, W.; Buch, R.; Hitzler, W.E. Effect of platelet concentration in platelet-rich plasma on peri-implant bone regeneration. *Bone* **2004**, *34*, 665–671. [CrossRef] [PubMed]
64. Woodell-May, J.E.; Ridderman, D.N.; Swift, M.J.; Higgins, J. Producing accurate platelet counts for platelet rich plasma: Validation of a hematology analyzer and preparation techniques for counting. *J. Craniofacial Surg.* **2005**, *16*, 749–756. [CrossRef] [PubMed]
65. Fitzpatrick, J.; Bulsara, M.K.; McCrory, P.R.; Richardson, M.D.; Zheng, M.H. Analysis of Platelet-Rich Plasma Extraction: Variations in Platelet and Blood Components Between 4 Common Commercial Kits. *Orthop. J. Sports Med.* **2017**, *5*, 2325967116675272. [CrossRef]
66. Melo, B.A.G.; Luzo, A.C.M.; Lana, J.; Santana, M.H.A. Centrifugation Conditions in the L-PRP Preparation Affect Soluble Factors Release and Mesenchymal Stem Cell Proliferation in Fibrin Nanofibers. *Molecules* **2019**, *24*. [CrossRef]
67. Anitua, E.; Zalduendo, M.; Troya, M.; Padilla, S.; Orive, G. Leukocyte inclusion within a platelet rich plasma-derived fibrin scaffold stimulates a more pro-inflammatory environment and alters fibrin properties. *PLoS ONE* **2015**, *10*, e0121713. [CrossRef]
68. Assirelli, E.; Filardo, G.; Mariani, E.; Kon, E.; Roffi, A.; Vaccaro, F.; Marcacci, M.; Facchini, A.; Pulsatelli, L. Effect of two different preparations of platelet-rich plasma on synoviocytes. *Knee Surg. Sports Traumatol. Arthrosc.* **2015**, *23*, 2690–2703. [CrossRef]
69. Cavallo, C.; Filardo, G.; Mariani, E.; Kon, E.; Marcacci, M.; Pereira Ruiz, M.T.; Facchini, A.; Grigolo, B. Comparison of platelet-rich plasma formulations for cartilage healing: An in vitro study. *J. Bone Joint Surg. Am.* **2014**, *96*, 423–429. [CrossRef]
70. McCarrel, T.M.; Minas, T.; Fortier, L.A. Optimization of leukocyte concentration in platelet-rich plasma for the treatment of tendinopathy. *J. Bone Jt. Surg. Am.* **2012**, *94*, e143. [CrossRef]
71. Pavlovic, V.; Ciric, M.; Jovanovic, V.; Stojanovic, P. Platelet Rich Plasma: A short overview of certain bioactive components. *Open Med. (Wars)* **2016**, *11*, 242–247. [CrossRef] [PubMed]
72. Anitua, E.; Sanchez, M.; Nurden, A.T.; Nurden, P.; Orive, G.; Andia, I. New insights into and novel applications for platelet-rich fibrin therapies. *Trends Biotechnol.* **2006**, *24*, 227–234. [CrossRef] [PubMed]
73. Li, T.; Ma, Y.; Wang, M.; Wang, T.; Wei, J.; Ren, R.; He, M.; Wang, G.; Boey, J.; Armstrong, D.G.; et al. Platelet-rich plasma plays an antibacterial, anti-inflammatory and cell proliferation-promoting role in an in vitro model for diabetic infected wounds. *Infect. Drug Resist.* **2019**, *12*, 297–309. [CrossRef] [PubMed]
74. Nurden, A.T. Platelets, inflammation and tissue regeneration. *Thromb. Haemost.* **2011**, *105* (Suppl. S1), S13–S33. [CrossRef]
75. Bielecki, T.; Dohan Ehrenfest, D.M.; Everts, P.A.; Wiczkowski, A. The role of leukocytes from L-PRP/L-PRF in wound healing and immune defense: New perspectives. *Curr. Pharm. Biotechnol.* **2012**, *13*, 1153–1162. [CrossRef]
76. Rane, D.; Patil, T.; More, V.; Patra, S.S.; Bodhale, N.; Dandapat, J.; Sarkar, A. Neutrophils: Interplay between host defense, cellular metabolism and intracellular infection. *Cytokine* **2018**, *112*, 44–51. [CrossRef]
77. Moojen, D.J.; Everts, P.A.; Schure, R.M.; Overdevest, E.P.; van Zundert, A.; Knape, J.T.; Castelein, R.M.; Creemers, L.B.; Dhert, W.J. Antimicrobial activity of platelet-leukocyte gel against Staphylococcus aureus. *J. Orthop. Res.* **2008**, *26*, 404–410. [CrossRef]
78. Magalon, J.; Bausset, O.; Serratrice, N.; Giraudo, L.; Aboudou, H.; Veran, J.; Magalon, G.; Dignat-Georges, F.; Sabatier, F. Characterization and comparison of 5 platelet-rich plasma preparations in a single-donor model. *Arthroscopy* **2014**, *30*, 629–638. [CrossRef]
79. Castillo, T.N.; Pouliot, M.A.; Kim, H.J.; Dragoo, J.L. Comparison of growth factor and platelet concentration from commercial platelet-rich plasma separation systems. *Am. J. Sports Med.* **2011**, *39*, 266–271. [CrossRef]
80. Kobayashi, Y.; Saita, Y.; Nishio, H.; Ikeda, H.; Takazawa, Y.; Nagao, M.; Takaku, T.; Komatsu, N.; Kaneko, K. Leukocyte concentration and composition in platelet-rich plasma (PRP) influences the growth factor and protease concentrations. *J. Orthop. Sci.* **2016**, *21*, 683–689. [CrossRef]
81. Celik, M.O.; Labuz, D.; Henning, K.; Busch-Dienstfertig, M.; Gaveriaux-Ruff, C.; Kieffer, B.L.; Zimmer, A.; Machelska, H. Leukocyte opioid receptors mediate analgesia via Ca(2+)-regulated release of opioid peptides. *Brain Behav. Immun.* **2016**, *57*, 227–242. [CrossRef] [PubMed]

82. Arnold, L.; Henry, A.; Poron, F.; Baba-Amer, Y.; van Rooijen, N.; Plonquet, A.; Gherardi, R.K.; Chazaud, B. Inflammatory monocytes recruited after skeletal muscle injury switch into antiinflammatory macrophages to support myogenesis. *J. Exp. Med.* **2007**, *204*, 1057–1069. [CrossRef] [PubMed]
83. Lana, J.F.; Macedo, A.; Ingrao, I.L.G.; Huber, S.C.; Santos, G.S.; Santana, M.H.A. Leukocyte-rich PRP for knee osteoarthritis: Current concepts. *J. Clin. Orthop. Trauma* **2019**, *10*, S179–S182. [CrossRef] [PubMed]
84. Eken, C.; Sadallah, S.; Martin, P.J.; Treves, S.; Schifferli, J.A. Ectosomes of polymorphonuclear neutrophils activate multiple signaling pathways in macrophages. *Immunobiology* **2013**, *218*, 382–392. [CrossRef] [PubMed]
85. Lawrence, T.; Natoli, G. Transcriptional regulation of macrophage polarization: Enabling diversity with identity. *Nat. Rev. Immunol.* **2011**, *11*, 750–761. [CrossRef] [PubMed]
86. Winterbourn, C.C.; Kettle, A.J.; Hampton, M.B. Reactive Oxygen Species and Neutrophil Function. *Annu. Rev. Biochem.* **2016**, *85*, 765–792. [CrossRef] [PubMed]
87. Parrish, W.R.; Roides, B.; Hwang, J.; Mafilios, M.; Story, B.; Bhattacharyya, S. Normal platelet function in platelet concentrates requires non-platelet cells: A comparative in vitro evaluation of leucocyte-rich (type 1a) and leucocyte-poor (type 3b) platelet concentrates. *BMJ Open Sport Exerc. Med.* **2016**, *2*, e000071. [CrossRef]
88. Kantarci, A.; Van Dyke, T.E. Lipoxins in chronic inflammation. *Crit. Rev. Oral Biol. Med.* **2003**, *14*, 4–12. [CrossRef]
89. Serhan, C.N. Resolution phase of inflammation: Novel endogenous anti-inflammatory and proresolving lipid mediators and pathways. *Annu. Rev. Immunol.* **2007**, *25*, 101–137. [CrossRef]
90. Everts, P.A.; Malanga, G.A.; Paul, R.V.; Rothenberg, J.B.; Stephens, N.; Mautner, K.R. Assessing clinical implications and perspectives of the pathophysiological effects of erythrocytes and plasma free hemoglobin in autologous biologics for use in musculoskeletal regenerative medicine therapies. A review. *Regen. Ther.* **2019**, *11*, 56–64. [CrossRef]
91. Karsten, E.; Hill, C.J.; Herbert, B.R. Red blood cells: The primary reservoir of macrophage migration inhibitory factor in whole blood. *Cytokine* **2018**, *102*, 34–40. [CrossRef] [PubMed]
92. Zhang, P.L.; Liu, J.; Xu, L.; Sun, Y.; Sun, X.C. Synovial Fluid Macrophage Migration Inhibitory Factor Levels Correlate with Severity of Self-Reported Pain in Knee Osteoarthritis Patients. *Med. Sci. Monit.* **2016**, *22*, 2182–2186. [CrossRef] [PubMed]
93. Liu, M.; Hu, C. Association of MIF in serum and synovial fluid with severity of knee osteoarthritis. *Clin. Biochem.* **2012**, *45*, 737–739. [CrossRef] [PubMed]
94. Braun, H.J.; Kim, H.J.; Chu, C.R.; Dragoo, J.L. The effect of platelet-rich plasma formulations and blood products on human synoviocytes: Implications for intra-articular injury and therapy. *Am. J. Sports Med.* **2014**, *42*, 1204–1210. [CrossRef]
95. Jansen, N.W.; Roosendaal, G.; Bijlsma, J.W.; DeGroot, J.; Theobald, M.; Lafeber, F.P. Degenerated and healthy cartilage are equally vulnerable to blood-induced damage. *Ann. Rheum. Dis.* **2008**, *67*, 1468–1473. [CrossRef]
96. Hooiveld, M.J.; Roosendaal, G.; van den Berg, H.M.; Bijlsma, J.W.; Lafeber, F.P. Haemoglobin-derived iron-dependent hydroxyl radical formation in blood-induced joint damage: An in vitro study. *Rheumatology* **2003**, *42*, 784–790. [CrossRef]
97. Hooiveld, M.; Roosendaal, G.; Wenting, M.; van den Berg, M.; Bijlsma, J.; Lafeber, F.P. Short-Term Exposure of Cartilage to Blood Results in Chondrocyte Apoptosis. *Am. J. Pathol.* **2003**, *162*, 943–951. [CrossRef]
98. Wagener, F.A.; Scharstuhl, A.; Tyrrell, R.M.; Von den Hoff, J.W.; Jozkowicz, A.; Dulak, J.; Russel, F.G.; Kuijpers-Jagtman, A.M. The heme-heme oxygenase system in wound healing; implications for scar formation. *Curr. Drug Targets* **2010**, *11*, 1571–1585. [CrossRef]
99. Lundvig, D.M.; Immenschuh, S.; Wagener, F.A. Heme oxygenase, inflammation, and fibrosis: The good, the bad, and the ugly? *Front. Pharmacol.* **2012**, *3*, 81. [CrossRef]
100. Vijayan, V.; Wagener, F.; Immenschuh, S. The macrophage heme-heme oxygenase-1 system and its role in inflammation. *Biochem. Pharmacol.* **2018**, *153*, 159–167. [CrossRef]
101. Coppinger, J.A.; Maguire, P.B. Insights into the platelet releasate. *Curr. Pharm. Des.* **2007**, *13*, 2640–2646. [CrossRef] [PubMed]
102. Mazzucco, L.; Borzini, P.; Gope, R. Platelet-derived factors involved in tissue repair-from signal to function. *Transfus. Med. Rev.* **2010**, *24*, 218–234. [CrossRef] [PubMed]
103. Le, A.D.K.; Enweze, L.; DeBaun, M.R.; Dragoo, J.L. Platelet-Rich Plasma. *Clin. Sports Med.* **2019**, *38*, 17–44. [CrossRef] [PubMed]

104. Foster, T.E.; Puskas, B.L.; Mandelbaum, B.R.; Gerhardt, M.B.; Rodeo, S.A. Platelet-rich plasma: From basic science to clinical applications. *Am. J. Sports Med.* **2009**, *37*, 2259–2272. [CrossRef]

105. Kobayashi, E.; Fluckiger, L.; Fujioka-Kobayashi, M.; Sawada, K.; Sculean, A.; Schaller, B.; Miron, R.J. Comparative release of growth factors from PRP, PRF, and advanced-PRF. *Clin. Oral Investig.* **2016**, *20*, 2353–2360. [CrossRef]

106. Mazzucco, L.; Balbo, V.; Cattana, E.; Guaschino, R.; Borzini, P. Not every PRP-gel is born equal. Evaluation of growth factor availability for tissues through four PRP-gel preparations: Fibrinet, RegenPRP-Kit, Plateltex and one manual procedure. *Vox Sang.* **2009**, *97*, 110–118. [CrossRef]

107. Mazzocca, A.D.; McCarthy, M.B.; Chowaniec, D.M.; Cote, M.P.; Romeo, A.A.; Bradley, J.P.; Arciero, R.A.; Beitzel, K. Platelet-rich plasma differs according to preparation method and human variability. *J. Bone Jt. Surg. Am.* **2012**, *94*, 308–316. [CrossRef]

108. Oh, J.H.; Kim, W.; Park, K.U.; Roh, Y.H. Comparison of the Cellular Composition and Cytokine-Release Kinetics of Various Platelet-Rich Plasma Preparations. *Am. J. Sports Med.* **2015**, *43*, 3062–3070. [CrossRef]

109. Ha, C.W.; Park, Y.B.; Jang, J.W.; Kim, M.; Kim, J.A.; Park, Y.G. Variability of the Composition of Growth Factors and Cytokines in Platelet-Rich Plasma From the Knee With Osteoarthritis. *Arthroscopy* **2019**, *35*, 2878–2884. [CrossRef]

110. Lane, N.E.; Brandt, K.; Hawker, G.; Peeva, E.; Schreyer, E.; Tsuji, W.; Hochberg, M.C. OARSI-FDA initiative: Defining the disease state of osteoarthritis. *Osteoarthr. Cartil.* **2011**, *19*, 478–482. [CrossRef]

111. Bannuru, R.R.; Osani, M.C.; Vaysbrot, E.E.; Arden, N.K.; Bennell, K.; Bierma-Zeinstra, S.M.A.; Kraus, V.B.; Lohmander, L.S.; Abbott, J.H.; Bhandari, M.; et al. OARSI guidelines for the non-surgical management of knee, hip, and polyarticular osteoarthritis. *Osteoarthr. Cartil.* **2019**, *27*, 1578–1589. [CrossRef]

112. Zhang, W.; Moskowitz, R.W.; Nuki, G.; Abramson, S.; Altman, R.D.; Arden, N.; Bierma-Zeinstra, S.; Brandt, K.D.; Croft, P.; Doherty, M.; et al. OARSI recommendations for the management of hip and knee osteoarthritis, Part II: OARSI evidence-based, expert consensus guidelines. *Osteoarthr. Cartil.* **2008**, *16*, 137–162. [CrossRef]

113. Sanchez, M.; Anitua, E.; Azofra, J.; Aguirre, J.J.; Andia, I. Intra-articular injection of an autologous preparation rich in growth factors for the treatment of knee OA: A retrospective cohort study. *Clin. Exp. Rheumatol.* **2008**, *26*, 910–913.

114. Sanchez, M.; Fiz, N.; Azofra, J.; Usabiaga, J.; Aduriz Recalde, E.; Garcia Gutierrez, A.; Albillos, J.; Garate, R.; Aguirre, J.J.; Padilla, S.; et al. A randomized clinical trial evaluating plasma rich in growth factors (PRGF-Endoret) versus hyaluronic acid in the short-term treatment of symptomatic knee osteoarthritis. *Arthroscopy* **2012**, *28*, 1070–1078. [CrossRef]

115. Vaquerizo, V.; Plasencia, M.A.; Arribas, I.; Seijas, R.; Padilla, S.; Orive, G.; Anitua, E. Comparison of intra-articular injections of plasma rich in growth factors (PRGF-Endoret) versus Durolane hyaluronic acid in the treatment of patients with symptomatic osteoarthritis: A randomized controlled trial. *Arthroscopy* **2013**, *29*, 1635–1643. [CrossRef]

116. Raeissadat, S.A.; Rayegani, S.M.; Ahangar, A.G.; Abadi, P.H.; Mojgani, P.; Ahangar, O.G. Efficacy of Intra-articular Injection of a Newly Developed Plasma Rich in Growth Factor (PRGF) Versus Hyaluronic Acid on Pain and Function of Patients with Knee Osteoarthritis: A Single-Blinded Randomized Clinical Trial. *Clin. Med. Insights Arthritis Musculoskelet. Disord.* **2017**, *10*, 1179544117733452. [CrossRef]

117. Wang-Saegusa, A.; Cugat, R.; Ares, O.; Seijas, R.; Cusco, X.; Garcia-Balletbo, M. Infiltration of plasma rich in growth factors for osteoarthritis of the knee short-term effects on function and quality of life. *Arch. Orthop. Trauma Surg.* **2011**, *131*, 311–317. [CrossRef]

118. Vaquerizo, V.; Padilla, S.; Aguirre, J.J.; Begona, L.; Orive, G.; Anitua, E. Two cycles of plasma rich in growth factors (PRGF-Endoret) intra-articular injections improve stiffness and activities of daily living but not pain compared to one cycle on patients with symptomatic knee osteoarthritis. *Knee Surg. Sports Traumatol. Arthrosc.* **2018**, *26*, 2615–2621. [CrossRef]

119. Kon, E.; Buda, R.; Filardo, G.; Di Martino, A.; Timoncini, A.; Cenacchi, A.; Fornasari, P.M.; Giannini, S.; Marcacci, M. Platelet-rich plasma: Intra-articular knee injections produced favorable results on degenerative cartilage lesions. *Knee Surg. Sports Traumatol. Arthrosc.* **2010**, *18*, 472–479. [CrossRef]

120. Filardo, G.; Kon, E.; Buda, R.; Timoncini, A.; Di Martino, A.; Cenacchi, A.; Fornasari, P.M.; Giannini, S.; Marcacci, M. Platelet-rich plasma intra-articular knee injections for the treatment of degenerative cartilage lesions and osteoarthritis. *Knee Surg. Sports Traumatol. Arthrosc.* **2011**, *19*, 528–535. [CrossRef]

121. Gobbi, A.; Karnatzikos, G.; Mahajan, V.; Malchira, S. Platelet-rich plasma treatment in symptomatic patients with knee osteoarthritis: Preliminary results in a group of active patients. *Sports Health* **2012**, *4*, 162–172. [CrossRef]
122. Halpern, B.; Chaudhury, S.; Rodeo, S.A.; Hayter, C.; Bogner, E.; Potter, H.G.; Nguyen, J. Clinical and MRI outcomes after platelet-rich plasma treatment for knee osteoarthritis. *Clin. J. Sport Med.* **2013**, *23*, 238–239. [CrossRef] [PubMed]
123. Gobbi, A.; Lad, D.; Karnatzikos, G. The effects of repeated intra-articular PRP injections on clinical outcomes of early osteoarthritis of the knee. *Knee Surg. Sports Traumatol. Arthrosc.* **2015**, *23*, 2170–2177. [CrossRef] [PubMed]
124. Hassan, A.S.; El-Shafey, A.M.; Ahmed, H.S.; Hamed, M.S. Effectiveness of the intra-articular injection of platelet rich plasma in the treatment of patients with primary knee osteoarthritis. *Egypt. Rheumatol.* **2015**, *37*, 119–124. [CrossRef]
125. Bottegoni, C.; Dei Giudici, L.; Salvemini, S.; Chiurazzi, E.; Bencivenga, R.; Gigante, A. Homologous platelet-rich plasma for the treatment of knee osteoarthritis in selected elderly patients: An open-label, uncontrolled, pilot study. *Ther. Adv. Musculoskelet. Dis.* **2016**, *8*, 35–41. [CrossRef] [PubMed]
126. Chen, C.P.C.; Cheng, C.H.; Hsu, C.C.; Lin, H.C.; Tsai, Y.R.; Chen, J.L. The influence of platelet rich plasma on synovial fluid volumes, protein concentrations, and severity of pain in patients with knee osteoarthritis. *Exp. Gerontol.* **2017**, *93*, 68–72. [CrossRef]
127. Huang, P.H.; Wang, C.J.; Chou, W.Y.; Wang, J.W.; Ko, J.Y. Short-term clinical results of intra-articular PRP injections for early osteoarthritis of the knee. *Int. J. Surg.* **2017**, *42*, 117–122. [CrossRef]
128. Fawzy, R.M.; Hashaad, N.I.; Mansour, A.I. Decrease of serum biomarker of type II Collagen degradation (Coll2-1) by intra-articular injection of an autologous plasma-rich-platelet in patients with unilateral primary knee osteoarthritis. *Eur. J. Rheumatol.* **2017**, *4*, 93–97. [CrossRef]
129. Taniguchi, Y.; Yoshioka, T.; Kanamori, A.; Aoto, K.; Sugaya, H.; Yamazaki, M. Intra-articular platelet-rich plasma (PRP) injections for treating knee pain associated with osteoarthritis of the knee in the Japanese population: A phase I and IIa clinical trial. *Nagoya J. Med. Sci.* **2018**, *80*, 39–51. [CrossRef]
130. Burchard, R.; Huflage, H.; Soost, C.; Richter, O.; Bouillon, B.; Graw, J.A. Efficiency of platelet-rich plasma therapy in knee osteoarthritis does not depend on level of cartilage damage. *J. Orthop. Surg. Res.* **2019**, *14*, 153. [CrossRef]
131. Sucuoglu, H.; Ustunsoy, S. The short-term effect of PRP on chronic pain in knee osteoarthritis. *Agri* **2019**, *31*, 63–69. [CrossRef] [PubMed]
132. Cerza, F.; Carni, S.; Carcangiu, A.; Di Vavo, I.; Schiavilla, V.; Pecora, A.; De Biasi, G.; Ciuffreda, M. Comparison between hyaluronic acid and platelet-rich plasma, intra-articular infiltration in the treatment of gonarthrosis. *Am. J. Sports Med.* **2012**, *40*, 2822–2827. [CrossRef] [PubMed]
133. Filardo, G.; Kon, E.; Di Martino, A.; Di Matteo, B.; Merli, M.L.; Cenacchi, A.; Fornasari, P.M.; Marcacci, M. Platelet-rich plasma vs hyaluronic acid to treat knee degenerative pathology: Study design and preliminary results of a randomized controlled trial. *BMC Musculoskelet. Disord.* **2012**, *13*, 229. [CrossRef] [PubMed]
134. Spakova, T.; Rosocha, J.; Lacko, M.; Harvanova, D.; Gharaibeh, A. Treatment of knee joint osteoarthritis with autologous platelet-rich plasma in comparison with hyaluronic acid. *Am. J. Phys. Med. Rehabil.* **2012**, *91*, 411–417. [CrossRef]
135. Say, F.; Gurler, D.; Yener, K.; Bulbul, M.; Malkoc, M. Platelet-rich plasma injection is more effective than hyaluronic acid in the treatment of knee osteoarthritis. *Acta Chir. Orthop. Traumatol. Cechoslov.* **2013**, *80*, 278–283.
136. Guler, O.; Mutlu, S.; Isyar, M.; Seker, A.; Kayaalp, M.E.; Mahirogullari, M. Comparison of short-term results of intraarticular platelet-rich plasma (PRP) and hyaluronic acid treatments in early-stage gonarthrosis patients. *Eur. J. Orthop. Surg. Traumatol.* **2015**, *25*, 509–513. [CrossRef]
137. Raeissadat, S.A.; Rayegani, S.M.; Hassanabadi, H.; Fathi, M.; Ghorbani, E.; Babaee, M.; Azma, K. Knee Osteoarthritis Injection Choices: Platelet- Rich Plasma (PRP) Versus Hyaluronic Acid (A one-year randomized clinical trial). *Clin. Med. Insights Arthritis Musculoskelet. Disord.* **2015**, *8*, 1–8. [CrossRef]
138. Montanez-Heredia, E.; Irizar, S.; Huertas, P.J.; Otero, E.; Del Valle, M.; Prat, I.; Diaz-Gallardo, M.S.; Peran, M.; Marchal, J.A.; Hernandez-Lamas Mdel, C. Intra-Articular Injections of Platelet-Rich Plasma versus Hyaluronic Acid in the Treatment of Osteoarthritic Knee Pain: A Randomized Clinical Trial in the Context of the Spanish National Health Care System. *Int. J. Mol. Sci.* **2016**, *17*. [CrossRef]

139. Ahmad, H.S.; Farrag, S.E.; Okasha, A.E.; Kadry, A.O.; Ata, T.B.; Monir, A.A.; Shady, I. Clinical outcomes are associated with changes in ultrasonographic structural appearance after platelet-rich plasma treatment for knee osteoarthritis. *Int. J. Rheum. Dis.* **2018**, *21*, 960–966. [CrossRef]
140. Louis, M.L.; Magalon, J.; Jouve, E.; Bornet, C.E.; Mattei, J.C.; Chagnaud, C.; Rochwerger, A.; Veran, J.; Sabatier, F. Growth Factors Levels Determine Efficacy of Platelets Rich Plasma Injection in Knee Osteoarthritis: A Randomized Double Blind Noninferiority Trial Compared with Viscosupplementation. *Arthroscopy* **2018**, *34*, 1530–1540. [CrossRef]
141. Filardo, G.; Di Matteo, B.; Di Martino, A.; Merli, M.L.; Cenacchi, A.; Fornasari, P.; Marcacci, M.; Kon, E. Platelet-Rich Plasma Intra-articular Knee Injections Show No Superiority Versus Viscosupplementation: A Randomized Controlled Trial. *Am. J. Sports Med.* **2015**, *43*, 1575–1582. [CrossRef] [PubMed]
142. Di Martino, A.; Di Matteo, B.; Papio, T.; Tentoni, F.; Selleri, F.; Cenacchi, A.; Kon, E.; Filardo, G. Platelet-Rich Plasma Versus Hyaluronic Acid Injections for the Treatment of Knee Osteoarthritis: Results at 5 Years of a Double-Blind, Randomized Controlled Trial. *Am. J. Sports Med.* **2019**, *47*, 347–354. [CrossRef] [PubMed]
143. Kon, E.; Mandelbaum, B.; Buda, R.; Filardo, G.; Delcogliano, M.; Timoncini, A.; Fornasari, P.M.; Giannini, S.; Marcacci, M. Platelet-rich plasma intra-articular injection versus hyaluronic acid viscosupplementation as treatments for cartilage pathology: From early degeneration to osteoarthritis. *Arthroscopy* **2011**, *27*, 1490–1501. [CrossRef] [PubMed]
144. Lana, J.F.; Weglein, A.; Sampson, S.E.; Vicente, E.F.; Huber, S.C.; Souza, C.V.; Ambach, M.A.; Vincent, H.; Urban-Paffaro, A.; Onodera, C.M.; et al. Randomized controlled trial comparing hyaluronic acid, platelet-rich plasma and the combination of both in the treatment of mild and moderate osteoarthritis of the knee. *J. Stem Cells Regen. Med.* **2016**, *12*, 69–78.
145. Yu, W.; Xu, P.; Huang, G.; Liu, L. Clinical therapy of hyaluronic acid combined with platelet-rich plasma for the treatment of knee osteoarthritis. *Exp. Ther. Med.* **2018**, *16*, 2119–2125. [CrossRef]
146. Lin, K.Y.; Yang, C.C.; Hsu, C.J.; Yeh, M.L.; Renn, J.H. Intra-articular Injection of Platelet-Rich Plasma Is Superior to Hyaluronic Acid or Saline Solution in the Treatment of Mild to Moderate Knee Osteoarthritis: A Randomized, Double-Blind, Triple-Parallel, Placebo-Controlled Clinical Trial. *Arthroscopy* **2019**, *35*, 106–117. [CrossRef]
147. Duymus, T.M.; Mutlu, S.; Dernek, B.; Komur, B.; Aydogmus, S.; Kesiktas, F.N. Choice of intra-articular injection in treatment of knee osteoarthritis: Platelet-rich plasma, hyaluronic acid or ozone options. *Knee Surg. Sports Traumatol. Arthrosc.* **2017**, *25*, 485–492. [CrossRef]
148. Dernek, B.; Kesiktas, F.N. Efficacy of combined ozone and platelet-rich-plasma treatment versus platelet-rich-plasma treatment alone in early stage knee osteoarthritis. *J. Back Musculoskelet. Rehabil.* **2019**, *32*, 305–311. [CrossRef]
149. Huang, Y.; Liu, X.; Xu, X.; Liu, J. Intra-articular injections of platelet-rich plasma, hyaluronic acid or corticosteroids for knee osteoarthritis: A prospective randomized controlled study. *Orthopade* **2019**, *48*, 239–247. [CrossRef]
150. Camurcu, Y.; Sofu, H.; Ucpunar, H.; Kockara, N.; Cobden, A.; Duman, S. Single-dose intra-articular corticosteroid injection prior to platelet-rich plasma injection resulted in better clinical outcomes in patients with knee osteoarthritis: A pilot study. *J. Back Musculoskelet. Rehabil.* **2018**, *31*, 603–610. [CrossRef]
151. Filardo, G.; Kon, E.; Pereira Ruiz, M.T.; Vaccaro, F.; Guitaldi, R.; Di Martino, A.; Cenacchi, A.; Fornasari, P.M.; Marcacci, M. Platelet-rich plasma intra-articular injections for cartilage degeneration and osteoarthritis: Single- versus double-spinning approach. *Knee Surg. Sports Traumatol. Arthrosc.* **2012**, *20*, 2082–2091. [CrossRef] [PubMed]
152. Hart, R.; Safi, A.; Komzak, M.; Jajtner, P.; Puskeiler, M.; Hartova, P. Platelet-rich plasma in patients with tibiofemoral cartilage degeneration. *Arch. Orthop. Trauma Surg.* **2013**, *133*, 1295–1301. [CrossRef] [PubMed]
153. Patel, S.; Dhillon, M.S.; Aggarwal, S.; Marwaha, N.; Jain, A. Treatment with platelet-rich plasma is more effective than placebo for knee osteoarthritis: A prospective, double-blind, randomized trial. *Am. J. Sports Med.* **2013**, *41*, 356–364. [CrossRef] [PubMed]
154. Gormeli, G.; Gormeli, C.A.; Ataoglu, B.; Colak, C.; Aslanturk, O.; Ertem, K. Multiple PRP injections are more effective than single injections and hyaluronic acid in knees with early osteoarthritis: A randomized, double-blind, placebo-controlled trial. *Knee Surg. Sports Traumatol. Arthrosc.* **2017**, *25*, 958–965. [CrossRef]

155. Guillibert, C.; Charpin, C.; Raffray, M.; Benmenni, A.; Dehaut, F.X.; El Ghobeira, G.; Giorgi, R.; Magalon, J.; Arniaud, D. Single Injection of High Volume of Autologous Pure PRP Provides a Significant Improvement in Knee Osteoarthritis: A Prospective Routine Care Study. *Int. J. Mol. Sci.* **2019**, *20*. [CrossRef]
156. Rayegani, S.M.; Raeissadat, S.A.; Taheri, M.S.; Babaee, M.; Bahrami, M.H.; Eliaspour, D.; Ghorbani, E. Does intra articular platelet rich plasma injection improve function, pain and quality of life in patients with osteoarthritis of the knee? A randomized clinical trial. *Orthop. Rev. (Pavia)* **2014**, *6*, 5405. [CrossRef]
157. Duif, C.; Vogel, T.; Topcuoglu, F.; Spyrou, G.; von Schulze Pellengahr, C.; Lahner, M. Does intraoperative application of leukocyte-poor platelet-rich plasma during arthroscopy for knee degeneration affect postoperative pain, function and quality of life? A 12-month randomized controlled double-blind trial. *Arch. Orthop. Trauma Surg.* **2015**, *135*, 971–977. [CrossRef]
158. Smith, P.A. Intra-articular Autologous Conditioned Plasma Injections Provide Safe and Efficacious Treatment for Knee Osteoarthritis: An FDA-Sanctioned, Randomized, Double-blind, Placebo-controlled Clinical Trial. *Am. J. Sports Med.* **2016**, *44*, 884–891. [CrossRef]
159. Simental-Mendia, M.; Vilchez-Cavazos, J.F.; Pena-Martinez, V.M.; Said-Fernandez, S.; Lara-Arias, J.; Martinez-Rodriguez, H.G. Leukocyte-poor platelet-rich plasma is more effective than the conventional therapy with acetaminophen for the treatment of early knee osteoarthritis. *Arch. Orthop. Trauma Surg.* **2016**, *136*, 1723–1732. [CrossRef]
160. Cole, B.J.; Karas, V.; Hussey, K.; Pilz, K.; Fortier, L.A. Hyaluronic Acid Versus Platelet-Rich Plasma: A Prospective, Double-Blind Randomized Controlled Trial Comparing Clinical Outcomes and Effects on Intra-articular Biology for the Treatment of Knee Osteoarthritis. *Am. J. Sports Med.* **2017**, *45*, 339–346. [CrossRef]
161. Buendia-Lopez, D.; Medina-Quiros, M.; Fernandez-Villacanas Marin, M.A. Clinical and radiographic comparison of a single LP-PRP injection, a single hyaluronic acid injection and daily NSAID administration with a 52-week follow-up: A randomized controlled trial. *J. Orthop. Traumatol.* **2018**, *19*, 3. [CrossRef] [PubMed]
162. Huang, G.; Hua, S.; Yang, T.; Ma, J.; Yu, W.; Chen, X. Platelet-rich plasma shows beneficial effects for patients with knee osteoarthritis by suppressing inflammatory factors. *Exp. Ther. Med.* **2018**, *15*, 3096–3102. [CrossRef]
163. Elik, H.; Dogu, B.; Yilmaz, F.; Begoglu, F.A.; Kuran, B. The Efficiency of Platelet Rich Plasma Treatment in patients with knee osteoarthritis. *J. Back Musculoskelet. Rehabil.* **2019**. [CrossRef] [PubMed]
164. Li, M.; Zhang, C.; Ai, Z.; Yuan, T.; Feng, Y.; Jia, W. [Therapeutic effectiveness of intra-knee-articular injection of platelet-rich plasma on knee articular cartilage degeneration]. *Zhongguo Xiu Fu Chong Jian Wai Ke Za Zhi* **2011**, *25*, 1192–1196. [PubMed]
165. Forogh, B.; Mianehsaz, E.; Shoaee, S.; Ahadi, T.; Raissi, G.R.; Sajadi, S. Effect of single injection of platelet-rich plasma in comparison with corticosteroid on knee osteoarthritis: A double-blind randomized clinical trial. *J. Sports Med. Phys. Fit.* **2016**, *56*, 901–908.
166. Rahimzadeh, P.; Imani, F.; Faiz, S.H.R.; Entezary, S.R.; Zamanabadi, M.N.; Alebouyeh, M.R. The effects of injecting intra-articular platelet-rich plasma or prolotherapy on pain score and function in knee osteoarthritis. *Clin. Interv. Aging* **2018**, *13*, 73–79. [CrossRef]
167. Paterson, K.L.; Nicholls, M.; Bennell, K.L.; Bates, D. Intra-articular injection of photo-activated platelet-rich plasma in patients with knee osteoarthritis: A double-blind, randomized controlled pilot study. *BMC Musculoskelet. Disord.* **2016**, *17*, 67. [CrossRef]
168. Bansal, H.; Comella, K.; Leon, J.; Verma, P.; Agrawal, D.; Koka, P.; Ichim, T. Intra-articular injection in the knee of adipose derived stromal cells (stromal vascular fraction) and platelet rich plasma for osteoarthritis. *J. Transl. Med.* **2017**, *15*, 141. [CrossRef]
169. Sanchez, M.; Delgado, D.; Sanchez, P.; Muinos-Lopez, E.; Paiva, B.; Granero-Molto, F.; Prosper, F.; Pompei, O.; Perez, J.C.; Azofra, J.; et al. Combination of Intra-Articular and Intraosseous Injections of Platelet Rich Plasma for Severe Knee Osteoarthritis: A Pilot Study. *Biomed. Res. Int.* **2016**, *2016*, 4868613. [CrossRef]
170. Sanchez, M.; Delgado, D.; Pompei, O.; Perez, J.C.; Sanchez, P.; Garate, A.; Bilbao, A.M.; Fiz, N.; Padilla, S. Treating Severe Knee Osteoarthritis with Combination of Intra-Osseous and Intra-Articular Infiltrations of Platelet-Rich Plasma: An Observational Study. *Cartilage* **2019**, *10*, 245–253. [CrossRef]

171. Su, K.; Bai, Y.; Wang, J.; Zhang, H.; Liu, H.; Ma, S. Comparison of hyaluronic acid and PRP intra-articular injection with combined intra-articular and intraosseous PRP injections to treat patients with knee osteoarthritis. *Clin. Rheumatol.* **2018**, *37*, 1341–1350. [CrossRef] [PubMed]
172. Patel, S.; Dhillon, M.S. The Anti-inflammatory and Matrix Restorative Mechanisms of Platelet-Rich Plasma in Osteoarthritis: Letter to the Editor. *Am. J. Sports Med.* **2014**, *42*. [CrossRef] [PubMed]
173. Anitua, E.; Sanchez, M.; De la Fuente, M.; Zalduendo, M.M.; Orive, G. Plasma rich in growth factors (PRGF-Endoret) stimulates tendon and synovial fibroblasts migration and improves the biological properties of hyaluronic acid. *Knee Surg. Sports Traumatol. Arthrosc.* **2012**, *20*, 1657–1665. [CrossRef] [PubMed]
174. Marmotti, A.; Bruzzone, M.; Bonasia, D.E.; Castoldi, F.; Rossi, R.; Piras, L.; Maiello, A.; Realmuto, C.; Peretti, G.M. One-step osteochondral repair with cartilage fragments in a composite scaffold. *Knee Surg. Sports Traumatol. Arthrosc.* **2012**, *20*, 2590–2601. [CrossRef]
175. Andia, I.; Abate, M. Knee osteoarthritis: Hyaluronic acid, platelet-rich plasma or both in association? *Expert Opin. Biol. Ther.* **2014**, *14*, 635–649. [CrossRef]
176. Chen, C.P.C.; Chen, J.L.; Hsu, C.C.; Pei, Y.C.; Chang, W.H.; Lu, H.C. Injecting autologous platelet rich plasma solely into the knee joint is not adequate in treating geriatric patients with moderate to severe knee osteoarthritis. *Exp. Gerontol.* **2019**, *119*, 1–6. [CrossRef]
177. Anitua, E.; Sanchez, M.; Aguirre, J.J.; Prado, R.; Padilla, S.; Orive, G. Efficacy and safety of plasma rich in growth factors intra-articular infiltrations in the treatment of knee osteoarthritis. *Arthroscopy* **2014**, *30*, 1006–1017. [CrossRef]
178. Bennell, K.L.; Hunter, D.J.; Paterson, K.L. Platelet-Rich Plasma for the Management of Hip and Knee Osteoarthritis. *Curr. Rheumatol. Rep.* **2017**, *19*, 24. [CrossRef]
179. Dai, W.L.; Zhou, A.G.; Zhang, H.; Zhang, J. Efficacy of Platelet-Rich Plasma in the Treatment of Knee Osteoarthritis: A Meta-analysis of Randomized Controlled Trials. *Arthroscopy* **2017**, *33*, 659–670. [CrossRef]
180. Han, Y.; Huang, H.; Pan, J.; Lin, J.; Zeng, L.; Liang, G.; Yang, W.; Liu, J. Meta-analysis Comparing Platelet-Rich Plasma vs. Hyaluronic Acid Injection in Patients with Knee Osteoarthritis. *Pain Med.* **2019**, *20*, 1418–1429. [CrossRef]
181. Di, Y.; Han, C.; Zhao, L.; Ren, Y. Is local platelet-rich plasma injection clinically superior to hyaluronic acid for treatment of knee osteoarthritis? A systematic review of randomized controlled trials. *Arthritis Res. Ther.* **2018**, *20*, 128. [CrossRef] [PubMed]
182. Sanchez, M.; Guadilla, J.; Fiz, N.; Andia, I. Ultrasound-guided platelet-rich plasma injections for the treatment of osteoarthritis of the hip. *Rheumatology* **2012**, *51*, 144–150. [CrossRef] [PubMed]
183. Singh, J.R.; Haffey, P.; Valimahomed, A.; Gellhorn, A. The Effectiveness of Autologous Platelet-Rich Plasma for Osteoarthritis of the Hip: A Retrospective Analysis. *Pain Med.* **2019**. [CrossRef] [PubMed]
184. Battaglia, M.; Guaraldi, F.; Vannini, F.; Rossi, G.; Timoncini, A.; Buda, R.; Giannini, S. Efficacy of ultrasound-guided intra-articular injections of platelet-rich plasma versus hyaluronic acid for hip osteoarthritis. *Orthopedics* **2013**, *36*, e1501–e1508. [CrossRef]
185. Di Sante, L.; Villani, C.; Santilli, V.; Valeo, M.; Bologna, E.; Imparato, L.; Paoloni, M.; Iagnocco, A. Intra-articular hyaluronic acid vs platelet-rich plasma in the treatment of hip osteoarthritis. *Med. Ultrason.* **2016**, *18*, 463–468. [CrossRef]
186. Doria, C.; Mosele, G.R.; Caggiari, G.; Puddu, L.; Ciurlia, E. Treatment of Early Hip Osteoarthritis: Ultrasound-Guided Platelet Rich Plasma versus Hyaluronic Acid Injections in a Randomized Clinical Trial. *Joints* **2017**, *5*, 152–155. [CrossRef]
187. Dallari, D.; Stagni, C.; Rani, N.; Sabbioni, G.; Pelotti, P.; Torricelli, P.; Tschon, M.; Giavaresi, G. Ultrasound-Guided Injection of Platelet-Rich Plasma and Hyaluronic Acid, Separately and in Combination, for Hip Osteoarthritis: A Randomized Controlled Study. *Am. J. Sports Med.* **2016**, *44*, 664–671. [CrossRef]
188. Fiz, N.; Perez, J.C.; Guadilla, J.; Garate, A.; Sanchez, P.; Padilla, S.; Delgado, D.; Sanchez, M. Intraosseous Infiltration of Platelet-Rich Plasma for Severe Hip Osteoarthritis. *Arthrosc. Tech.* **2017**, *6*, e821–e825. [CrossRef]
189. Ye, Y.; Zhou, X.; Mao, S.; Zhang, J.; Lin, B. Platelet rich plasma versus hyaluronic acid in patients with hip osteoarthritis: A meta-analysis of randomized controlled trials. *Int. J. Surg.* **2018**, *53*, 279–287. [CrossRef]
190. Ali, M.; Mohamed, A.; Ahmed, H.E.; Malviya, A.; Atchia, I. The use of ultrasound-guided platelet-rich plasma injections in the treatment of hip osteoarthritis: A systematic review of the literature. *J. Ultrason.* **2018**, *18*, 332–337. [CrossRef]

191. Repetto, I.; Biti, B.; Cerruti, P.; Trentini, R.; Felli, L. Conservative Treatment of Ankle Osteoarthritis: Can Platelet-Rich Plasma Effectively Postpone Surgery? *J. Foot Ankle Surg.* **2017**, *56*, 362–365. [CrossRef] [PubMed]
192. Fukawa, T.; Yamaguchi, S.; Akatsu, Y.; Yamamoto, Y.; Akagi, R.; Sasho, T. Safety and Efficacy of Intra-articular Injection of Platelet-Rich Plasma in Patients With Ankle Osteoarthritis. *Foot Ankle Int.* **2017**, *38*, 596–604. [CrossRef] [PubMed]
193. Rajabi, H.; Sheikhani Shahin, H.; Norouzian, M.; Mehrabani, D.; Dehghani Nazhvani, S. The Healing Effects of Aquatic Activities and Allogenic Injection of Platelet-Rich Plasma (PRP) on Injuries of Achilles Tendon in Experimental Rat. *World J. Plast. Surg.* **2015**, *4*, 66–73. [PubMed]
194. Sinnott, C.; White, H.M.; Cuchna, J.W.; Van Lunen, B.L. Autologous Blood and Platelet-Rich Plasma Injections in the Treatment of Achilles Tendinopathy: A Critically Appraised Topic. *J. Sport Rehabil.* **2017**, *26*, 279–285. [CrossRef]
195. Gaweda, K.; Tarczynska, M.; Krzyzanowski, W. Treatment of Achilles tendinopathy with platelet-rich plasma. *Int. J. Sports Med.* **2010**, *31*, 577–583. [CrossRef]
196. Volpi, P.; Quaglia, A.; Schoenhuber, H.; Melegati, G.; Corsi, M.M.; Banfi, G.; de Girolamo, L. Growth factors in the management of sport-induced tendinopathies: Results after 24 months from treatment. A pilot study. *J. Sports Med. Phys. Fit.* **2010**, *50*, 494–500.
197. Finnoff, J.T.; Fowler, S.P.; Lai, J.K.; Santrach, P.J.; Willis, E.A.; Sayeed, Y.A.; Smith, J. Treatment of chronic tendinopathy with ultrasound-guided needle tenotomy and platelet-rich plasma injection. *PM R* **2011**, *3*, 900–911. [CrossRef]
198. Deans, V.M.; Miller, A.; Ramos, J. A prospective series of patients with chronic Achilles tendinopathy treated with autologous-conditioned plasma injections combined with exercise and therapeutic ultrasonography. *J. Foot Ankle Surg.* **2012**, *51*, 706–710. [CrossRef]
199. Ferrero, G.; Fabbro, E.; Orlandi, D.; Martini, C.; Lacelli, F.; Serafini, G.; Silvestri, E.; Sconfienza, L.M. Ultrasound-guided injection of platelet-rich plasma in chronic Achilles and patellar tendinopathy. *J. Ultrasound* **2012**, *15*, 260–266. [CrossRef]
200. Monto, R.R. Platelet rich plasma treatment for chronic Achilles tendinosis. *Foot Ankle Int.* **2012**, *33*, 379–385. [CrossRef]
201. Murawski, C.D.; Smyth, N.A.; Newman, H.; Kennedy, J.G. A single platelet-rich plasma injection for chronic midsubstance achilles tendinopathy: A retrospective preliminary analysis. *Foot Ankle Spec.* **2014**, *7*, 372–376. [CrossRef]
202. Guelfi, M.; Pantalone, A.; Vanni, D.; Abate, M.; Guelfi, M.G.; Salini, V. Long-term beneficial effects of platelet-rich plasma for non-insertional Achilles tendinopathy. *Foot Ankle Surg.* **2015**, *21*, 178–181. [CrossRef]
203. Salini, V.; Vanni, D.; Pantalone, A.; Abate, M. Platelet Rich Plasma Therapy in Non-insertional Achilles Tendinopathy: The Efficacy is Reduced in 60-years Old People Compared to Young and Middle-Age Individuals. *Front. Aging Neurosci.* **2015**, *7*, 228. [CrossRef] [PubMed]
204. Owens, R.F., Jr.; Ginnetti, J.; Conti, S.F.; Latona, C. Clinical and magnetic resonance imaging outcomes following platelet rich plasma injection for chronic midsubstance Achilles tendinopathy. *Foot Ankle Int.* **2011**, *32*, 1032–1039. [CrossRef] [PubMed]
205. Filardo, G.; Kon, E.; Di Matteo, B.; Di Martino, A.; Tesei, G.; Pelotti, P.; Cenacchi, A.; Marcacci, M. Platelet-rich plasma injections for the treatment of refractory Achilles tendinopathy: Results at 4 years. *Blood Transfus.* **2014**, *12*, 533–540. [CrossRef] [PubMed]
206. De Jonge, S.; de Vos, R.J.; Weir, A.; van Schie, H.T.; Bierma-Zeinstra, S.M.; Verhaar, J.A.; Weinans, H.; Tol, J.L. One-year follow-up of platelet-rich plasma treatment in chronic Achilles tendinopathy: A double-blind randomized placebo-controlled trial. *Am. J. Sports Med.* **2011**, *39*, 1623–1629. [CrossRef]
207. Krogh, T.P.; Ellingsen, T.; Christensen, R.; Jensen, P.; Fredberg, U. Ultrasound-Guided Injection Therapy of Achilles Tendinopathy With Platelet-Rich Plasma or Saline: A Randomized, Blinded, Placebo-Controlled Trial. *Am. J. Sports Med.* **2016**, *44*, 1990–1997. [CrossRef]
208. Hanisch, K.; Wedderkopp, N. Platelet-rich plasma (PRP) treatment of noninsertional Achilles tendinopathy in a two case series: No significant difference in effect between leukocyte-rich and leukocyte-poor PRP. *Orthop. Res. Rev.* **2019**, *11*, 55–60. [CrossRef]
209. Gentile, P.; De Angelis, B.; Agovino, A.; Orlandi, F.; Migner, A.; Di Pasquali, C.; Cervelli, V. Use of Platelet Rich Plasma and Hyaluronic Acid in the Treatment of Complications of Achilles Tendon Reconstruction. *World J. Plast. Surg.* **2016**, *5*, 124–132.

210. Erroi, D.; Sigona, M.; Suarez, T.; Trischitta, D.; Pavan, A.; Vulpiani, M.C.; Vetrano, M. Conservative treatment for Insertional Achilles Tendinopathy: Platelet-rich plasma and focused shock waves. A retrospective study. *Muscles Ligaments Tendons J.* **2017**, *7*, 98–106. [CrossRef]
211. Oloff, L.; Elmi, E.; Nelson, J.; Crain, J. Retrospective Analysis of the Effectiveness of Platelet-Rich Plasma in the Treatment of Achilles Tendinopathy: Pretreatment and Posttreatment Correlation of Magnetic Resonance Imaging and Clinical Assessment. *Foot Ankle Spec.* **2015**, *8*, 490–497. [CrossRef] [PubMed]
212. Kearney, R.S.; Parsons, N.; Costa, M.L. Achilles tendinopathy management: A pilot randomised controlled trial comparing platelet-richplasma injection with an eccentric loading programme. *Bone Jt. Res.* **2013**, *2*, 227–232. [CrossRef] [PubMed]
213. De Vos, R.J.; Weir, A.; van Schie, H.T.; Bierma-Zeinstra, S.M.; Verhaar, J.A.; Weinans, H.; Tol, J.L. Platelet-rich plasma injection for chronic Achilles tendinopathy: A randomized controlled trial. *JAMA* **2010**, *303*, 144–149. [CrossRef] [PubMed]
214. De Vos, R.J.; Weir, A.; Tol, J.L.; Verhaar, J.A.; Weinans, H.; van Schie, H.T. No effects of PRP on ultrasonographic tendon structure and neovascularisation in chronic midportion Achilles tendinopathy. *Br. J. Sports Med.* **2011**, *45*, 387–392. [CrossRef] [PubMed]
215. Boesen, A.P.; Hansen, R.; Boesen, M.I.; Malliaras, P.; Langberg, H. Effect of High-Volume Injection, Platelet-Rich Plasma, and Sham Treatment in Chronic Midportion Achilles Tendinopathy: A Randomized Double-Blinded Prospective Study. *Am. J. Sports Med.* **2017**, *45*, 2034–2043. [CrossRef]
216. Zhang, Y.J.; Xu, S.Z.; Gu, P.C.; Du, J.Y.; Cai, Y.Z.; Zhang, C.; Lin, X.J. Is Platelet-rich Plasma Injection Effective for Chronic Achilles Tendinopathy? A Meta-analysis. *Clin. Orthop. Relat. Res.* **2018**, *476*, 1633–1641. [CrossRef]
217. Chen, X.; Jones, I.A.; Park, C.; Vangsness, C.T., Jr. The Efficacy of Platelet-Rich Plasma on Tendon and Ligament Healing: A Systematic Review and Meta-analysis With Bias Assessment. *Am. J. Sports Med.* **2018**, *46*, 2020–2032. [CrossRef]
218. Liu, C.J.; Yu, K.L.; Bai, J.B.; Tian, D.H.; Liu, G.L. Platelet-rich plasma injection for the treatment of chronic Achilles tendinopathy: A meta-analysis. *Medicine* **2019**, *98*, e15278. [CrossRef]
219. Wang, Y.; Han, C.; Hao, J.; Ren, Y.; Wang, J. Efficacy of platelet-rich plasma injections for treating Achilles tendonitis: Systematic review of high-quality randomized controlled trials. *Orthopade* **2019**, *48*, 784–791. [CrossRef]
220. Dragoo, J.L.; Meadows, M.C. The use of biologics for the elbow: A critical analysis review. *J. Shoulder Elbow Surg.* **2019**, *28*, 2053–2060. [CrossRef]
221. Mishra, A.; Pavelko, T. Treatment of chronic elbow tendinosis with buffered platelet-rich plasma. *Am. J. Sports Med.* **2006**, *34*, 1774–1778. [CrossRef] [PubMed]
222. Hechtman, K.S.; Uribe, J.W.; Botto-vanDemden, A.; Kiebzak, G.M. Platelet-rich plasma injection reduces pain in patients with recalcitrant epicondylitis. *Orthopedics* **2011**, *34*, 92. [CrossRef] [PubMed]
223. Brkljac, M.; Kumar, S.; Kalloo, D.; Hirehal, K. The effect of platelet-rich plasma injection on lateral epicondylitis following failed conservative management. *J. Orthop.* **2015**, *12*, S166–S170. [CrossRef] [PubMed]
224. Brkljac, M.; Conville, J.; Sonar, U.; Kumar, S. Long-term follow-up of platelet-rich plasma injections for refractory lateral epicondylitis. *J. Orthop.* **2019**, *16*, 496–499. [CrossRef] [PubMed]
225. Creaney, L.; Wallace, A.; Curtis, M.; Connell, D. Growth factor-based therapies provide additional benefit beyond physical therapy in resistant elbow tendinopathy: A prospective, single-blind, randomised trial of autologous blood injections versus platelet-rich plasma injections. *Br. J. Sports Med.* **2011**, *45*, 966–971. [CrossRef] [PubMed]
226. Thanasas, C.; Papadimitriou, G.; Charalambidis, C.; Paraskevopoulos, I.; Papanikolaou, A. Platelet-rich plasma versus autologous whole blood for the treatment of chronic lateral elbow epicondylitis: A randomized controlled clinical trial. *Am. J. Sports Med.* **2011**, *39*, 2130–2134. [CrossRef] [PubMed]
227. Raeissadat, S.A.; Rayegani, S.M.; Hassanabadi, H.; Rahimi, R.; Sedighipour, L.; Rostami, K. Is Platelet-rich plasma superior to whole blood in the management of chronic tennis elbow: One year randomized clinical trial. *BMC Sports Sci. Med. Rehabil.* **2014**, *6*, 12. [CrossRef]
228. Montalvan, B.; Le Goux, P.; Klouche, S.; Borgel, D.; Hardy, P.; Breban, M. Inefficacy of ultrasound-guided local injections of autologous conditioned plasma for recent epicondylitis: Results of a double-blind placebo-controlled randomized clinical trial with one-year follow-up. *Rheumatology* **2016**, *55*, 279–285. [CrossRef]

229. Schoffl, V.; Willauschus, W.; Sauer, F.; Kupper, T.; Schoffl, I.; Lutter, C.; Gelse, K.; Dickschas, J. Autologous Conditioned Plasma Versus Placebo Injection Therapy in Lateral Epicondylitis of the Elbow: A Double Blind, Randomized Study. *Sportverletz Sportschaden* **2017**, *31*, 31–36. [CrossRef]
230. Yerlikaya, M.; Talay Calis, H.; Tomruk Sutbeyaz, S.; Sayan, H.; Ibis, N.; Koc, A.; Karakukcu, C. Comparison of Effects of Leukocyte-Rich and Leukocyte-Poor Platelet-Rich Plasma on Pain and Functionality in Patients With Lateral Epicondylitis. *Arch. Rheumatol.* **2018**, *33*, 73–79. [CrossRef]
231. Mishra, A.K.; Skrepnik, N.V.; Edwards, S.G.; Jones, G.L.; Sampson, S.; Vermillion, D.A.; Ramsey, M.L.; Karli, D.C.; Rettig, A.C. Efficacy of platelet-rich plasma for chronic tennis elbow: A double-blind, prospective, multicenter, randomized controlled trial of 230 patients. *Am. J. Sports Med.* **2014**, *42*, 463–471. [CrossRef] [PubMed]
232. Peerbooms, J.C.; Sluimer, J.; Bruijn, D.J.; Gosens, T. Positive effect of an autologous platelet concentrate in lateral epicondylitis in a double-blind randomized controlled trial: Platelet-rich plasma versus corticosteroid injection with a 1-year follow-up. *Am. J. Sports Med.* **2010**, *38*, 255–262. [CrossRef] [PubMed]
233. Gosens, T.; Peerbooms, J.C.; van Laar, W.; den Oudsten, B.L. Ongoing positive effect of platelet-rich plasma versus corticosteroid injection in lateral epicondylitis: A double-blind randomized controlled trial with 2-year follow-up. *Am. J. Sports Med.* **2011**, *39*, 1200–1208. [CrossRef] [PubMed]
234. Gautam, V.K.; Verma, S.; Batra, S.; Bhatnagar, N.; Arora, S. Platelet-rich plasma versus corticosteroid injection for recalcitrant lateral epicondylitis: Clinical and ultrasonographic evaluation. *J. Orthop. Surg.* **2015**, *23*, 1–5. [CrossRef] [PubMed]
235. Khaliq, A.; Khan, I.; Inam, M.; Saeed, M.; Khan, H.; Iqbal, M.J. Effectiveness of platelets rich plasma versus corticosteroids in lateral epicondylitis. *J. Pak. Med. Assoc.* **2015**, *65*, S100–S104. [PubMed]
236. Varshney, A.; Maheshwari, R.; Juyal, A.; Agrawal, A.; Hayer, P. Autologous Platelet-rich Plasma versus Corticosteroid in the Management of Elbow Epicondylitis: A Randomized Study. *Int. J. Appl. Basic Med. Res.* **2017**, *7*, 125–128. [CrossRef] [PubMed]
237. Gupta, P.K.; Acharya, A.; Khanna, V.; Roy, S.; Khillan, K.; Sambandam, S.N. PRP versus steroids in a deadlock for efficacy: Long-term stability versus short-term intensity-results from a randomised trial. *Musculoskelet. Surg.* **2019**. [CrossRef]
238. Lebiedzinski, R.; Synder, M.; Buchcic, P.; Polguj, M.; Grzegorzewski, A.; Sibinski, M. A randomized study of autologous conditioned plasma and steroid injections in the treatment of lateral epicondylitis. *Int. Orthop.* **2015**, *39*, 2199–2203. [CrossRef]
239. Palacio, E.P.; Schiavetti, R.R.; Kanematsu, M.; Ikeda, T.M.; Mizobuchi, R.R.; Galbiatti, J.A. Effects of platelet-rich plasma on lateral epicondylitis of the elbow: Prospective randomized controlled trial. *Rev. Bras. Ortop.* **2016**, *51*, 90–95. [CrossRef]
240. Yadav, R.; Kothari, S.Y.; Borah, D. Comparison of Local Injection of Platelet Rich Plasma and Corticosteroids in the Treatment of Lateral Epicondylitis of Humerus. *J. Clin. Diagn. Res.* **2015**, *9*, RC05–RC07. [CrossRef]
241. Behera, P.; Dhillon, M.; Aggarwal, S.; Marwaha, N.; Prakash, M. Leukocyte-poor platelet-rich plasma versus bupivacaine for recalcitrant lateral epicondylar tendinopathy. *J. Orthop. Surg.* **2015**, *23*, 6–10. [CrossRef] [PubMed]
242. Tonk, G.; Kumar, A.; Gupta, A. Platelet rich plasma versus laser therapy in lateral epicondylitis of elbow. *Indian J. Orthop.* **2014**, *48*, 390–393. [CrossRef] [PubMed]
243. Alessio-Mazzola, M.; Repetto, I.; Biti, B.; Trentini, R.; Formica, M.; Felli, L. Autologous US-guided PRP injection versus US-guided focal extracorporeal shock wave therapy for chronic lateral epicondylitis: A minimum of 2-year follow-up retrospective comparative study. *J. Orthop. Surg.* **2018**, *26*, 2309499017749986. [CrossRef] [PubMed]
244. Seetharamaiah, V.B.; Gantaguru, A.; Basavarajanna, S. A comparative study to evaluate the efficacy of platelet-rich plasma and triamcinolone to treat tennis elbow. *Indian J. Orthop.* **2017**, *51*, 304–311. [CrossRef]
245. Krogh, T.P.; Fredberg, U.; Stengaard-Pedersen, K.; Christensen, R.; Jensen, P.; Ellingsen, T. Treatment of lateral epicondylitis with platelet-rich plasma, glucocorticoid, or saline: A randomized, double-blind, placebo-controlled trial. *Am. J. Sports Med.* **2013**, *41*, 625–635. [CrossRef]
246. Stenhouse, G.; Sookur, P.; Watson, M. Do blood growth factors offer additional benefit in refractory lateral epicondylitis? A prospective, randomized pilot trial of dry needling as a stand-alone procedure versus dry needling and autologous conditioned plasma. *Skelet. Radiol.* **2013**, *42*, 1515–1520. [CrossRef]

247. Merolla, G.; Dellabiancia, F.; Ricci, A.; Mussoni, M.P.; Nucci, S.; Zanoli, G.; Paladini, P.; Porcellini, G. Arthroscopic Debridement Versus Platelet-Rich Plasma Injection: A Prospective, Randomized, Comparative Study of Chronic Lateral Epicondylitis With a Nearly 2-Year Follow-Up. *Arthroscopy* **2017**, *33*, 1320–1329. [CrossRef]
248. Boden, A.L.; Scott, M.T.; Dalwadi, P.P.; Mautner, K.; Mason, R.A.; Gottschalk, M.B. Platelet-rich plasma versus Tenex in the treatment of medial and lateral epicondylitis. *J. Shoulder Elbow Surg.* **2019**, *28*, 112–119. [CrossRef]
249. Chiavaras, M.M.; Jacobson, J.A.; Carlos, R.; Maida, E.; Bentley, T.; Simunovic, N.; Swinton, M.; Bhandari, M. IMpact of Platelet Rich plasma OVer alternative therapies in patients with lateral Epicondylitis (IMPROVE): Protocol for a multicenter randomized controlled study: A multicenter, randomized trial comparing autologous platelet-rich plasma, autologous whole blood, dry needle tendon fenestration, and physical therapy exercises alone on pain and quality of life in patients with lateral epicondylitis. *Acad. Radiol.* **2014**, *21*, 1144–1155. [CrossRef]
250. Hastie, G.; Soufi, M.; Wilson, J.; Roy, B. Platelet rich plasma injections for lateral epicondylitis of the elbow reduce the need for surgical intervention. *J. Orthop.* **2018**, *15*, 239–241. [CrossRef]
251. Arirachakaran, A.; Sukthuayat, A.; Sisayanarane, T.; Laoratanavoraphong, S.; Kanchanatawan, W.; Kongtharvonskul, J. Platelet-rich plasma versus autologous blood versus steroid injection in lateral epicondylitis: Systematic review and network meta-analysis. *J. Orthop. Traumatol.* **2016**, *17*, 101–112. [CrossRef]
252. Mi, B.; Liu, G.; Zhou, W.; Lv, H.; Liu, Y.; Wu, Q.; Liu, J. Platelet rich plasma versus steroid on lateral epicondylitis: Meta-analysis of randomized clinical trials. *Phys. Sportsmed.* **2017**, *45*, 97–104. [CrossRef]
253. Ben-Nafa, W.; Munro, W. The effect of corticosteroid versus platelet-rich plasma injection therapies for the management of lateral epicondylitis: A systematic review. *SICOT J.* **2018**, *4*, 11. [CrossRef] [PubMed]
254. Houck, D.A.; Kraeutler, M.J.; Thornton, L.B.; McCarty, E.C.; Bravman, J.T. Treatment of Lateral Epicondylitis With Autologous Blood, Platelet-Rich Plasma, or Corticosteroid Injections: A Systematic Review of Overlapping Meta-analyses. *Orthop. J. Sports Med.* **2019**, *7*, 2325967119831052. [CrossRef] [PubMed]
255. Barnett, J.; Bernacki, M.N.; Kainer, J.L.; Smith, H.N.; Zaharoff, A.M.; Subramanian, S.K. The effects of regenerative injection therapy compared to corticosteroids for the treatment of lateral Epicondylitis: A systematic review and meta-analysis. *Arch. Physiother.* **2019**, *9*, 12. [CrossRef] [PubMed]
256. Monteagudo, M.; de Albornoz, P.M.; Gutierrez, B.; Tabuenca, J.; Alvarez, I. Plantar fasciopathy: A current concepts review. *EFORT Open Rev.* **2018**, *3*, 485–493. [CrossRef]
257. Henning, P.R.; Grear, B.J. Platelet-rich plasma in the foot and ankle. *Curr. Rev. Musculoskelet. Med.* **2018**, *11*, 616–623. [CrossRef]
258. Ragab, E.M.; Othman, A.M. Platelets rich plasma for treatment of chronic plantar fasciitis. *Arch. Orthop. Trauma Surg.* **2012**, *132*, 1065–1070. [CrossRef]
259. Kumar, V.; Millar, T.; Murphy, P.N.; Clough, T. The treatment of intractable plantar fasciitis with platelet-rich plasma injection. *Foot* **2013**, *23*, 74–77. [CrossRef]
260. Martinelli, N.; Marinozzi, A.; Carni, S.; Trovato, U.; Bianchi, A.; Denaro, V. Platelet-rich plasma injections for chronic plantar fasciitis. *Int. Orthop.* **2013**, *37*, 839–842. [CrossRef]
261. O'Malley, M.J.; Vosseller, J.T.; Gu, Y. Successful use of platelet-rich plasma for chronic plantar fasciitis. *HSS J.* **2013**, *9*, 129–133. [CrossRef] [PubMed]
262. Wilson, J.J.; Lee, K.S.; Miller, A.T.; Wang, S. Platelet-rich plasma for the treatment of chronic plantar fasciopathy in adults: A case series. *Foot Ankle Spec.* **2014**, *7*, 61–67. [CrossRef] [PubMed]
263. Malahias, M.A.; Mavrogenis, A.F.; Nikolaou, V.S.; Megaloikonomos, P.D.; Kazas, S.T.; Chronopoulos, E.; Babis, G.C. Similar effect of ultrasound-guided platelet-rich plasma versus platelet-poor plasma injections for chronic plantar fasciitis. *Foot* **2019**, *38*, 30–33. [CrossRef] [PubMed]
264. Johnson-Lynn, S.; Cooney, A.; Ferguson, D.; Bunn, D.; Gray, W.; Coorsh, J.; Kakwani, R.; Townshend, D. A Feasibility Study Comparing Platelet-Rich Plasma Injection With Saline for the Treatment of Plantar Fasciitis Using a Prospective, Randomized Trial Design. *Foot Ankle Spec.* **2019**, *12*, 153–158. [CrossRef] [PubMed]
265. Aksahin, E.; Dogruyol, D.; Yuksel, H.Y.; Hapa, O.; Dogan, O.; Celebi, L.; Bicimoglu, A. The comparison of the effect of corticosteroids and platelet-rich plasma (PRP) for the treatment of plantar fasciitis. *Arch. Orthop. Trauma Surg.* **2012**, *132*, 781–785. [CrossRef] [PubMed]

266. Tiwari, M.; Bhargava, R. Platelet rich plasma therapy: A comparative effective therapy with promising results in plantar fasciitis. *J. Clin. Orthop. Trauma* **2013**, *4*, 31–35. [CrossRef] [PubMed]
267. Omar, A.S.; Ibrahim, M.E.; Ahmed, A.S.; Said, M. Local injection of autologous platelet rich plasma and corticosteroid in treatment of lateral epicondylitis and plantar fasciitis: Randomized clinical trial. *Egypt. Rheumatol.* **2012**, *34*, 43–49. [CrossRef]
268. Shetty, V.D.; Dhillon, M.; Hegde, C.; Jagtap, P.; Shetty, S. A study to compare the efficacy of corticosteroid therapy with platelet-rich plasma therapy in recalcitrant plantar fasciitis: A preliminary report. *Foot Ankle Surg.* **2014**, *20*, 10–13. [CrossRef]
269. Jain, K.; Murphy, P.N.; Clough, T.M. Platelet rich plasma versus corticosteroid injection for plantar fasciitis: A comparative study. *Foot* **2015**, *25*, 235–237. [CrossRef]
270. Sherpy, N.A.; Hammad, M.A.; Hagrass, H.A.; Samir, H.S.; Abu-ElMaaty, S.E.; Mortadaa, M.A. Local injection of autologous platelet rich plasma compared to corticosteroid treatment of chronic plantar fasciitis patients: A clinical and ultrasonographic follow-up study. *Egypt. Rheumatol.* **2016**, *38*, 247–252. [CrossRef]
271. Vahdatpour, B.; Kianimehr, L.; Moradi, A.; Haghighat, S. Beneficial effects of platelet-rich plasma on improvement of pain severity and physical disability in patients with plantar fasciitis: A randomized trial. *Adv. Biomed. Res.* **2016**, *5*, 179. [CrossRef] [PubMed]
272. Acosta-Olivo, C.; Elizondo-Rodriguez, J.; Lopez-Cavazos, R.; Vilchez-Cavazos, F.; Simental-Mendia, M.; Mendoza-Lemus, O. Plantar Fasciitis-A Comparison of Treatment with Intralesional Steroids versus Platelet-Rich Plasma A Randomized, Blinded Study. *J. Am. Podiatr. Med. Assoc.* **2017**, *107*, 490–496. [CrossRef] [PubMed]
273. Jain, S.K.; Suprashant, K.; Kumar, S.; Yadav, A.; Kearns, S.R. Comparison of Plantar Fasciitis Injected With Platelet-Rich Plasma vs Corticosteroids. *Foot Ankle Int.* **2018**, *39*, 780–786. [CrossRef] [PubMed]
274. Monto, R.R. Platelet-rich plasma efficacy versus corticosteroid injection treatment for chronic severe plantar fasciitis. *Foot Ankle Int.* **2014**, *35*, 313–318. [CrossRef] [PubMed]
275. Say, F.; Gurler, D.; Inkaya, E.; Bulbul, M. Comparison of platelet-rich plasma and steroid injection in the treatment of plantar fasciitis. *Acta Orthop. Traumatol. Turc.* **2014**, *48*, 667–672. [CrossRef]
276. Peerbooms, J.C.; Lodder, P.; den Oudsten, B.L.; Doorgeest, K.; Schuller, H.M.; Gosens, T. Positive Effect of Platelet-Rich Plasma on Pain in Plantar Fasciitis: A Double-Blind Multicenter Randomized Controlled Trial. *Am. J. Sports Med.* **2019**, *47*, 3238–3246. [CrossRef]
277. Jimenez-Perez, A.E.; Gonzalez-Arabio, D.; Diaz, A.S.; Maderuelo, J.A.; Ramos-Pascua, L.R. Clinical and imaging effects of corticosteroids and platelet-rich plasma for the treatment of chronic plantar fasciitis: A comparative non randomized prospective study. *Foot Ankle Surg.* **2019**, *25*, 354–360. [CrossRef]
278. Mahindra, P.; Yamin, M.; Selhi, H.S.; Singla, S.; Soni, A. Chronic Plantar Fasciitis: Effect of Platelet-Rich Plasma, Corticosteroid, and Placebo. *Orthopedics* **2016**, *39*, e285–e289. [CrossRef]
279. Shetty, S.H.; Dhond, A.; Arora, M.; Deore, S. Platelet-Rich Plasma Has Better Long-Term Results Than Corticosteroids or Placebo for Chronic Plantar Fasciitis: Randomized Control Trial. *J. Foot Ankle Surg.* **2019**, *58*, 42–46. [CrossRef]
280. Chew, K.T.; Leong, D.; Lin, C.Y.; Lim, K.K.; Tan, B. Comparison of autologous conditioned plasma injection, extracorporeal shockwave therapy, and conventional treatment for plantar fasciitis: A randomized trial. *PM R* **2013**, *5*, 1035–1043. [CrossRef]
281. Kim, E.; Lee, J.H. Autologous platelet-rich plasma versus dextrose prolotherapy for the treatment of chronic recalcitrant plantar fasciitis. *PM R* **2014**, *6*, 152–158. [CrossRef] [PubMed]
282. Gonnade, N.; Bajpayee, A.; Elhence, A.; Lokhande, V.; Mehta, N.; Mishra, M.; Kaur, A. Regenerative efficacy of therapeutic quality platelet-rich plasma injections versus phonophoresis with kinesiotaping for the treatment of chronic plantar fasciitis: A prospective randomized pilot study. *Asian J. Transfus. Sci.* **2018**, *12*, 105–111. [CrossRef] [PubMed]
283. Gogna, P.; Gaba, S.; Mukhopadhyay, R.; Gupta, R.; Rohilla, R.; Yadav, L. Plantar fasciitis: A randomized comparative study of platelet rich plasma and low dose radiation in sportspersons. *Foot* **2016**, *28*, 16–19. [CrossRef] [PubMed]
284. Chiew, S.K.; Ramasamy, T.S.; Amini, F. Effectiveness and relevant factors of platelet-rich plasma treatment in managing plantar fasciitis: A systematic review. *J. Res. Med. Sci.* **2016**, *21*, 38. [CrossRef]
285. Yang, W.Y.; Han, Y.H.; Cao, X.W.; Pan, J.K.; Zeng, L.F.; Lin, J.T.; Liu, J. Platelet-rich plasma as a treatment for plantar fasciitis: A meta-analysis of randomized controlled trials. *Medicine* **2017**, *96*, e8475. [CrossRef]

286. Singh, P.; Madanipour, S.; Bhamra, J.S.; Gill, I. A systematic review and meta-analysis of platelet-rich plasma versus corticosteroid injections for plantar fasciopathy. *Int. Orthop.* **2017**, *41*, 1169–1181. [CrossRef]
287. Zwerver, J.; Bredeweg, S.W.; van den Akker-Scheek, I. Prevalence of Jumper's knee among nonelite athletes from different sports: A cross-sectional survey. *Am. J. Sports Med.* **2011**, *39*, 1984–1988. [CrossRef]
288. Mautner, K.; Colberg, R.E.; Malanga, G.; Borg-Stein, J.P.; Harmon, K.G.; Dharamsi, A.S.; Chu, S.; Homer, P. Outcomes after ultrasound-guided platelet-rich plasma injections for chronic tendinopathy: A multicenter, retrospective review. *PM R* **2013**, *5*, 169–175. [CrossRef]
289. Crescibene, A.; Napolitano, M.; Sbano, R.; Costabile, E.; Almolla, H. Infiltration of Autologous Growth Factors in Chronic Tendinopathies. *J. Blood Transfus.* **2015**, *2015*, 924380. [CrossRef]
290. Kaux, J.F.; Bruyere, O.; Croisier, J.L.; Forthomme, B.; Le Goff, C.; Crielaard, J.M. One-year follow-up of platelet-rich plasma infiltration to treat chronic proximal patellar tendinopathies. *Acta Orthop. Belg.* **2015**, *81*, 251–256.
291. Bowman, K.F., Jr.; Muller, B.; Middleton, K.; Fink, C.; Harner, C.D.; Fu, F.H. Progression of patellar tendinitis following treatment with platelet-rich plasma: Case reports. *Knee Surg. Sports Traumatol. Arthrosc.* **2013**, *21*, 2035–2039. [CrossRef]
292. Manfreda, F.; Palmieri, D.; Antinolfi, P.; Rinonapoli, G.; Caraffa, A. Can platelet-rich plasma be an alternative to surgery for resistant chronic patellar tendinopathy in sportive people? Poor clinical results at 1-year follow-up. *J. Orthop. Surg.* **2019**, *27*, 2309499019842424. [CrossRef]
293. Filardo, G.; Kon, E.; Di Matteo, B.; Pelotti, P.; Di Martino, A.; Marcacci, M. Platelet-rich plasma for the treatment of patellar tendinopathy: Clinical and imaging findings at medium-term follow-up. *Int. Orthop.* **2013**, *37*, 1583–1589. [CrossRef]
294. Charousset, C.; Zaoui, A.; Bellaiche, L.; Bouyer, B. Are multiple platelet-rich plasma injections useful for treatment of chronic patellar tendinopathy in athletes? a prospective study. *Am. J. Sports Med.* **2014**, *42*, 906–911. [CrossRef]
295. Zayni, R.; Thaunat, M.; Fayard, J.M.; Hager, J.P.; Carrillon, Y.; Clechet, J.; Gadea, F.; Archbold, P.; Sonnery Cottet, B. Platelet-rich plasma as a treatment for chronic patellar tendinopathy: Comparison of a single versus two consecutive injections. *Muscles Ligaments Tendons J.* **2015**, *5*, 92–98. [CrossRef]
296. Kaux, J.F.; Croisier, J.L.; Forthomme, B.; Le Goff, C.; Buhler, F.; Savanier, B.; Delcour, S.; Gothot, A.; Crielaard, J.M. Using platelet-rich plasma to treat jumper's knees: Exploring the effect of a second closely-timed infiltration. *J. Sci. Med. Sport* **2016**, *19*, 200–204. [CrossRef]
297. Filardo, G.; Kon, E.; Della Villa, S.; Vincentelli, F.; Fornasari, P.M.; Marcacci, M. Use of platelet-rich plasma for the treatment of refractory jumper's knee. *Int. Orthop.* **2010**, *34*, 909–915. [CrossRef]
298. Gosens, T.; Den Oudsten, B.L.; Fievez, E.; van 't Spijker, P.; Fievez, A. Pain and activity levels before and after platelet-rich plasma injection treatment of patellar tendinopathy: A prospective cohort study and the influence of previous treatments. *Int. Orthop.* **2012**, *36*, 1941–1946. [CrossRef]
299. Vetrano, M.; Castorina, A.; Vulpiani, M.C.; Baldini, R.; Pavan, A.; Ferretti, A. Platelet-rich plasma versus focused shock waves in the treatment of jumper's knee in athletes. *Am. J. Sports Med.* **2013**, *41*, 795–803. [CrossRef]
300. Abate, M.; Di Carlo, L.; Verna, S.; Di Gregorio, P.; Schiavone, C.; Salini, V. Synergistic activity of platelet rich plasma and high volume image guided injection for patellar tendinopathy. *Knee Surg. Sports Traumatol. Arthrosc.* **2018**, *26*, 3645–3651. [CrossRef]
301. Dragoo, J.L.; Wasterlain, A.S.; Braun, H.J.; Nead, K.T. Platelet-rich plasma as a treatment for patellar tendinopathy: A double-blind, randomized controlled trial. *Am. J. Sports Med.* **2014**, *42*, 610–618. [CrossRef] [PubMed]
302. Scott, A.; LaPrade, R.F.; Harmon, K.G.; Filardo, G.; Kon, E.; Della Villa, S.; Bahr, R.; Moksnes, H.; Torgalsen, T.; Lee, J.; et al. Platelet-Rich Plasma for Patellar Tendinopathy: A Randomized Controlled Trial of Leukocyte-Rich PRP or Leukocyte-Poor PRP Versus Saline. *Am. J. Sports Med.* **2019**, *47*, 1654–1661. [CrossRef] [PubMed]
303. Filardo, G.; Di Matteo, B.; Kon, E.; Merli, G.; Marcacci, M. Platelet-rich plasma in tendon-related disorders: Results and indications. *Knee Surg. Sports Traumatol. Arthrosc.* **2018**, *26*, 1984–1999. [CrossRef] [PubMed]
304. Liddle, A.D.; Rodriguez-Merchan, E.C. Platelet-Rich Plasma in the Treatment of Patellar Tendinopathy: A Systematic Review. *Am. J. Sports Med.* **2015**, *43*, 2583–2590. [CrossRef] [PubMed]

305. Jeong, D.U.; Lee, C.R.; Lee, J.H.; Pak, J.; Kang, L.W.; Jeong, B.C.; Lee, S.H. Clinical applications of platelet-rich plasma in patellar tendinopathy. *Biomed. Res. Int.* **2014**, *2014*, 249498. [CrossRef] [PubMed]
306. Dupley, L.; Charalambous, C.P. Platelet-Rich Plasma Injections as a Treatment for Refractory Patellar Tendinosis: A Meta-Analysis of Randomised Trials. *Knee Surg. Relat. Res.* **2017**, *29*, 165–171. [CrossRef] [PubMed]
307. Vander Doelen, T.; Jelley, W. Non-surgical treatment of patellar tendinopathy: A systematic review of randomized controlled trials. *J. Sci. Med. Sport* **2020**, *23*, 118–124. [CrossRef]
308. Andriolo, L.; Altamura, S.A.; Reale, D.; Candrian, C.; Zaffagnini, S.; Filardo, G. Nonsurgical Treatments of Patellar Tendinopathy: Multiple Injections of Platelet-Rich Plasma Are a Suitable Option: A Systematic Review and Meta-analysis. *Am. J. Sports Med.* **2019**, *47*, 1001–1018. [CrossRef]
309. Ekstrand, J.; Healy, J.C.; Walden, M.; Lee, J.C.; English, B.; Hagglund, M. Hamstring muscle injuries in professional football: The correlation of MRI findings with return to play. *Br. J. Sports Med.* **2012**, *46*, 112–117. [CrossRef]
310. Orchard, J.W.; Seward, H.; Orchard, J.J. Results of 2 decades of injury surveillance and public release of data in the Australian Football League. *Am. J. Sports Med.* **2013**, *41*, 734–741. [CrossRef]
311. Bernuzzi, G.; Petraglia, F.; Pedrini, M.F.; De Filippo, M.; Pogliacomi, F.; Verdano, M.A.; Costantino, C. Use of platelet-rich plasma in the care of sports injuries: Our experience with ultrasound-guided injection. *Blood Transfus.* **2014**, *12* (Suppl. S1), s229–s234. [CrossRef] [PubMed]
312. Zanon, G.; Combi, F.; Combi, A.; Perticarini, L.; Sammarchi, L.; Benazzo, F. Platelet-rich plasma in the treatment of acute hamstring injuries in professional football players. *Joints* **2016**, *4*, 17–23. [CrossRef] [PubMed]
313. Reurink, G.; Goudswaard, G.J.; Moen, M.H.; Weir, A.; Verhaar, J.A.; Bierma-Zeinstra, S.M.; Maas, M.; Tol, J.L.; Dutch Hamstring Injection Therapy Study, I. Platelet-rich plasma injections in acute muscle injury. *N. E. J. Med.* **2014**, *370*, 2546–2547. [CrossRef] [PubMed]
314. Reurink, G.; Goudswaard, G.J.; Moen, M.H.; Weir, A.; Verhaar, J.A.; Bierma-Zeinstra, S.M.; Maas, M.; Tol, J.L.; Dutch, H.I.T.s.I. Rationale, secondary outcome scores and 1-year follow-up of a randomised trial of platelet-rich plasma injections in acute hamstring muscle injury: The Dutch Hamstring Injection Therapy study. *Br. J. Sports Med.* **2015**, *49*, 1206–1212. [CrossRef] [PubMed]
315. Punduk, Z.; Oral, O.; Ozkayin, N.; Rahman, K.; Varol, R. Single dose of intra-muscular platelet rich plasma reverses the increase in plasma iron levels in exercise-induced muscle damage: A pilot study. *J. Sport Health Sci.* **2016**, *5*, 109–114. [CrossRef]
316. Martinez-Zapata, M.J.; Orozco, L.; Balius, R.; Soler, R.; Bosch, A.; Rodas, G.; Til, L.; Peirau, X.; Urrutia, G.; Gich, I.; et al. Efficacy of autologous platelet-rich plasma for the treatment of muscle rupture with haematoma: A multicentre, randomised, double-blind, placebo-controlled clinical trial. *Blood Transfus.* **2016**, *14*, 245–254. [CrossRef]
317. Bubnov, R.; Yevseenko, V.; Semeniv, I. Ultrasound guided injections of platelets rich plasma for muscle injury in professional athletes. Comparative study. *Med. Ultrason.* **2013**, *15*, 101–105. [CrossRef]
318. Wetzel, R.J.; Patel, R.M.; Terry, M.A. Platelet-rich plasma as an effective treatment for proximal hamstring injuries. *Orthopedics* **2013**, *36*, e64–e70. [CrossRef]
319. Park, P.Y.S.; Cai, C.; Bawa, P.; Kumaravel, M. Platelet-rich plasma vs. steroid injections for hamstring injury-is there really a choice? *Skelet. Radiol.* **2019**, *48*, 577–582. [CrossRef]
320. A Hamid, M.S.; Mohamed Ali, M.R.; Yusof, A.; George, J.; Lee, L.P. Platelet-rich plasma injections for the treatment of hamstring injuries: A randomized controlled trial. *Am. J. Sports Med.* **2014**, *42*, 2410–2418. [CrossRef]
321. Rossi, L.A.; Molina Romoli, A.R.; Bertona Altieri, B.A.; Burgos Flor, J.A.; Scordo, W.E.; Elizondo, C.M. Does platelet-rich plasma decrease time to return to sports in acute muscle tear? A randomized controlled trial. *Knee Surg. Sports Traumatol. Arthrosc.* **2017**, *25*, 3319–3325. [CrossRef] [PubMed]
322. Borrione, P.; Fossati, C.; Pereira, M.T.; Giannini, S.; Davico, M.; Minganti, C.; Pigozzi, F. The use of platelet-rich plasma (PRP) in the treatment of gastrocnemius strains: A retrospective observational study. *Platelets* **2018**, *29*, 596–601. [CrossRef] [PubMed]
323. Guillodo, Y.; Madouas, G.; Simon, T.; Le Dauphin, H.; Saraux, A. Platelet-rich plasma (PRP) treatment of sports-related severe acute hamstring injuries. *Muscles Ligaments Tendons J.* **2015**, *5*, 284–288. [CrossRef]

324. Hamilton, B.; Tol, J.L.; Almusa, E.; Boukarroum, S.; Eirale, C.; Farooq, A.; Whiteley, R.; Chalabi, H. Platelet-rich plasma does not enhance return to play in hamstring injuries: A randomised controlled trial. *Br. J. Sports Med.* **2015**, *49*, 943–950. [CrossRef] [PubMed]
325. Grassi, A.; Napoli, F.; Romandini, I.; Samuelsson, K.; Zaffagnini, S.; Candrian, C.; Filardo, G. Is Platelet-Rich Plasma (PRP) Effective in the Treatment of Acute Muscle Injuries? A Systematic Review and Meta-Analysis. *Sports Med.* **2018**, *48*, 971–989. [CrossRef] [PubMed]
326. Sheth, U.; Dwyer, T.; Smith, I.; Wasserstein, D.; Theodoropoulos, J.; Takhar, S.; Chahal, J. Does Platelet-Rich Plasma Lead to Earlier Return to Sport When Compared With Conservative Treatment in Acute Muscle Injuries? A Systematic Review and Meta-analysis. *Arthroscopy* **2018**, *34*, 281–288. [CrossRef] [PubMed]
327. Andia, I.; Abate, M. Platelet-rich plasma: Combinational treatment modalities for musculoskeletal conditions. *Front. Med.* **2018**, *12*, 139–152. [CrossRef]
328. Miroshnychenko, O.; Chang, W.T.; Dragoo, J.L. The Use of Platelet-Rich and Platelet-Poor Plasma to Enhance Differentiation of Skeletal Myoblasts: Implications for the Use of Autologous Blood Products for Muscle Regeneration. *Am. J. Sports Med.* **2017**, *45*, 945–953. [CrossRef]
329. Scully, D.; Naseem, K.M.; Matsakas, A. Platelet biology in regenerative medicine of skeletal muscle. *Acta Physiol. (Oxf.)* **2018**, *223*, e13071. [CrossRef]
330. Hammond, J.W.; Hinton, R.Y.; Curl, L.A.; Muriel, J.M.; Lovering, R.M. Use of autologous platelet-rich plasma to treat muscle strain injuries. *Am. J. Sports Med.* **2009**, *37*, 1135–1142. [CrossRef]
331. Chellini, F.; Tani, A.; Zecchi-Orlandini, S.; Sassoli, C. Influence of Platelet-Rich and Platelet-Poor Plasma on Endogenous Mechanisms of Skeletal Muscle Repair/Regeneration. *Int. J. Mol. Sci.* **2019**, *20*. [CrossRef] [PubMed]
332. Zhang, X.; Wang, J.; Chen, Z.; Hu, Q.; Wang, C.; Yan, J.; Dotti, G.; Huang, P.; Gu, Z. Engineering PD-1-Presenting Platelets for Cancer Immunotherapy. *Nano Lett.* **2018**, *18*, 5716–5725. [CrossRef] [PubMed]
333. Son, S.R.; Sarkar, S.K.; Nguyen-Thuy, B.L.; Padalhin, A.R.; Kim, B.R.; Jung, H.I.; Lee, B.T. Platelet-rich plasma encapsulation in hyaluronic acid/gelatin-BCP hydrogel for growth factor delivery in BCP sponge scaffold for bone regeneration. *J. Biomater. Appl.* **2015**, *29*, 988–1002. [CrossRef] [PubMed]
334. Garcia, J.P.; Stein, J.; Cai, Y.; Riemers, F.; Wexselblatt, E.; Wengel, J.; Tryfonidou, M.; Yayon, A.; Howard, K.A.; Creemers, L.B. Fibrin-hyaluronic acid hydrogel-based delivery of antisense oligonucleotides for ADAMTS5 inhibition in co-delivered and resident joint cells in osteoarthritis. *J. Control. Release* **2019**, *294*, 247–258. [CrossRef]
335. Lolli, A.; Sivasubramaniyan, K.; Vainieri, M.L.; Oieni, J.; Kops, N.; Yayon, A.; van Osch, G. Hydrogel-based delivery of antimiR-221 enhances cartilage regeneration by endogenous cells. *J. Control. Release* **2019**, *309*, 220–230. [CrossRef]
336. Dhillon, M.S.; Patel, S.; Bansal, T. Improvising PRP for use in osteoarthritis knee- upcoming trends and futuristic view. *J. Clin. Orthop. Trauma* **2019**, *10*, 32–35. [CrossRef]

5

Platelet Proteasome Activity and Metabolism is Upregulated during Bacterial Sepsis

Katharina Grundler Groterhorst [1,2,†], Hanna Mannell [2,3,†], Joachim Pircher [1,4] and Bjoern F Kraemer [1,*]

1 Medizinische Klinik und Poliklinik I, Klinikum der Universität München, Marchioninistrasse 15, 81377 Munich, Germany; gruendler.k@gmail.com (K.G.G.); Joachim.Pircher@med.uni-muenchen.de (J.P.)
2 Walter Brendel Centre of Experimental Medicine, University Hospital, Ludwig-Maximilians-University, Marchioninistr. 27, 81377 Munich, Germany; Hanna.Mannell@med.uni-muenchen.de
3 Biomedical Center, Ludwig-Maximilians-University, Großhaderner Str. 9, 82152 Planegg, Germany
4 DZHK (German Center for Cardiovascular Research), Partner Site Munich Heart Alliance, 80802 Munich, Germany
* Correspondence: bjoern.kraemer@klinik-ebe.de
† These authors contributed equally to this work.

Abstract: Dysregulation of platelet function can contribute to the disease progression in sepsis. The proteasome represents a critical and vital element of cellular protein metabolism in platelets and its proteolytic activity has been associated with platelet function. However, the role of the platelet proteasome as well as its response to infection under conditions of sepsis have not been studied so far. We measured platelet proteasome activity by fluorescent substrates, degradation of poly-ubiquitinated proteins and cleavage of the proteasome substrate Talin-1 in the presence of living *E. coli* strains and in platelets isolated from sepsis patients. Upregulation of the proteasome activator PA28 (PSME1) was assessed by quantitative real-time PCR in platelets from sepsis patients. We show that co-incubation of platelets with living *E. coli* (UTI89) results in increased degradation of poly-ubiquitinated proteins and cleavage of Talin-1 by the proteasome. Proteasome activity and cleavage of Talin-1 was significantly increased in α-hemolysin (HlyA)-positive *E. coli* strains. Supporting these findings, proteasome activity was also increased in platelets of patients with sepsis. Finally, the proteasome activator PA28 (PSME1) was upregulated in this group of patients. In this study we demonstrate for the first time that the proteasome in platelets is activated in the septic milieu.

Keywords: proteasome activity; platelets; sepsis; bacteria

---

1. Introduction

Sepsis is a life-threatening disease caused by a dysregulated inflammatory host response, often induced upon a systemic bacterial infection. In this context, *E. coli* is a frequent cause of sepsis. Apart from an exaggerated systemic inflammatory response, a pro-coagulant and pro-thrombotic state is present during sepsis and the uncontrolled activation of platelets can contribute to the progression of the disease [1]. Recent work has demonstrated that platelets also possess a functional proteasome [2] and others and we have shown that platelet function is associated with proteolytic regulation of proteins by the proteasome [3,4]. Moreover, the expression as well as the proteolytic activity of the proteasome were shown to be increased in muscle tissue during sepsis [5–7]. However, proteasome activity has not yet been studied in platelets during sepsis.

The proteasome represents a critical element for protein processing in human cells and is crucial for protein degradation, turnover and antigen presentation [8]. Proteins designated for proteasome processing are tagged with ubiquitin to be unfolded and identified by the proteasomal catalytic subunits [9]. Especially, in platelets, as anucleate cells, the proteolytic cleavage of proteins is an important mechanism for regulation of their cellular functions [10]. Indeed, proteasomal activity was shown to be important for platelet aggregation and thrombosis formation in vitro and in vivo and interestingly, these physiological processes could be efficiently prevented by proteasome inhibition [2–4,10,11]. Moreover, by studying the proteasomal cleavage of proteins involved in cytoskeletal regulation, such as Filamin A and Talin-1, our group was able to identify a link between the proteasome and NFκB in the regulation of collagen-induced platelet aggregation [3]. During inflammatory conditions, additional proteasomal subunits (PSME1 and PSME2) are expressed and form an immunoproteasome together with subunits of the conventional proteasome [12]. Apart from its important role in antigen presentation by MHC class I molecules, the immunoproteasome has been shown to exhibit a higher proteolytic activity and to prevent cellular damage during inflammation [13]. Of note, a functional immunoproteasome as well as the capacity to process and present antigens is present also in platelets [14,15]. Malfunction of the proteasome has been associated with several disease processes [16]. However, our knowledge about its role and function in platelets, especially under disease conditions, is still scarce.

In this study, we therefore investigated the activity of the proteasome in platelets in the septic milieu using living *E. coli* in vitro and in sepsis patients. We observed an upregulation of the immuno-proteasome subunit and activator PA28 (PSME1) in platelets from sepsis patients and increased processing of polyubiquitinated proteins as well as the proteasome substrate Talin-1 under conditions of sepsis. Proteasome activation was more pronounced when platelets were exposed to pathogenic *E. coli* (UTI89) expressing the exotoxin α-hemolysin compared to toxin-negative *E. coli* strains. Our novel data demonstrate that the proteasome in platelets responds to the septic environment and is upregulated in patients with sepsis.

2. Results

2.1. Platelet Proteasome Activity and Protein Metabolism is Increased in the Septic Milieu

As systemic *E. coli* infection is a frequent cause of sepsis, we were first interested in whether *E. coli* affects platelet proteasome activity. Incubation of isolated human platelets with the pathogenic *E. coli* strain UTI89 led to increased proteasome activity in vitro. This effect was specific, as it was effectively inhibited by the proteasome inhibitor epoxomicin (Figure 1A). Poly-ubiquitinated proteins, which represent proteins marked for proteasomal processing, were excessively degraded over time during coincubation with *E. coli* UTI89. This process was equally inhibited by treatment with epoxomicin (Figure 1B).

2.2. E. coli Exotoxin α-Hemolysin (hlyA) may be a Contributing Factor to Increased Proteasome Activity in Platelets

E. coli α-hemolysin is a potent exotoxin, which can activate proteases in a calcium-dependent fashion [17,18]. Platelet proteasome activity, assessed by fluorescent substrate cleavage, was significantly enhanced in the presence of *E. coli* UTI89 (Figure 2A). Effects were less pronounced during coincubation with *E. coli* strains, where the α-hemolysin gene was knocked-out (UTI89 ΔhlyA). Typical cleavage of the proteasome substrate Talin-1 by the proteasome from a 235 kDa fragment to a 190 kDa fragment was also increased in the presence of *E. coli* UTI89 (ratio 190/235 kDa). This effect was again weaker when platelets were coincubated with the hlyA-negative *E. coli* strain UTI89 ΔhlyA (Figure 2B).

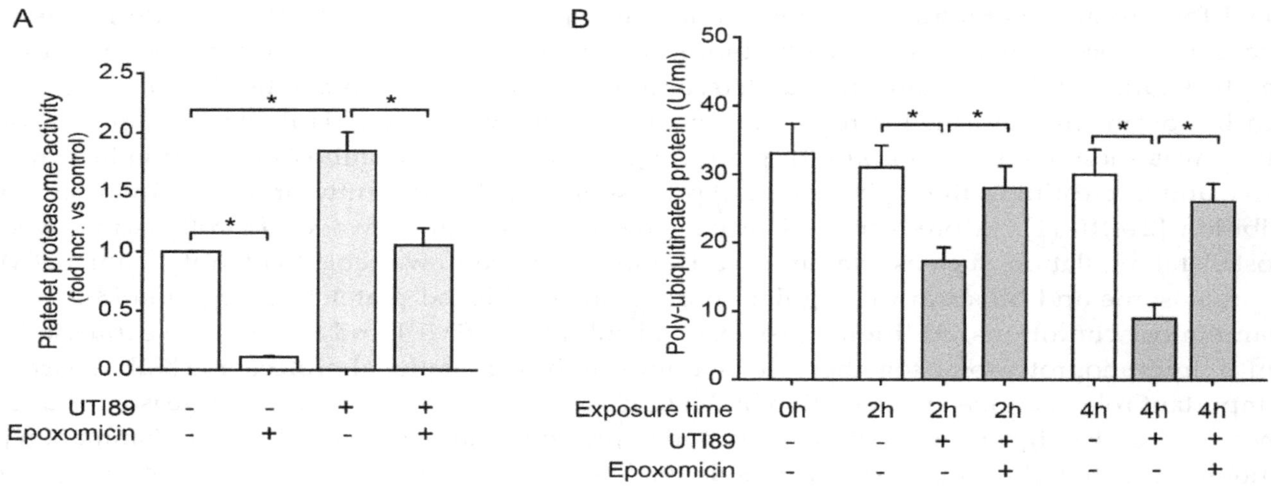

Figure 1. Bacteria induce proteasome activation and increased protein degradation in human platelets. (**A**) Co-incubation of platelets with living *E. coli* (UTI89) for 4 h induced a significant increase in platelet proteasome activity measured by fluorescent substrate cleavage, compared to control platelets (*$p < 0.05$, $n = 3$). Increased activation of the proteasome was effectively reversed with the proteasome inhibitor epoxomicin (10 μM; *$p < 0.05$, $n = 3$). (**B**) Platelet coincubation with living *E. coli* (UTI89) led to accelerated degradation of polyubiquitinated proteins (U/mL) in platelets after 2 and 4 h of incubation, as assessed by ELISA. This was effectively inhibited by proteasome inhibition (epoxomicin, 10 μM; *$p < 0.05$, $n = 3$).

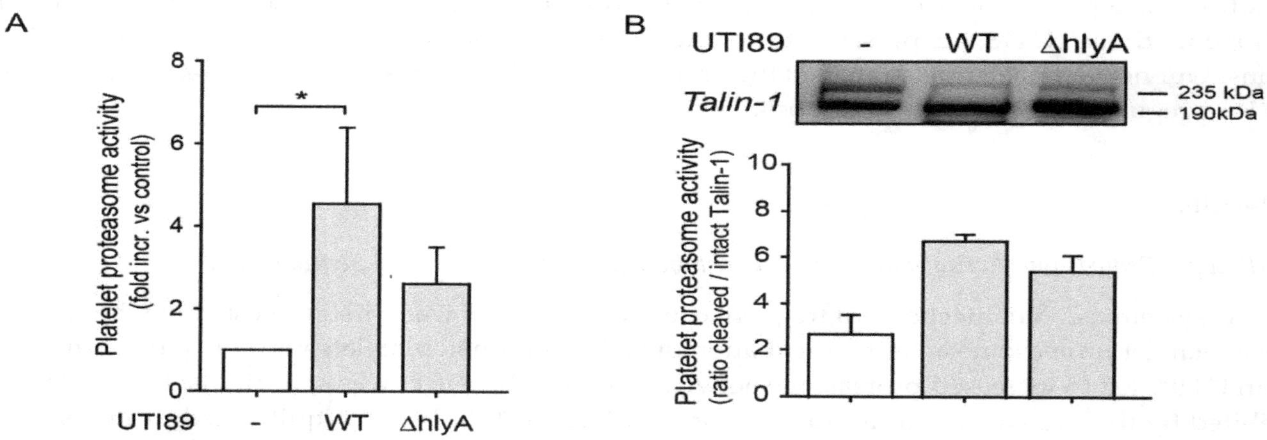

Figure 2. α-hemolysin expression in *E. coli* enhances platelet proteasome activation. (**A**) Platelet proteasome activity was enhanced upon incubation (4 h) with α-hemolysin expressing *E. coli* (UTI89 WT), whereas incubation with *E. coli* lacking functional α-hemolysin (UTI89 ΔhlyA) showed a trend towards less proteasome activation, as assessed by fluorescent peptide cleavage (*$p < 0.05$, $n = 5$). (**B**) The proteasome substrate Talin-1 was increasingly cleaved during coincubation with *E. coli* (UTI89) (ratio of 190 kDa to 235 kDa fragments) compared to control platelets. Increase of cleavage was less with the α-hemolysin negative strain UTI89 ΔhlyA compared to platelet control ($n = 2$).

2.3. Platelet Proteasome Activity is Increased in Platelets from Sepsis Patients

Having observed that pathogenic *E. coli* enhances platelet proteasome activity, we next studied the platelet proteasome in patients with bacterial sepsis. Clinical information on sepsis patients is shown in Table 1. As seen in Figure 3A, proteasome activity in platelets from sepsis patients was significantly upregulated compared to healthy controls. In addition, cleavage of the proteasome substrate Talin-1, expressed as a 190 to 235 kDa ratio, was also significantly increased in platelets of sepsis patients compared to controls (Figure 3B). As the immunoproteasome has been shown to be upregulated upon inflammatory conditions, we next investigated immunoproteasome activator PA28 (PSME1) expression in platelets from sepsis patients. mRNA expression analysis using real-time PCR revealed an upregulation of PA28 (PSME1) in platelets of sepsis patients compared to controls (Figure 3C).

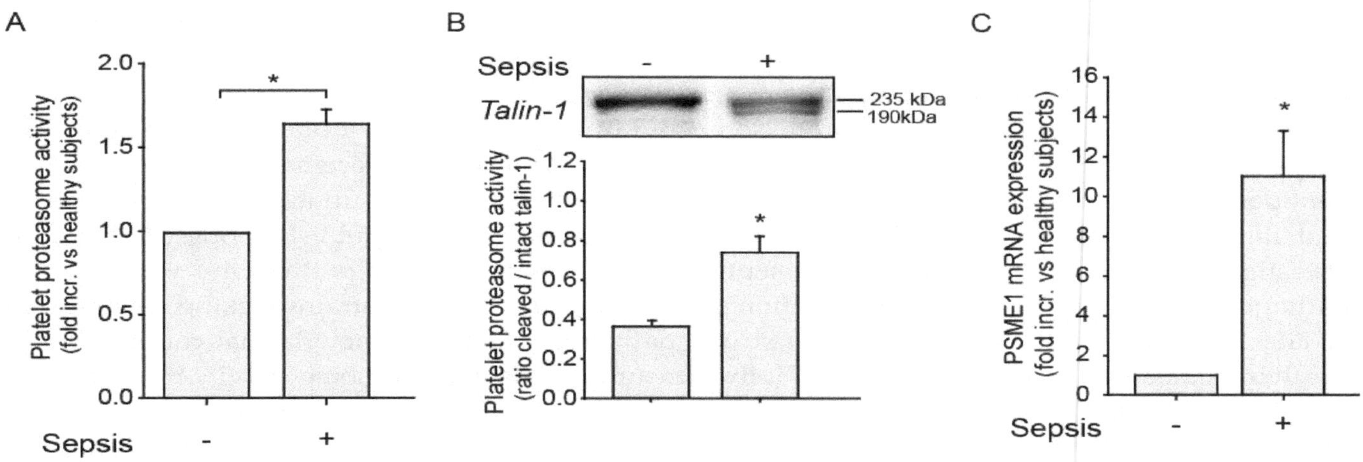

Figure 3. Platelets proteasome activity is increased in platelets of sepsis patients. (**A**) Platelet proteasome activity is increased in patients with sepsis (*$p < 0.05$, $n = 7$) compared to healthy controls ($n = 9$). (**B**) Cleavage of the proteasome substrate Talin-1 from a 235 to a 190 kDa fragment is significantly increased in platelets of the sepsis population ($n = 5$) compared to controls ($n = 3$; *$p < 0.05$). Proteasome activity was expressed as a ratio of cleaved 190 kDa to intact 235 kDa Talin-1 products by immunoblotting. The protein bands from a healthy control and a sepsis patient in the western blot image are derived from the same blot but cropped out and placed next to each other (white dotted line) for illustrative purposes. (**C**) mRNA expression analysis by real-time PCR revealed that the proteasome activator PA28 (PSME1) was overexpressed in platelets of sepsis patients compared to healthy individuals ($n = 4$, *$p < 0.05$).

2.4. Clinical Characteristics of Sepsis Patients

We have summarized clinical characteristics of sepsis patients in Table 1.

Table 1. Clinical characteristics of seven patients with bacterial sepsis are shown. Patients were diagnosed with bacterial sepsis based on clinical presentation and positive bacterial culture results. Markers of inflammation including leukocyte count, C-reactive protein (CRP) and procalcitonin were markedly elevated. SOFA score (sequential organ failure assessment) showed multi-organ dysfunction and all patients were treated with antibiotics. Data is presented as median and (interquartile range (IQR)), n = 7.

Clinical Characteristics of Seven Patients with Bacterial Sepsis	
Age	63 [IQR 49; 69]
Leukocyte count (1000/μL)	16 [IQR 11.0; 21.0]
CRP level (mg/dl)	25 [IQR 19.5; 28.5]
Procalcitonin (ng/mL)	5.5 [IQR 4.0; 9.8]
Positive bacterial culture	7 of 7
Antibiotic treatment	7 of 7
SOFA score	9 [IQR 6; 11]

3. Discussion

In the present study, we demonstrated for the first time that the proteasome of platelets was not static but responded to environmental stress conditions, such as a septic milieu, by increasing proteasome activity and protein metabolism. We know that the proteasome is activated in muscle tissue and in muscle wasting in sepsis [6,7], but in platelets its activity and response under septic conditions has not been studied so far. In vitro, we demonstrated that exposure of platelets with living *E. coli*, which is frequently identified in blood cultures of sepsis patients, increased platelet proteasome activity and subsequently resulted in increased degradation of poly-ubiquitinated proteins in platelets. This metabolic activation was effectively inhibited with the proteasome inhibitor epoxomicin. Polyubiquitination is a characteristic mechanism for labeling proteins for proteasomal processing and thus levels of polyubiquitinated proteins represent a good marker for proteasomal activity. More precisely, we showed that the proteasome substrate Talin-1 is excessively cleaved by the proteasome from a 235 to a 190 kDa fragment after co-incubation with *E. coli* (UTI89). Talin-1 has been shown to be a classical target of the proteasome and functional aspects of platelet aggregation and shape change have been associated with proteolytic regulation by the proteasome [3,4]. Talin-1-derived peptides were found to be presented on HLA-1 molecules on the platelet surface, which also confirms that Talin-1 is processed by the (immuno)-proteasome of platelets (unpublished observations). To investigate underlying mechanisms how septic processes may affect the proteasome, we compared pathogenic *E. coli* strains expressing functional α-hemolysin (UTI89) to strains lacking α-hemolysin (UTI89 ΔhlyA) [19,20]. α-hemolysin is a strong pathogenic factor of bacteria that can induce cell death in platelets as previously shown [17]. By measuring platelet proteasome activity by fluorescent substrate and Talin-1 cleavage during coincubation with living *E. coli*, we observed that the increase of proteasome activity seemed to be more pronounced in wild-type α-hemolysin positive UTI89 strains compared to strains not expressing α-hemolysin (UTI89 ΔhlyA). Although this needs further confirmation, the results suggest that α-hemolysin may be one of the pathogenic factors mediating the activation of the platelet proteasome. As previously shown for calpain [17], this process likely involves calcium-dependent activation of proteasomal subunits to increase proteolytic activity, although the precise mechanism remains to be elucidated in future studies. Finally, to study platelet proteasome activity in a complex sepsis environment, we compared platelets from patients with bacterial sepsis to healthy patients on an mRNA expression level ex vivo. Using real-time PCR we observed upregulation of the proteasome activator PA28 (PSME1) in platelets specifically in the sepsis population in line with increased proteasome activity. This finding was of particular interest, as PA28 (PSME1) is primarily an activator of the immunoproteasome that processes peptides for antigen presentation and which is activated during inflammation [5]. The presented data support our concept that the proteasome in platelets is involved in cellular adaptation to inflammatory environmental stress, such as sepsis, on different levels. PA28 (PSME1) encodes for the 11S proteasomal subunit that replaces the 19S regulatory subunit when the immunoproteasome is assembled. PA28 (PMSE1) is expressed in platelets at the mRNA level [15] and protein level [21] as previously shown. Interestingly, the complete machinery to process and present antigens on the cell surface via MHC-I is present in platelets, which underscores that platelets play an active role in the immune defense [14,15]. As an example, previous work shows that HLA-1 expression in platelets is altered during infection with dengue virus [22]. Supporting our in vitro findings, we observed increased platelet proteasome activity in platelets of patients with sepsis by fluorescent substrate cleavage as well as significantly increased cleavage of the proteasome substrate Talin-1 in the sepsis population ex vivo. Although this finding supports the observation that *E. coli* induces proteasomal proteolytic activity, we cannot exclude that the inflammatory environment per se additionally contributes to increased proteasome activity in sepsis patients. It also still remains to be investigated, which specific effects an increased proteasome activity has on platelet function in sepsis. Nevertheless, our collected data implied that platelet proteasome activity was affected both in terms of activity and function but also on an mRNA expression level during sepsis.

4. Materials and Methods

4.1. *Materials*

Talin-1 antibody was purchased from Cell Signaling, Danvers, USA, and horseradish peroxidase-conjugated secondary antibody was from Merck Millipore (Billerica, MA, USA). Epoxomicin was purchased from EMD Biosciences (LaJolla, CA, USA). All other chemicals were purchased from Sigma Aldrich (Darmstadt, Germany).

4.2. Bacteria

E. coli bacteria of the α-hemolytic strain (hlyA) UTI89 [19] as well as the *E. coli* strain with deletion of hlyA (UTI89 ΔhlyA) [20] were a kind gift from Dr. Matthew Mulvey, Department of Pathology at the University of Utah (Salt Lake City, UT, USA). Similar to recent work [17], bacteria were expanded on blood agar plates overnight at 37 °C until they reached a stationary growth phase. Bacteria were then resuspended in phosphate-buffered saline (PBS) and their concentration was determined by colorimetry (VITEK Colorimeter, bioMerieux, Inc., Durham, NC, USA). Prior to experiments, *E. coli* bacteria were resuspended in M199 culture. For every experiment, the soluble agonists or bacteria were incubated in the presence of freshly isolated platelets (1×10^8 platelets at bacteria to platelet ratio of 1:30) for 4 h in M199 culture media under cell culture conditions.

4.3. Platelet Isolation and Preparation

As previously described [3,17,23], washed platelets were freshly isolated from healthy human subjects and patients with bacterial sepsis, who consented to participate in the study. All studies involving patients and human cells were performed in accordance with the declaration of Helsinki and were approved by the local ethics committee (No. 290-11, approved at 12 July 2011). For inhibitor studies of the proteasome, platelets were incubated with epoxomicin (10 μM) for the indicated time in warm M199 media under cell culture conditions (5% CO_2, 37 °C).

4.4. RNA Isolation

Platelets (1×10^9) were isolated from healthy donors and patients with bacterial sepsis and were exposed to CD45 positive selection, which effectively depletes contaminating leukocytes as previously described [24]. Cells were lysed in Trizol (Invitrogen, Carlsbad, CA, USA) and glycogen was added to the aqueous phase before precipitation with isopropanol to optimize RNA yields. The RNA was treated with DNAse (DNA free Kit, Ambion, Austin, TX, USA), precipitated with ethanol and dissolved in 12 μL of RNAse-free water. RNA (1 μg) was used to generate cDNA to characterize the expression of PSME1. Integrin αIIb was used as a positive control for platelet-specific RNA. The relative abundance of the PSME1 RNA was measured by real-time PCR. Primer sets for these studies were as follows: PSME1, forward-GAAGCCAACTTGAGCAATCTGA, reverse-AGCCTTCTAGCTTGGTGTGGAG; and β2-Microglobulin, forward-ACACTATTCTAGCAGGAGGGTTGG, reverse- CAGGGCTCAGTCTCTTTATTAGGC.

4.5. Detection of Platelet Proteasome Activity

Proteasome activity was measured using a proteasome activity assay with fluorescent substrate according to the manufacturer's instructions (Millipore, Billerica, MA, USA) and as previously described [3]. For experiments where platelets were exposed to living *E. coli* bacteria (see above), cells were pelleted and lysed after 4 h of incubation. As a negative control equal amounts of bacteria were incubated under identical conditions, pelleted, lysed and incubated with the proteasome substrate. Platelet proteasome activity (relative fluorescent units, RFU) was detected at 4 h under cell culture conditions at 450 nm using a fluorescence microplate reader (Tecan Group, Mannedorf, Switzerland).

4.6. Quantification of Polyubiquitinated Protein in Human Platelets

After incubation with living *E. coli* or the proteasome inhibitor epoxomicin, cells were pelleted and lysed in 200 μL of ubiquitinated protein lysis buffer according to the manufacturer's protocol (human poly-ubiquitinated protein ELISA, MBL International, Woburn, MA, USA). Platelet lysates were diluted 1:200 in the provided dilution buffer and then added to the ELISA plate. ELISA was carried out according to the manufacturer's instructions. Quantitative changes of total poly-ubiquitinated protein (U/mL) in platelets were measured at 450 nm with an ELISA plate reader (BMG LabTec, Worchester, MA, USA).

4.7. Immunoblotting

Immunoblotting for Talin-1 was performed as previously described [3]. In brief, pellets were lysed in cell lysis buffer (Cell Signaling, Danvers, MA, USA) containing 1 mM PMSF. After centrifugation at 10,000 g for 5 min at 4 °C, protein quantification was performed through the bicinchoninic acid assay (ThermoScientific, Waltham, MA, USA) according to the manufacturer's protocol. Equal volumes of sample were separated by SDS-PAGE using 10% gels and blotted onto PVDF membrane. Membranes were blocked in 5% milk and dissolved in Tris buffered saline with 0.1% Tween and were subsequently incubated with the primary antibody against Talin-1 (1:1000 dilution) at 4 °C overnight. Membranes were washed and incubated with horseradish peroxidase-conjugated secondary antibodies at room temperature for 1 h. Enzymatic activity was detected with a chemiluminescence detection kit according to the supplier's protocol and recorded with a digital camera (Hamamatsu Photonics, Hamamatsu City, Japan). Protein band density measurements for Talin-1 235 kDa and 190 kDa fragments were performed with Image J software. For evaluation of proteasomal cleavage of Talin-1, the ratio 235 kDa/190 kDa was calculated.

4.8. Statistical Analysis

All data are presented as means ± SEM. Statistical analyses were performed with Sigma Plot 10.0. For comparisons between two groups of normal distributed data, the student's *t*-test was used. For the comparison of two groups without normal distributed data, a rank-sum test was performed. For multiple comparisons between groups of normal distributed data, the one-way analysis of variance (1-way ANOVA) was used. Differences were considered significant at an error probability level of $p < 0.05$.

5. Conclusions

In summary, our study provided novel evidence that the proteasome of platelets responded to the septic environment by increased proteasome activity, increased proteolytic cleavage of proteasome substrates such as Talin-1 and upregulation of (immuno)proteasome activator PA28 (PSME1). Although we still know little about the precise function of the proteasome in platelets, it appeared that the proteasome was an important element in the response to severe cellular stress situations such as sepsis. Future studies will have to show if this response is simply a matter of increased metabolic protein turnover required during systemic inflammation or if it serves the purpose of anti-pathogenic defense strategies, antigen processing, hemostasis or all of them. This study, however, provides a basis to further characterize the precise role and regulation of the proteasome in platelets in complex disease processes, such as sepsis.

Author Contributions: Conceptualization and project administration: B.F.K.; Supervision: H.M., B.F.K.; Investigation: K.G.G., J.P., B.F.K.; Methodology: K.G.G., J.P., B.F.K.; Visualization: K.G.G., H.M.; Writing—original draft preparation: H.M., B.F.K.

Acknowledgments: We thank the group of Professor Matthew Mulvey, Department of Pathology at the University of Utah (Salt Lake City, USA) for generously providing the E. coli strains used in the study. We are grateful to the group of Professor Andrew Weyrich, Eccles Institute of Human Genetics at the University of Utah (Salt Lake City, USA) for their support in establishing some of the methods used in the study and to the group of Professor Stefan Stevanovic, Department of Immunology at the University of Tuebingen, Germany for sharing their expertise with antigen presentation.

References

1. Assinger, A.; Schrottmaier, W.C.; Salzmann, M.; Rayes, J. Platelets in Sepsis: An Update on Experimental Models and Clinical Data. *Front. Immunol.* **2019**, *10*, 1687. [CrossRef] [PubMed]
2. Nayak, M.K.; Kumar, K.; Dash, D. Regulation of proteasome activity in activated human platelets. *Cell Calcium* **2011**, *49*, 226–232. [CrossRef] [PubMed]
3. Grundler, K.; Rotter, R.; Tilley, S.; Pircher, J.; Czermak, T.; Yakac, M.; Gaitzsch, E.; Massberg, S.; Krotz, F.; Sohn, H.Y.; et al. The proteasome regulates collagen-induced platelet aggregation via nuclear-factor-kappa-B (NFkB) activation. *Thromb. Res.* **2016**, *148*, 15–22. [CrossRef] [PubMed]
4. Gupta, N.; Li, W.; Willard, B.; Silverstein, R.L.; McIntyre, T.M. Proteasome proteolysis supports stimulated platelet function and thrombosis. *Arterioscler. Thromb. Vasc. Biol.* **2014**, *34*, 160–168. [CrossRef] [PubMed]
5. Ferrington, D.A.; Gregerson, D.S. Immunoproteasomes: Structure, function, and antigen presentation. *Prog. Mol. Biol. Transl. Sci.* **2012**, *109*, 75–112. [PubMed]
6. Hobler, S.C.; Williams, A.; Fischer, D.; Wang, J.J.; Sun, X.; Fischer, J.E.; Monaco, J.J.; Hasselgren, P.O. Activity and expression of the 20S proteasome are increased in skeletal muscle during sepsis. *Am. J. Physiol.* **1999**, *277*, R434–R440. [CrossRef] [PubMed]
7. Klaude, M.; Fredriksson, K.; Tjader, I.; Hammarqvist, F.; Ahlman, B.; Rooyackers, O.; Wernerman, J. Proteasome proteolytic activity in skeletal muscle is increased in patients with sepsis. *Clin. Sci. (Lond.)* **2007**, *112*, 499–506. [CrossRef]
8. Tanaka, K. The proteasome: Overview of structure and functions. *Proc. Jpn. Acad. Ser. B Phys. Biol. Sci.* **2009**, *85*, 12–36. [CrossRef]
9. Hershko, A.; Ciechanover, A. The ubiquitin system. *Annu. Rev. Biochem.* **1998**, *67*, 425–479. [CrossRef]
10. Kraemer, B.F.; Weyrich, A.S.; Lindemann, S. Protein degradation systems in platelets. *Thromb. Haemost.* **2013**, *110*, 920–924. [CrossRef]
11. Avcu, F.; Ural, A.U.; Cetin, T.; Nevruz, O. Effects of bortezomib on platelet aggregation and ATP release in human platelets, in vitro. *Thromb. Res.* **2008**, *121*, 567–571. [CrossRef] [PubMed]
12. Kimura, H.; Caturegli, P.; Takahashi, M.; Suzuki, K. New Insights into the Function of the Immunoproteasome in Immune and Nonimmune Cells. *J. Immunol. Res.* **2015**, *2015*, 541984. [CrossRef] [PubMed]
13. Seifert, U.; Bialy, L.P.; Ebstein, F.; Bech-Otschir, D.; Voigt, A.; Schroter, F.; Prozorovski, T.; Lange, N.; Steffen, J.; Rieger, M.; et al. Immunoproteasomes preserve protein homeostasis upon interferon-induced oxidative stress. *Cell* **2010**, *142*, 613–624. [CrossRef] [PubMed]
14. Chapman, L.M.; Aggrey, A.A.; Field, D.J.; Srivastava, K.; Ture, S.; Yui, K.; Topham, D.J.; Baldwin, W.M., 3rd; Morrell, C.N. Platelets present antigen in the context of MHC class I. *J. Immunol.* **2012**, *189*, 916–923. [CrossRef]
15. Klockenbusch, C.; Walsh, G.M.; Brown, L.M.; Hoffman, M.D.; Ignatchenko, V.; Kislinger, T.; Kast, J. Global proteome analysis identifies active immunoproteasome subunits in human platelets. *Mol. Cell. Proteomics* **2014**, *13*, 3308–3319. [CrossRef]
16. Dahlmann, B. Role of proteasomes in disease. *BMC Biochem.* **2007**, *8* (Suppl. 1), S3. [CrossRef]
17. Kraemer, B.F.; Campbell, R.A.; Schwertz, H.; Franks, Z.G.; Vieira de Abreu, A.; Grundler, K.; Kile, B.T.; Dhakal, B.K.; Rondina, M.T.; Kahr, W.H.; et al. Bacteria differentially induce degradation of Bcl-xL, a survival protein, by human platelets. *Blood* **2012**, *120*, 5014–5020. [CrossRef]
18. Ristow, L.C.; Welch, R.A. Hemolysin of uropathogenic Escherichia coli: A cloak or a dagger? *Biochim. Biophys. Acta* **2016**, *1858*, 538–545. [CrossRef]
19. Mulvey, M.A.; Schilling, J.D.; Hultgren, S.J. Establishment of a persistent Escherichia coli reservoir during the acute phase of a bladder infection. *Infect. Immun.* **2001**, *69*, 4572–4579. [CrossRef]
20. Wiles, T.J.; Dhakal, B.K.; Eto, D.S.; Mulvey, M.A. Inactivation of host Akt/protein kinase B signaling by bacterial pore-forming toxins. *Mol. Biol. Cell* **2008**, *19*, 1427–1438. [CrossRef]
21. Ostrowska, H.; Ostrowska, J.K.; Worowski, K.; Radziwon, P. Human platelet 20S proteasome: Inhibition of its chymotrypsin-like activity and identification of the proteasome activator PA28. A preliminary report. *Platelets* **2003**, *14*, 151–157. [CrossRef] [PubMed]
22. Trugilho, M.R.O.; Hottz, E.D.; Brunoro, G.V.F.; Teixeira-Ferreira, A.; Carvalho, P.C.; Salazar, G.A.; Zimmerman, G.A.; Bozza, F.A.; Bozza, P.T.; Perales, J. Platelet proteome reveals novel pathways of platelet activation and platelet-mediated immunoregulation in dengue. *PLoS Pathog.* **2017**, *13*, e1006385. [CrossRef] [PubMed]

23. Weyrich, A.S.; Elstad, M.R.; McEver, R.P.; McIntyre, T.M.; Moore, K.L.; Morrissey, J.H.; Prescott, S.M.; Zimmerman, G.A. Activated platelets signal chemokine synthesis by human monocytes. *J. Clin. Invest.* **1996**, *97*, 1525–1534. [CrossRef] [PubMed]
24. Kraemer, B.F.; Campbell, R.A.; Schwertz, H.; Cody, M.J.; Franks, Z.; Tolley, N.D.; Kahr, W.H.; Lindemann, S.; Seizer, P.; Yost, C.C.; et al. Novel anti-bacterial activities of beta-defensin 1 in human platelets: Suppression of pathogen growth and signaling of neutrophil extracellular trap formation. *PLoS Pathog.* **2011**, *7*, e1002355. [CrossRef] [PubMed]

6

The MICELI (MICrofluidic, ELectrical, Impedance): Prototyping a Point-of-Care Impedance Platelet Aggregometer

Yana Roka-Moiia [1,†], Silvia Bozzi [2,†], Chiara Ferrari [2], Gabriele Mantica [2], Annalisa Dimasi [2], Marco Rasponi [2], Andrea Santoleri [2], Mariangela Scavone [3], Filippo Consolo [4], Marco Cattaneo [3], Marvin J. Slepian [1,*,‡] and Alberto Redaelli [2,‡]

1 Departments of Medicine and Biomedical Engineering, College of Medicine, University of Arizona, Tucson, AZ 85724, USA; rokamoiia@email.arizona.edu
2 Department of Electronics, Information and Bioengineering, Politecnico di Milano, Milan 20133, Italy; silvia.bozzi@polimi.it (S.B.); chiara6.ferrari@mail.polimi.it (C.F.); gabriele.mantica@mail.polimi.it (G.M.); annalisa.dimasi@polimi.it (A.D.); marco.rasponi@polimi.it (M.R.); andrea.santoleri@gmail.com (A.S.); alberto.redaelli@polimi.it (A.R.)
3 Divisione di Medicina, Ospedale San Paolo, Dipartimento di Scienze della Salute, Università degli Studi di Milano, Milan 20122, Italy; mariangela.scavone@guest.unimi.it (M.S.); marco.cattaneo@unimi.it (M.C.)
4 Università Vita Salute, San Raffaele, Milan 20132, Italy; consolo.filippo@hsr.it
* Correspondence: slepian@email.arizona.edu
† Both first authors contributed equally to this manuscript.
‡ Both authors served as senior authors for this work.

Abstract: As key cellular elements of hemostasis, platelets represent a primary target for thrombosis and bleeding management. Currently, therapeutic manipulations of platelet function (antithrombotic drugs) and count (platelet transfusion) are performed with limited or no real-time monitoring of the desired outcome at the point-of-care. To address the need, we have designed and fabricated an easy-to-use, accurate, and portable impedance aggregometer called "MICELI" (MICrofluidic, ELectrical, Impedance). It improves on current platelet aggregation technology by decreasing footprint, assay complexity, and time to obtain results. The current study aimed to optimize the MICELI protocol; validate sensitivity to aggregation agonists and key blood parameters, i.e., platelet count and hematocrit; and verify the MICELI operational performance as compared to commercial impedance aggregometry. We demonstrated that the MICELI aggregometer could detect platelet aggregation in 250 μL of whole blood or platelet-rich plasma, stimulated by ADP, TRAP-6, collagen, epinephrine, and calcium ionophore. Using hirudin as blood anticoagulant allowed higher aggregation values. Aggregation values obtained by the MICELI strongly correlated with platelet count and were not affected by hematocrit. The operational performance comparison of the MICELI and the Multiplate® Analyzer demonstrated strong correlation and similar interdonor distribution of aggregation values obtained between these devices. With the proven reliability of the data obtained by the MICELI aggregometer, it can be further translated into a point-of-care diagnostic device aimed at monitoring platelet function in order to guide pharmacological hemostasis management and platelet transfusions.

Keywords: platelet aggregation; electrical impedance; whole blood aggregometry; point-of-care; platelet function testing

---

1. Introduction

Platelets are vital blood cells critical for maintenance of vascular integrity and tissue repair [1]. Unfortunately, in many disease states platelets become activated with resultant pathologic thrombosis, reducing organ blood flow, leading to ischemia, infarction, and death [2]. To prevent inadvertent platelet activation, coagulation, and thrombosis, an increasing percentage of the U.S. population are prescribed a range of antithrombotic medications. However, many of these medications have varying interpatient pharmacodynamic efficacy [3]. For example, while aspirin has been proven to reduce thrombotic risk by 25%, still 10%–20% of aspirin-treated patients experience thrombotic events [4]. Conversely, many patients on appropriate dose therapy experience untoward bleeding consequences, with a net increase in bleeding risk [5,6]. This is particularly problematic when patients present with trauma or acute illness, requiring emergent surgery and intervention.

For patients experiencing massive trauma, falls, intestinal ischemia, and acute arterial insufficiency, an immediate surgical intervention is needed. For many patients encountered in emergency care situations, platelet function is compromised due to long-term use of antithrombotic therapy [7]. To date, there are no specific agents able to reverse the antiplatelet effect of contemporary antithrombotic drugs. Platelet transfusion has previously shown its efficacy in preventing perioperative hemorrhage, though the dosing and administration time remain unclear [8–10]. Therefore, careful monitoring of platelet function at the point-of-care (POC) is recommended by recent surgical guidelines to control risk of bleeding in pre-, peri-, and postoperative settings [7]. However, rapid platelet function testing is often not available at the POC, and surgeons rely only on platelet count and coagulation tests for bleeding risk assessment and platelet transfusion management [11]. The result is elevated procedure-related bleeding with significant morbidity and mortality [12].

Platelet aggregometry is the "gold standard" of platelet function monitoring [13,14]. The magnitude and rate of platelet aggregation are crucial indicators of platelet function and overall blood hemostasis state. To quantify platelet aggregation, benchtop optical and electronic platelet aggregometers are utilized. Light transmission aggregometry (LTA) detects platelet aggregation induced by chemical agonists, e.g., adenosine diphosphate (ADP), collagen, and arachidonic acid, in platelet-rich plasma (PRP) [15]. As light passes through a PRP sample, transmission is recorded via photometry; as platelets aggregate, PRP turbidity decreases, with increased light transmittance detected [16,17]. While LTA provides useful information about platelet function, its utility in clinical laboratory or POC environments is limited due to a series of significant limitations: the need for blood sample processing (to obtain PRP), a large blood volume requirement (6–10 mL) and significant time (>1 h) for a single test, expensive bulky equipment, and the need for trained personnel to run the test and interpret obtained results [3,18].

Electrical impedance aggregometry (EIA) affords an improvement on several limitations of LTA. EIA works via measuring resistance between two electrodes submerged in whole blood or PRP (Figure A3) [19]. The addition of an agonist causes platelets to aggregate around the electrodes, corresponding to an increase in sample resistance [20,21]. EIA allows measurement of platelet aggregation in whole blood, thus eliminating the need for blood sample processing, but currently requires expensive semiautomated equipment and trained personnel to operate it. Despite the reliability and high value of the information provided by both LTA and EIA, their shared limitations minimize their clinical use. Based on the FDA premarket approval, LTA and EIA instruments are generally conducted as "research use only" devices or with restricted "intended use" considerations restraining their use for POC diagnostics [22,23].

To overcome the limitations of commercial aggregometers, we designed and fabricated a portable, accurate, and easy-to-use platelet aggregometer able to measure platelet aggregation in whole blood and PRP. Our platform, called "MICELI" (MICrofluidic, ELectrical, Impedance), is an impedance aggregometer recording the change in resistance resulting from platelet aggregation on electrodes within a miniature polymeric cartridge. Impedance data are converted into aggregation values: maximum amplitude (Amax) and area under the curve (AUC) [24]. Only a small sample volume (250 μL) is required for a single test, and aggregation values indicating platelet function are obtained in under 10 min. The initial testing of the MICELI performed with PRP demonstrated its accuracy and reproducibility [25]. As a portable, accurate, easy-to-use, and low-cost technology, the MICELI aggregometer has potential for further clinical translation as a POC diagnostic device for platelet function testing. Therefore, our current study aimed to: (1) optimize the MICELI assay protocol, i.e., blood sample volume, storage time, and anticoagulation regimen; (2) validate the device's sensitivity to aggregation agonists and crucial blood characteristics, i.e., platelet count and hematocrit; and (3) verify the MICELI operational performance as compared with the Multiplate® Analyzer, a commercial impedance aggregometer.

2. Results

The current study was dedicated to the optimization and internal validation of the MICELI system, a miniature impedance aggregometer designed and fabricated by our team. Optimization of the MICELI aggregometer and assay protocol aimed at addressing the following points: (1) defining the optimal blood sample volume, allowing consistent data acquisition (as indicated by high AUC and Amax, along with low lag time (LT) and noise); (2) establishing the maximum blood sample storage time (as a time window when valid and consistent aggregation data could be collected); and (3) identifying optimal blood anticoagulant, allowing maximal and consistent aggregation measurements. On the other hand, internal validation aimed at evaluating the crucial characteristics of the MICELI system included: (1) sensitivity to classic biochemical agonists and their concentrations, (2) sensitivity to platelet count and hematocrit, and (3) operational performance as compared with the Multiplate® Analyzer, a commercial impedance aggregometer.

2.1. Optimization of the MICELI aggregometer and assay protocol

2.1.1. Sample Volume

Uniform platelet aggregation on the electrodes of the MICELI cartridge largely relies on the sample volume. Sufficient blood volume is required to fully submerge electrodes and minimize stirring flow forces that might disturb formation of platelet aggregates on the electrode surface. Low sample volume at comparably high angular speed of stirring (up to 1000 rpm) recommended for aggregation assays may result in high noise, affecting the consistency of impedance data acquisition and low aggregation values [26,27]. Therefore, the sample volume optimization, with a specific focus on the sample to stir bar volume ratio (VR), is usually recommended for the specific device in use.

In our study, we evaluated the effect of three sample volumes (200, 250, and 300 μL) on platelet aggregation induced by ADP (20 μM). The VR values for every sample volume were calculated as shown in Figure 1A. The effect of VRs on platelet aggregation parameters, i.e., AUC, Amax, and LT, as well as reproducibility of data acquisition (noise level) were evaluated. We found that VR 20 was characterized by the highest values of AUC and Amax, yet the measurements were quite noisy and resulted in higher standard deviation of aggregation parameters as compared to other VR tests were registered (Figure 1B). VR 25 showed comparable values of aggregation parameters as those of VR 20, however, results were more reproducible, as indicated by low SD and noise. The VR 30 was characterized by significantly lower AUC and Amax, with the same degree of reproducibility as VR 25. Interestingly, among all aggregation parameters LT was not affected by the VR.

Statistical analysis of the results confirmed that the Amax and AUC values obtained for VR 25 were significantly higher than for VR 30 (11.8 ± 2.8 Ohm and 32.6 ± 5.6 Ohm*min vs. 7.9 ± 2.3 Ohm and 24.7 ± 7.4 Ohm*min, $p < 0.05$). Aggregation levels obtained with VR 20 did not significantly differ from VR 25, but were affected by a significantly higher level of noise (0.21 ± 0.05 Ohm vs. 0.41 ± 0.17 Ohm, $p < 0.01$). Therefore, VR 25, which corresponds to a blood sample volume of 250 μL, was selected as the optimal sample volume for the MICELI aggregometer, allowing acquisition of high aggregation values with good reproducibly and low levels of noise.

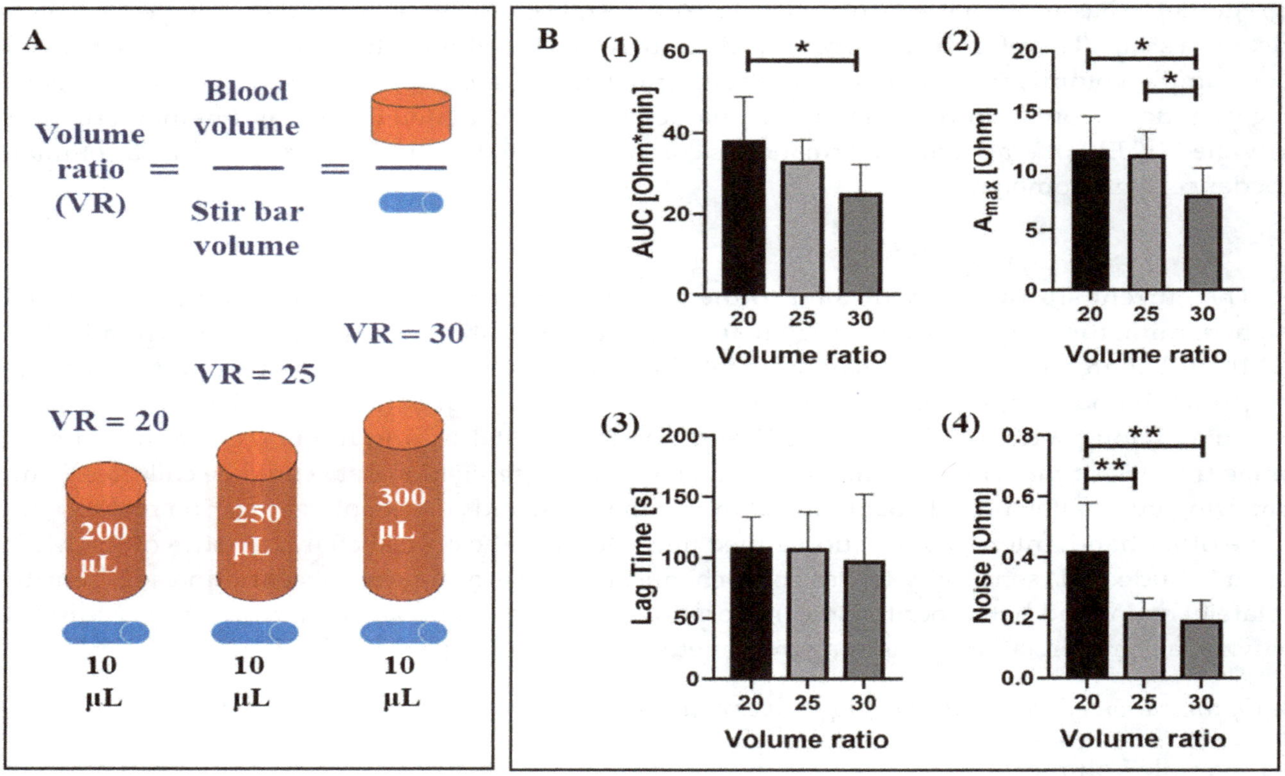

Figure 1. The effect of the sample to stir bar volume ratio (VR) on aggregation parameters collected by the MICELI (microfluidic, electrical, impedance) system: (**A**) calculation of VR for three sample volume tested, i.e., 200, 250, and 300 µL, and siliconized stir bar (#311, Chrono-log Corporation, Havertown, PA); (**B**) the effect of different VRs on ADP-induced platelet aggregation in ACD-anticoagulated whole blood: 1—area under the curve (AUC), 2—maximum aggregation (Amax), 3—lag time, and 4—noise, i.e., the amplitude of the impedance fluctuation over assay time. Data from 4 independent experiments using blood sample from 4 different individuals are reported. Mean ± SD as error bars of aggregation parameters are indicated, ANOVA: * $p < 0.05$, ** $p < 0.01$.

2.1.2. Blood Storage Time

For LTA, it is well documented that platelet aggregation in PRP could be affected by the sample storage time [28,29]. The recommended time interval for assay performance is within 3 h of blood draw [29]. Yet, the effect of storage time on platelet aggregation in whole blood when tested via EIA remained to be established. We evaluated the effect of the blood storage time, as the time interval between blood collection and aggregation test, on platelet aggregation parameters, i.e., AUC, Amax, and LT, recorded by the MICELI aggregometer.

We demonstrated that ADP-induced platelet aggregation was sensitive to the time elapsed from blood collection and tended to decrease over the storage time (Figure 2). The maximum aggregation decreased significantly within four hours, as compared with the first and second hour (3.5 ± 1 Ohm vs. 5.6 ± 2.2 Ohm and 5.5 ± 2.2 Ohm, respectively; ANOVA: $p < 0.05$). The area under the curve also tended to decrease over time, in particular between the second and third hour, although the decrease did not reach a level of significance. Lag time was not affected by the storage time. Thus, our results indicate that storage time has a significant impact on ADP-induced platelet aggregation in whole blood, and a delay of greater than two hours should be avoided.

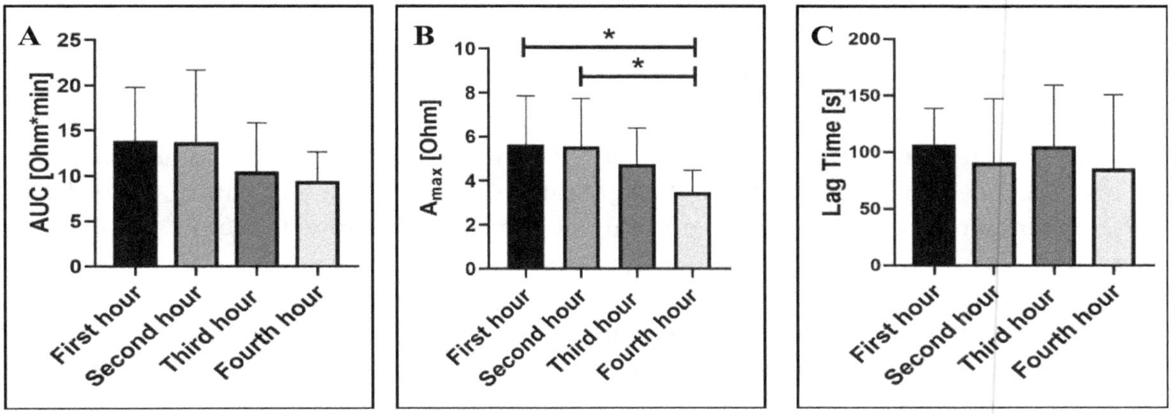

Figure 2. The effect of storage time on ADP-induced platelet aggregation in ACD-anticoagulated whole blood: (**A**) area under the curve (AUC); (**B**) maximum aggregation (Amax); and (**C**) lag time. Data from 6 independent experiments using blood samples from 6 different individuals are reported. Mean ± SD as error bars of aggregation parameters are indicated, ANOVA: no asterisk—$p > 0.05$; * $p < 0.05$.

2.1.3. Blood Anticoagulation Regimen

The anticoagulation of blood samples with calcium chelating agents, i.e., sodium citrate, ACD, and EDTA, as well as thrombin inhibitors, i.e., hirudin and heparin, is well established as to agent efficacy in blocking coagulation. Nevertheless, these agents have been demonstrated to affect platelet aggregation in whole blood [30–32]. In our study, we tested four anticoagulants commonly used for platelet aggregometry, i.e., sodium citrate, ACD, hirudin, and heparin. We demonstrated that all three aggregation parameters (AUC, Amax, and LT) were largely affected by the anticoagulation regimen selected (Figure 3). As such, hirudin showed the highest values for area under the curve and maximum aggregation as combined with low lag time. Statistically significant differences were observed between citrate and hirudin for both AUC (15.8 ± 7.8 Ohm*min vs. 27.3 ± 6.7 Ohm*min, $p < 0.05$) and Amax (5.9 ± 2.8 Ohm vs. 10.4 ± 2.2 Ohm, $p < 0.05$). For lag time, a statistically significant difference was demonstrated between ACD and hirudin (99 ± 30.9 s vs. 62.6 ± 26.1 s, respectively; $p < 0.05$). In conclusion, amongst the four anticoagulants tested, hirudin was shown to be the most appropriate option for measuring platelet aggregation in whole blood using the MICELI aggregometer. The MICELI validation was then performed with the primary focus on hirudinized whole blood samples.

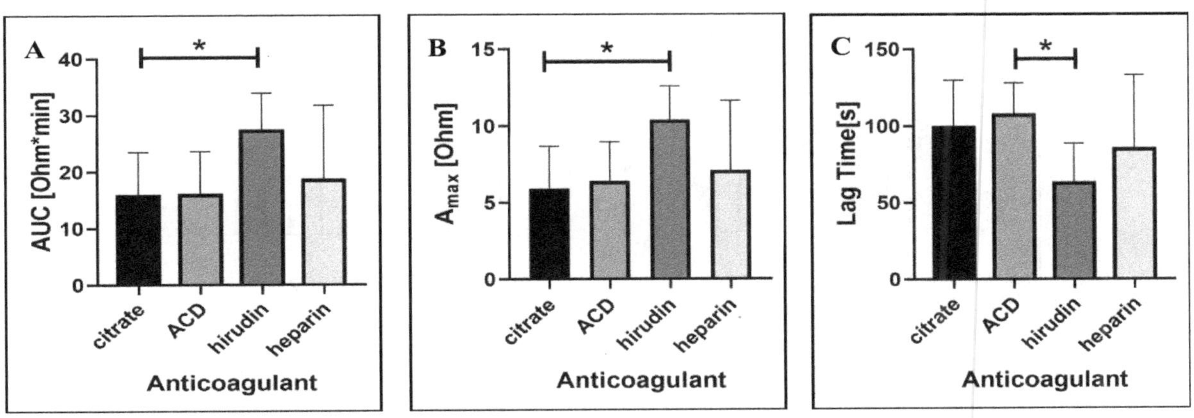

Figure 3. The effect of anticoagulant regimen on ADP-induced platelet aggregation in ACD-anticoagulated whole blood: (**A**) area under the curve (AUC); (**B**) maximum aggregation (Amax); (**C**) lag time. Data from 4 independent experiments using blood samples from 4 different individuals are reported. Mean ± SD as error bars of aggregation parameters are indicated, ANOVA: no asterisk—$p > 0.05$; * $p < 0.05$.

2.2. *Validation of the MICELI aggregometer*

2.2.1. Sensitivity and Precision with Different Aggregation Agonists and Their Concentration

In vivo platelet aggregation occurs as a result of platelet exposure to various biochemical agonists, e.g., ADP, collagen, epinephrine, and thrombin [1]. These agents, as well as their synthetic mimetics, are also used to promote platelet aggregation in vitro [33]. For internal validation of the MICELI aggregometer, four commonly used platelet aggregation agonists were used: ADP, collagen, epinephrine, and thrombin receptor-activating peptide 6 (TRAP-6)—a thrombin mimetic activating platelets via PAR-1-dependent pathway. Additionally, calcium ionophore A23187, a powerful platelet agonist facilitating a rapid increase of intracellular calcium concentration and inducing maximum platelet activation and aggregation response, was used as a positive control [34]. Platelet aggregation in both hirudin-anticoagulated blood and PRP was tested.

We showed that platelet aggregation response was registered for all agonists tested, yet the extent of platelet aggregation significantly varied. For the vast majority of biochemical agonists, platelet aggregation curves appear as steady increase of impedance with a tendency to saturation more evident in PRP than in whole blood (Figure 4a,b). Also, platelet aggregation values recorded in whole blood were higher than those in PRP. As expected, the highest platelet aggregation was detected for calcium ionophore in both whole blood and PRP samples, while epinephrine showed considerably lower aggregation values. In whole blood, significant differences were observed between platelet aggregation induced by epinephrine and calcium ionophore (17.3 ± 8.4 Ohm*min vs. 37.6 ± 13.3 Ohm*min, ANOVA: $p < 0.05$), as well as between epinephrine and TRAP-6 (17.3 ± 8.4 Ohm*min vs. 28.1 ± 11.7 Ohm*min, ANOVA: $p < 0.001$) (Figure 4c). The difference did not reach statistical significance in PRP samples (Figure 4d, ANOVA: $p < 0.05$). ADP induced consistent high-amplitude aggregation in both whole blood and PRP, when collagen was more potent in blood and showed much lower aggregation response in PRP.

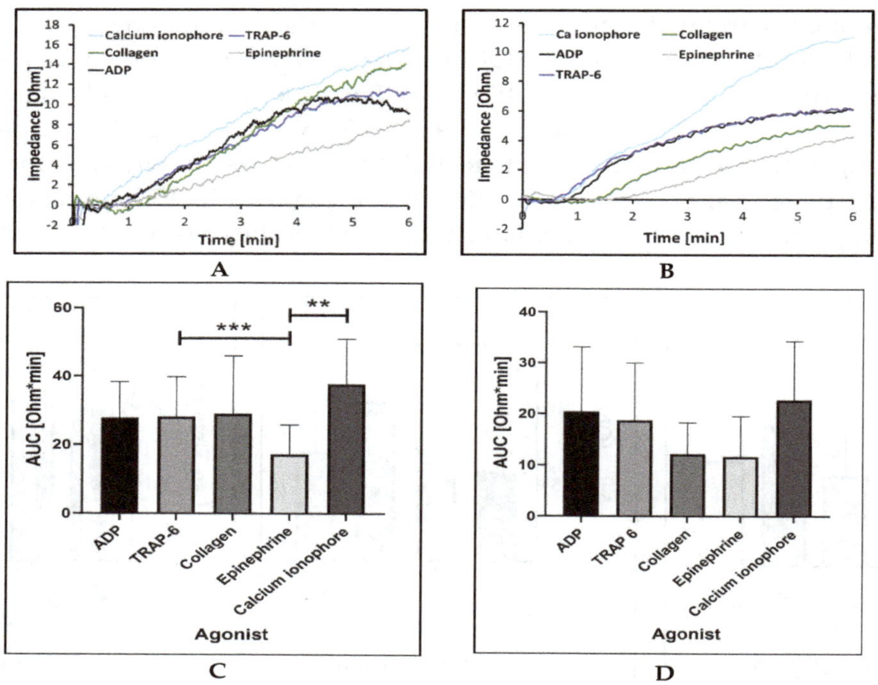

Figure 4. Platelet aggregation induced by different agonists in hirudin-anticoagulated whole blood (**A**,**C**) and platelet-rich plasma (**B**,**D**): AUC—area under the curve. (**A**,**B**) Illustrative aggregation curves recorded by the MICELI in whole blood and PRP, correspondingly. Data from 5 independent experiments using blood samples from 5 different individuals are reported. Mean ± SD as error bars of AUC are indicated, ANOVA: no asterisk–$p > 0.05$; ** $p < 0.01$, *** $p < 0.005$.

To evaluate precision of the MICELI aggregometer, we calculated coefficient of variation (CV) of aggregation values obtained in whole blood and PRP using platelet agonists inducing consistent platelet aggregation, i.e., ADP, TRAP-6, collagen, and calcium ionophore. Thus, interdonor variability was evaluated as CV of aggregation parameters measured across three different donors having similar platelet count and hematocrit levels. Intradonor variability was calculated as CV of aggregation parameters measured for the same donor within the same experiments. The MICELI device demonstrated decent precision as indicated by inter- and intradonor variability of AUC values (Table 1). Both inter- and intradonor variability of the aggregation parameters was significantly lower in PRP than in whole blood for all agonists tested. ADP and TRAP-6 showed lower CV for both inter- and intradonor distribution in PRP and whole blood, while collagen exhibited higher level of variability in whole blood than in PRP. Calcium ionophore-induced aggregation was characterized by good reproducibility regardless of the platelet sample type across all donors tested.

Table 1. Precision of the MICELI aggregometer using different platelet agonists as indicated by intra- and interdonor variability of area under the curve (CV, %).

	Intradonor Variability [1]		**Interdonor Variability** [2]	
Agonist Type	**PRP**	**Whole blood**	**PRP**	**Whole blood**
ADP	16	24	4	18
TRAP-6	6	23	11	11
Collagen	4	26	14	22
Ca ionophore	23	21	8	6

[1] Aggregation tested in one donor, platelet count: 280,000 platelets/μL, hematocrit: 39%; [2] Aggregation tested in 3 donors, platelet count: 250–290,000 platelets/μL, hematocrit: 37–39%.

To test the sensitivity of the MICELI device to agonist concentrations in whole blood, platelet aggregation was induced with ADP and collagen. Threshold agonists' concentrations were chosen based on recommendations of the previous studies aimed to validate platelet aggregometry assays for clinical use [35]. The resulting curves showing a linear, dose-dependent increase of platelet aggregation with the increase of agonist concentration are reported in Figure 5. Platelet aggregation induced by ADP appeared to be less sensitive to increments of agonist concentration than collagen-induced aggregation, as indicated by the notable increase of the Amax in response to the increase of collagen but not ADP concentration (Figure 5a,b). For ADP-induced aggregation, significant differences were found for 1 μM vs. 10 μM vs. 20 μM ADP (16.4 ± 11.7 Ohm*min, 29.3 ± 16.9 Ohm*min, and 33.3 ± 17.7, ANOVA: $p < 0.05$) with $R^2 = 0.9845$, indicating dissent linear correlation of AUC and ADP concentration in whole blood. Platelet aggregation response to collagen showed significant differences between final collagen concentrations of 1 μM and 10 μM (12.5 ± 6.3 Ohm*min vs. 31.9 ± 17.8 Ohm*min, ANOVA: $p < 0.01$) with $R^2 = 0.998$, indicating very strong linear correlation between AUC and collagen concentration. The obtained results prove that the MICELI aggregometer was sensitive to changes in the concentration of both ADP and collagen in whole blood anticoagulated with hirudin.

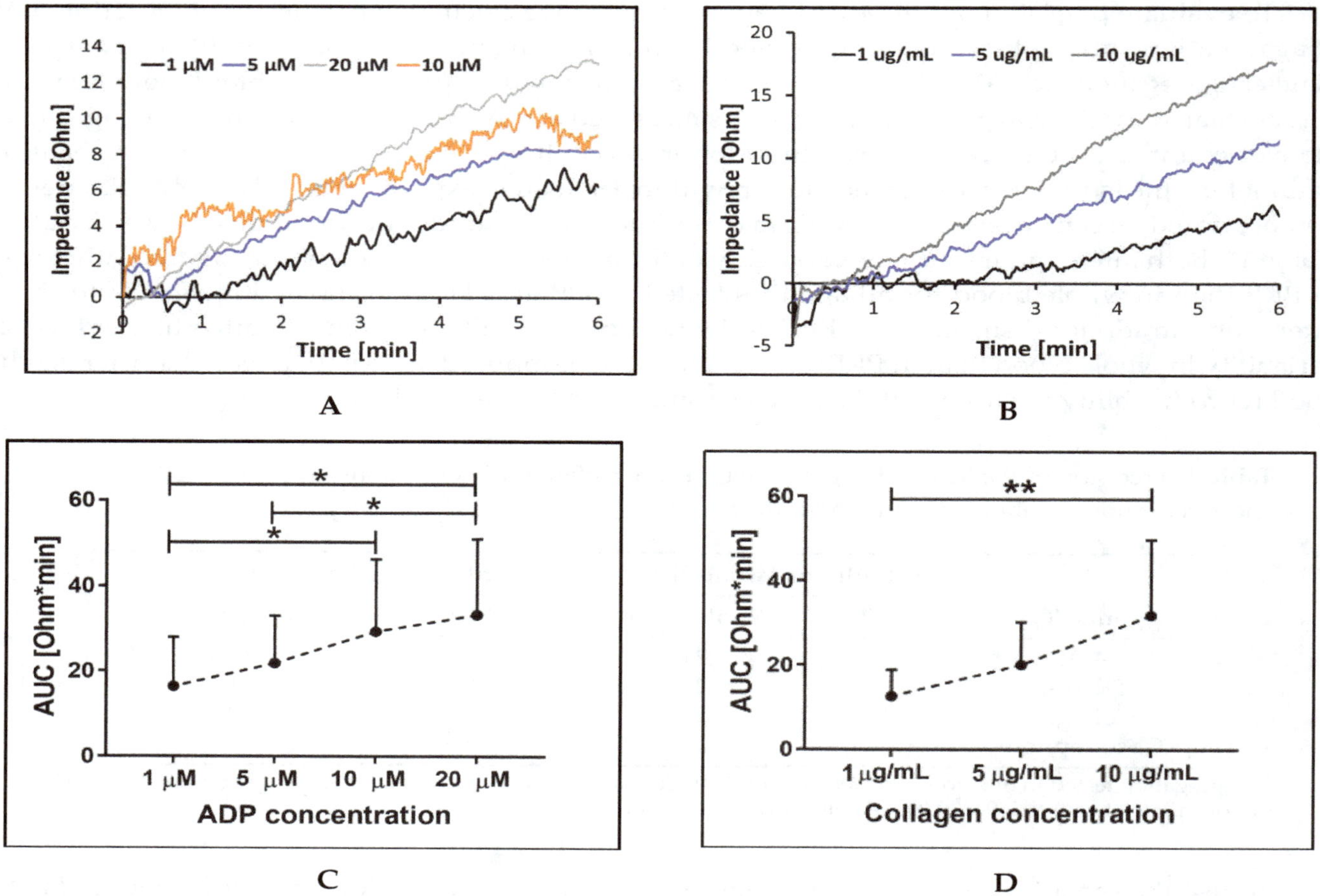

Figure 5. Platelet aggregation induced by different concentrations of ADP (**A**,**C**) and collagen (**B**,**D**) in hirudin-anticoagulated whole blood: AUC—area under the curve. (**A**,**B**)—illustrative curves of ADP- and collagen-induced aggregation recorded by the MICELI. Data from 4 independent experiments using blood samples from 4 different individuals are reported. Mean ± SD as error bars of AUC are indicated, ANOVA: no asterisk—$p > 0.05$; *—$p < 0.05$; **—$p < 0.01$.

2.2.2. Sensitivity to platelet count and hematocrit

The influence of platelet count and red blood cell volume (hematocrit) on aggregation parameters has been previously reported for both LTA and EIA [32,36–38]. Thus, we tested the MICELI sensitivity to platelet count and hematocrit in order to establish a lower detection limit for platelet count and optimal operating conditions. To evaluate the effect of platelet count on platelet aggregation parameters, we employed two strategies: (1) platelet count in hirudin-anticoagulated whole blood was defined for every donor, ADP-induced platelet aggregation was recorded, and correlation of platelet count with aggregation parameters was evaluated; (2) platelet count in PRP was manually adjusted with PPP, ADP-induced platelet aggregation was recorded, and correlation of platelet count with aggregation parameters was evaluated. As shown in Figure 6a, the amplitude of platelet aggregation in hirudinized blood positively correlated with platelet count, within the count range of 150,000–400,000 platelets/µL with a modest correlation coefficient (R^2 = 0.7661). Similarly, the extent of platelet aggregation steeply increased with the elevation of platelet count in PRP within the count range of 100,000–370,000 platelets/µL, showing even stronger linear correlation (R^2 = 0.9862, Figure 6b). No aggregation was detected in PRP samples containing 50,000 platelets/µL. To summarize, the MICELI aggregometer showed high sensitivity to platelet count in hirudin-anticoagulated whole blood and PRP, with the lower detection limit of 100,000 platelets/µL.

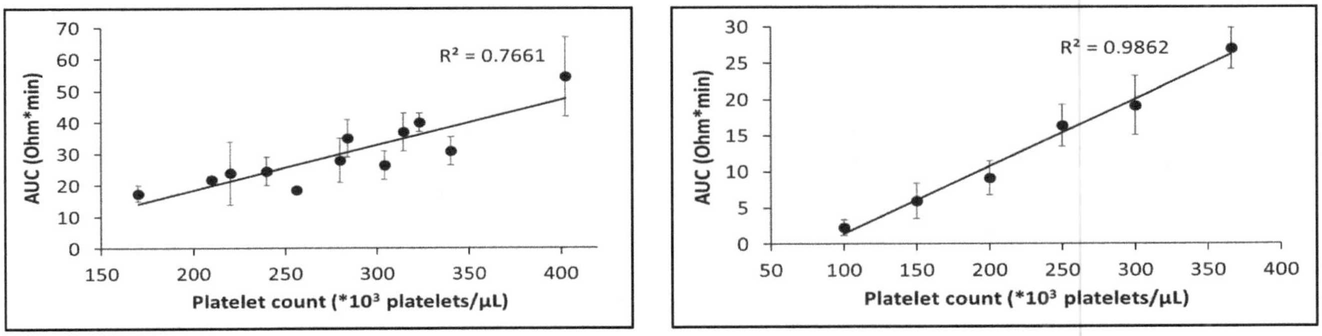

(a) (b)

Figure 6. The extent of platelet aggregation linearly corelates with platelet count in hirudin-anticoagulated whole blood (**a**) and platelet-rich plasma (**b**): AUC—area under the curve. In whole blood, the actual platelet count was not adjusted. In PRP, platelet count was manually adjusted by adding platelet-poor plasma. Then, platelet aggregation was induced by 20 µM ADP. Data from 12 (**a**) or 6 (**b**) independent experiments using blood samples from different donors are reported. Mean ± SD as error bars of AUC are indicated.

Platelet aggregation values showed no significant correlation with hematocrit in hirudin-anticoagulated whole blood, although a slight negative correlation could be noticed ($R^2 = 0.463$, Figure 7a). Our observations revealed that individuals with high hematocrit and lower platelet count tended to show lower platelet aggregation levels than those with low platelet count and moderate or low hematocrit. Yet, another interesting finding emerged; we found that the baseline level of impedance recorded by the MICELI (a "starting point" of the aggregation curve prior to agonist introduction) positively correlated with hematocrit value in hirudinized blood ($R^2 = 0.7393$, Figure 7b). Very strong linear correlation of the baseline impedance with the hematocrit values was further confirmed in the system where hematocrit was manually adjusted with PRP within hematocrit values of 35–50% ($R^2 = 0.9662$). Thus, we concluded that hematocrit did not significantly affect platelet aggregation values tested with the MICELI aggregometer, yet hematocrit values indeed predefined the baseline impedance of the whole blood sample.

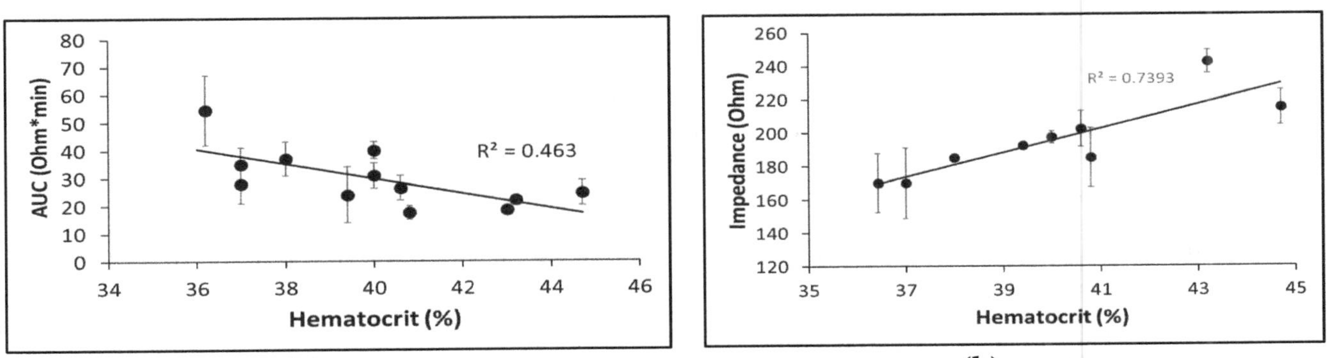

(a) (b)

Figure 7. The influence of hematocrit on ADP-induced platelet aggregation in hirudin-anticoagulated whole blood. (**a**) Platelet aggregation is not significantly affected by hematocrit within physiological range 36%–45%; (**b**) The baseline impedance strongly corelates with hematocrit: AUC—area under the curve. A total of 20 µM ADP was used to induce platelet aggregation. Data from 12 (**a**) or 9 (**b**) independent experiments using blood samples from different individuals are reported. Mean ± SD as error bars are indicated.

2.2.3. Operational Performance of the MICELI Aggregometer as Compared with Multiplate® Analyzer

The Multiplate® Analyzer is a semiautomated commercial impedance aggregometer capable of measuring platelet aggregation in whole blood and PRP. In our study, it was used as a reference device for comparison with the MICELI aggregometer. Aggregation response was tested in hirudin-anticoagulated blood samples of six individuals, using each of these devices. Platelet aggregation was induced with different concentrations of ADP (1, 5, or 10 μM). The dose-dependent increase of aggregation parameters with the increase of agonist concentration was reported for both aggregometers (Figure 8). Aggregation parameters obtained by the MICELI and Multiplate® showed good correlation, with Pearson coefficients equal to 0.92 and 0.98, for AUC and Amax, respectively. Both aggregometers demonstrated the same sensitivity to ADP concentration. For the MICELI, AUC values were significantly different for 1 and 5 μM ADP (4.3 ± 2.5 Ohm*min vs. 8.6 ± 2.1 Ohm*min, $p < 0.05$) and for 1 and 10 μM ADP (4.3 ± 2.5 Ohm*min vs. 13.2 ± 5.9 Ohm*min, $p < 0.05$). Similarly, for the Multiplate® Analyzer, AUC significantly differed between 1 and 5 μM ADP (299 ± 67 AU*min vs. 515 ± 90 AU*min, $p < 0.05$) and between 1 and 10 μM ADP (299 ± 67 AU*min vs. 555 ± 168 AU*min, $p < 0.005$). We also have noticed that when aggregation parameters reported by the MICELI linearly increased with the increase of ADP concentration from 1 to 10 μM, with the Multiplate® Analyzer, the saturation of aggregation levels observed after 10 μM ADP was added (for both AUC and Amax).

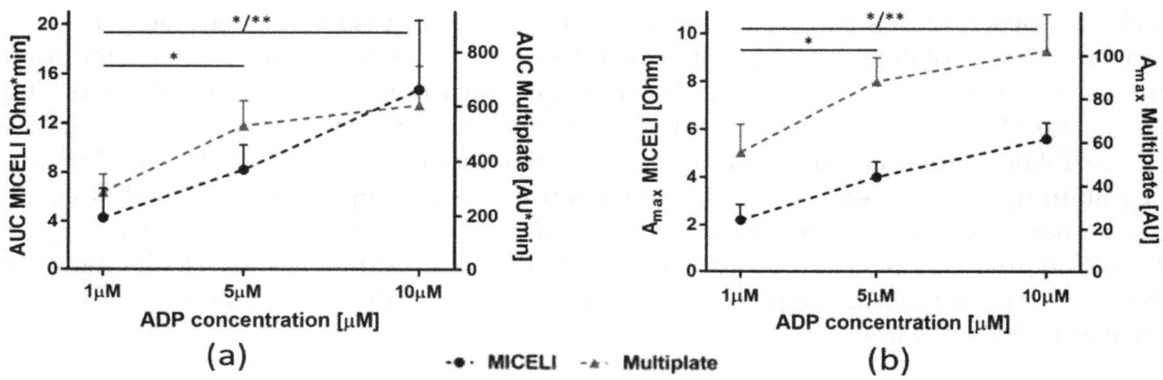

Figure 8. Comparison of aggregation parameters of ADP-induced platelet aggregation recorded by the MICELI aggregometer and Multiplate® Analyzer. Platelet aggregation in hirudin-anticoagulated whole blood was induced by 1, 5, or 10 μM ADP: (**a**) area under the curve (AUC); (**b**) maximum aggregation (Amax). Data from 6 independent experiments using blood samples from 6 different individuals are reported. Mean ± SD as error bars of aggregation parameters are indicated, ANOVA: no asterisk—$p > 0.05$; * $p < 0.05$; ** $p < 0.01$.

In Table A1, the interdonor distribution of AUC values obtained by the MICELI and the Multiplate® Analyzer aggregometers in our lab, as well as AUC values provided in the Multiplate® user manual, are reported. The 5°–50° and the 50°–95° percentiles obtained for the Multiplate® in our study were significantly lower than those reported in the instrument manual. It could be explained by a high interdonor variability and the low number of donors tested. The Multiplate® percentile values were very close to the MICELI ones, indicating that the statistical distributions of data obtained with the MICELI and Multiplate® are similar.

3. Discussion

In the current study, we optimized and validated the MICELI, a miniature impedance aggregometer designed and fabricated by our team. As a result of this optimization and validation effort, we developed a standard protocol of aggregometry testing using the MICELI; defined the instrument sensitivity to classic biochemical agonists and key blood characteristics, i.e., platelet count and hematocrit;

and compared the MICELI operational performance with the Multiplate® Analyzer, a commercial impedance aggregometer.

In the protocol optimization step, three critical settings of the experimental protocol were evaluated: the sample volume, the reliability of the aggregation measurements over blood storage time, and blood anticoagulation regimen. The optimal blood sample volume for MICELI aggregometry was defined as 250 μL, which allows acquisition of maximum platelet aggregation parameters with high reproducibility and minimum measurement noise. We showed that prolonged sample storage resulted in a steady decline of platelet aggregation. Thus, the optimal time gap for data collection using MICELI system should not exceed 2 h following blood sampling. This finding is in agreement with previous reports showing that impedance aggregometry in whole blood is very sensitive to blood storage time. Dézsi et al. observed a significant drop of ADP-induced aggregation as measured by the Multiplate® Analyzer from 1 to 4 h post blood collection [39], while both Jilma-Stohlawetz et al. and Johnson et al. reported a significant decline of platelet function and the increase of result variability, even after 2–3 h post blood sampling [40,41].

Testing the blood anticoagulation regimen, we showed a significantly weaker platelet aggregation response in blood anticoagulated with sodium citrate and ACD, as compared to hirudin-anticoagulated samples. The anticoagulant preference can be ordered as: hirudin > heparin > ACD >/= sodium citrate. The same tendency was observed by other researchers and could be explained by the fact that citrate inhibits coagulation by chelating calcium from the blood sample, while hirudin directly inhibits thrombin maintaining physiologic calcium levels required for platelet aggregation [30,42,43]. Johnson et al. showed that recalcification of citrate-anticoagulated blood resulted in aggregation values similar to those obtained for hirudin [41]. Therefore, our study clearly demonstrated that hirudin is the most suitable anticoagulant for MICELI aggregometry testing, which is in agreement with the current recommendations for other platelet aggregometry assays, such as LTA and Multiplate® [41,43].

After assessing the impact of the preanalytical variables, the standard protocol for the MICELI aggregometry was devised as follows: 1) a sample volume of 250 μL, 2) hirudin anticoagulated whole blood, and 3) an assay time window of 2 h from blood collection. This protocol was further used for the MICELI validation. Within the validation efforts, a series of experiments were performed to investigate the device's capability to measure platelet aggregation induced by different agonists acting via different platelet signaling pathways. For this purpose, four commonly utilized platelet agonists (ADP, TRAP-6, collagen, and epinephrine) were tested as compared with calcium ionophore, a powerful platelet stimulator nonspecifically inducing the increase of intracellular calcium and facilitating a maximum platelet response. We found that the MICELI aggregometer was able to record platelet aggregation induced by all agonists tested, in whole blood and PRP, with high consistency and reproducibility. The platelet agonists' capability to promote platelet aggregation in whole blood is as follows: calcium ionophore > collagen > ADP > TRAP-6 > epinephrine. A significantly lower platelet response to epinephrine detected by the MICELI is in agreement with previous reports showing that platelet response to epinephrine, as measured by impedance aggregometry, is typically absent or very weak [44,45]. Inter- and intradonor variability of the aggregation values assessed by the MICELI was also comparable to those previously reported for the Multiplate® aggregometry system [30].

We then explored MICELI sensitivity to different agonist concentrations, using the most potent physiologically relevant agonists, i.e., ADP and collagen. A strong linear correlation of aggregation parameters with agonist concentration revealed good sensitivity of the MICELI aggregometer to agonist concentration increments. Significant differences were observed between agonist doses, suggesting that the device can detect even micromolar changes in ADP and collagen concentrations in whole blood. Low-level aggregation was detected even at minimal agonist concentrations, 1 μM ADP and 1 μg/mL collagen, recommended for clinical application of EIA [35]. These experiments showed that the MICELI aggregometer could be used to evaluate different pathways of platelet activation and provide comprehensive information on platelet function and the effectiveness of antiplatelet therapy.

The effect of key blood parameters on MICELI results was investigated by assessing the influence of platelet count and hematocrit on platelet aggregation. Platelet aggregation parameters recorded by the MICELI strongly correlated with platelet number in whole blood and PRP. The minimum detection limit of the MICELI, defined as minimal platelet count when aggregation could be recorded, was identified as 100,000 platelets/μL. Therefore, the MICELI aggregometer could be successfully employed to detect platelet aggregation in donors within the normal platelet range, while platelet function testing in thrombocytopenic or anemic patents might be somewhat challenging with the current MICELI fabrication. Hematocrit did not significantly affect platelet aggregation values, though a slight negative correlation was noticed ($R^2 = 0.463$). Similarly, aggregation values obtained by the Multiplate® Analyzer were reported to depend on both platelet count and hematocrit, even within the normal range of these parameters [32,36,38]. Muller et al. specifically reported that the presence of large number of red blood cells was associated with low levels of aggregation detected via impedance aggregometry [38], which corresponds to our observations.

Analyzing aggregation curves obtained by the MICELI, we discovered that the baseline impedance value recorded prior to agonist introduction (see Figure A3a) strongly correlated with the hematocrit value of the sample tested (Figure 7b). In other impedance aggregometers, i.e., Multiplate® and CHRONO-LOG®, the baseline impedance value is not displayed to a user. The device's software normalizes the increment of impedance towards the baseline value of the sample, and the typical starting point of the aggregation curve equals to 0 Ohm (Figure A3c). Nevertheless, our finding clearly demonstrates that the baseline impedance value of a blood sample could be used to collect additional data from the donor blood sample using the same hardware of the MICELI aggregometer. In other words, the MICELI system could be equipped with the capability of electronic measurement of the hematocrit, a crucial blood parameter. The successful use of the Multiplate® Analyzer in the hemostatic management of patients with a history of antiplatelet medication use, with concomitant emergent need for neurosurgical therapy, was previously demonstrated in several small-cohort studies [46,47]. Thus, the MICELI aggregometer, when equipped with additional capabilities and translated to a POC diagnostic device, will provide platelet function and hematocrit data, both critical diagnostic parameters for bleeding management and verification of hemostatic transfusion outcomes [47–49].

The last step of the MICELI validation aimed to compare its operational performance with the Multiplate® Analyzer. Good correlation has been reported between aggregation parameters of ADP-induced platelet aggregation detected by the MICELI and Multiplate® Analyzer (with Pearson's coefficients for AUC and Amax equal to 0.92 and 0.98, correspondingly). The interdonor distribution of the MICELI measurements was also similar to the Multiplate®, as detected in our lab (Table A1) and reported by the Multiplate® user manual and other studies [30,41]. The sensitivity of both devices to ADP concentrations were similar. The MICELI and Multiplate® detected platelet aggregation in response to stimulation with 1 μM ADP (minimal concentration tested). However, the Multiplate® reported lower aggregation values induced by 10 μM ADP (maximum concentration tested). The last observation might be explained by the fact that the Multiplate® protocol requires dilution of a blood sample with saline (1:1), which results in the decrease of platelet count as compared with the nondiluted blood sample used for the MICELI data acquisition. So, the MICELI has a higher ratio between platelet count and electrode surface available. Therefore, when the ADP concentration increases from 5 to 10 μM in the MICELI sample, there are more platelets available for aggregation, while in the Multiplate®, there are no more platelets available and maximum impedance has almost reached its saturation point. This difference in the MICELI vs. Multiplate® protocol might provide a critical benefit when the measurement of platelet function is required in thrombocytopenic or anemic patients with low platelet count. Additional dilution of the blood sample may lead to the Multiplate® being unable to record aggregation, while the MICELI, given its equal reliability and operational performance, will still provide valid aggregation data.

The results of the MICELI aggregometer validation underscore the efficacy of this approach and facilitate further design advance towards its translation. The novelty of the prototype lies in the miniature form factor as compared with bulky competitors as well as the original design of the MICELI cartridge. As utilized in this study the single cartridge prototype was proven as a fully functional device enabling acquisition of real-time aggregation data from a 250 uL blood aliquot. The cartridge represents a single-well unit with two silver electrodes crossing the well horizontally and is secured from both ends within the walls of the well. Unlike the Multiplate® analyzer, the MICELI protocol does not require blood dilution, which decreases the number of pipetting steps and correspondent user error. Analyzing rough impedance data of a donor sample, the MICELI is capable of acquiring other important blood parameters, e.g., hematocrit (given the good positive correlation between blood impedance values and hematocrit).

Based on the results of the MICELI validation and feedback from its in-house utilization, the next immediate steps will include development of the second-generation prototype overcoming limitations of the described system. Thus, the effort was focused on the design and manufacturing of the self-filling cartridge with strict volume control; an agonist in biostable form preloaded in the cartridge well; multi-well cartridge to increase device turnaround; real-time analysis of impedance data and quick readout of the aggregation parameters. These advances will allow the evolution of the present system to one that is effective in the clinic and eventually in the point-of-care.

Limitations. The MICELI aggregometer prototype described in the manuscript shares the limitations of the current impedance aggregometers, e.g., comparably low reproducibility of aggregation data; high intra- and interdonor variability; and a multistep protocol requiring highly accurate pipetting of whole blood samples and agonists. Given the early stage of technology development, the prototype has a fairly primitive look and multicomponent setup.

4. Materials and Methods

4.1. Blood Collection and Anticoagulation

Blood was obtained from 42 healthy adult volunteers (19 females and 23 males) who provided their informed consent for inclusion before they participated in the study and claimed to abstain from medications affecting platelet function for two weeks prior to the experiment. The study was conducted in accordance with the Declaration of Helsinki, and the protocol was approved by the Institutional Review Board of the University of Arizona (Study Protocol #1810013264). The donor pool approximated the ethnic, racial, and gender categories distribution per 100 enrollees according to the geographic location of the University of Arizona (Pima County, AZ) [50]. Blood was drawn by venipuncture via a 21-gauge needle and anticoagulated with either acid citrate dextrose solution (ACD, 85 mM trisodium citrate, 78 mM citric acid, 111 mM glucose), blood: ACD ratio—10:1, or hirudin (Aniara, West Chester, OH, USA) in final concentration 525 ATU/mL. Alternatively, blood was collected into Vacutainer® tubes containing sodium citrate (final concentration—3.2%) or heparin (final concentration—17 IU/mL). Platelet-rich plasma (PRP) was obtained by centrifugation of ACD-anticoagulated blood at 300× *g* for 15 min at room temperature. The remaining blood was recentrifuged at 1200× *g* for 15 min at room temperature to obtain the platelet-poor plasma (PPP) required for platelet count standardization. Whole blood and PRP were stored and handled at room temperature if not otherwise indicated. Platelet count was quantified with Z1 Coulter Particle Counter (Beckman Coulter Inc., Indianapolis, IN, USA). Hematocrit was assessed using routine microhematocrit method [51].

4.2. The MICELI Impedance Aggregometry System Setup

The MICELI system is a multicomponent prototype of a miniature impedance aggregometer capable of evaluation of platelet aggregation in whole blood and PRP. The system consists of a miniature polymeric cartridge, magnetic stirrer, thermostatic chamber, and impedance analyzer communicating with an operator's laptop via Wi-Fi (Figure 9a,c). The MICELI cartridge, a crucial element of the system, is composed from two superimposed polydimethylsiloxane layers (13 mm × 20 mm) bonded to form a cylindrical reaction well (d = 8 mm, h = 6.88 mm, V = 0.35 mL) (Figure 9b,c). The drawing of the fully assembled cartridge with its geometric dimensions is reported in Figure A1. Two silver wire electrodes (d = 0.25 mm, ALFA AESAR, Haverhill, MA, USA) bridge the reaction well and are connected to the impedance analyzer via small alligator clamps. A siliconized microstir bar (1.64 mm × 4.78 mm, CHRONO-LOG Corporation, Havertown, PA, USA) was placed on the bottom of the reaction well when cartridge was fully assembled. The magnetic stirrer maintained continuous sample mixing at 250 rpm. Our in-house manufactured thermostatic chamber with a temperature controller allowed us to maintain reaction temperature at 37 °C (Figure A2). The impedance analyzer used in this MICELI setup was a STEMlab 125-10 single board computer (RedPitaya, Solkan, Slovenia) operated by the web-based application provided by the manufacturer. The impedance data acquisition and numerical analysis was performed using MATLAB software (Mathworks, Natick, MA, USA).

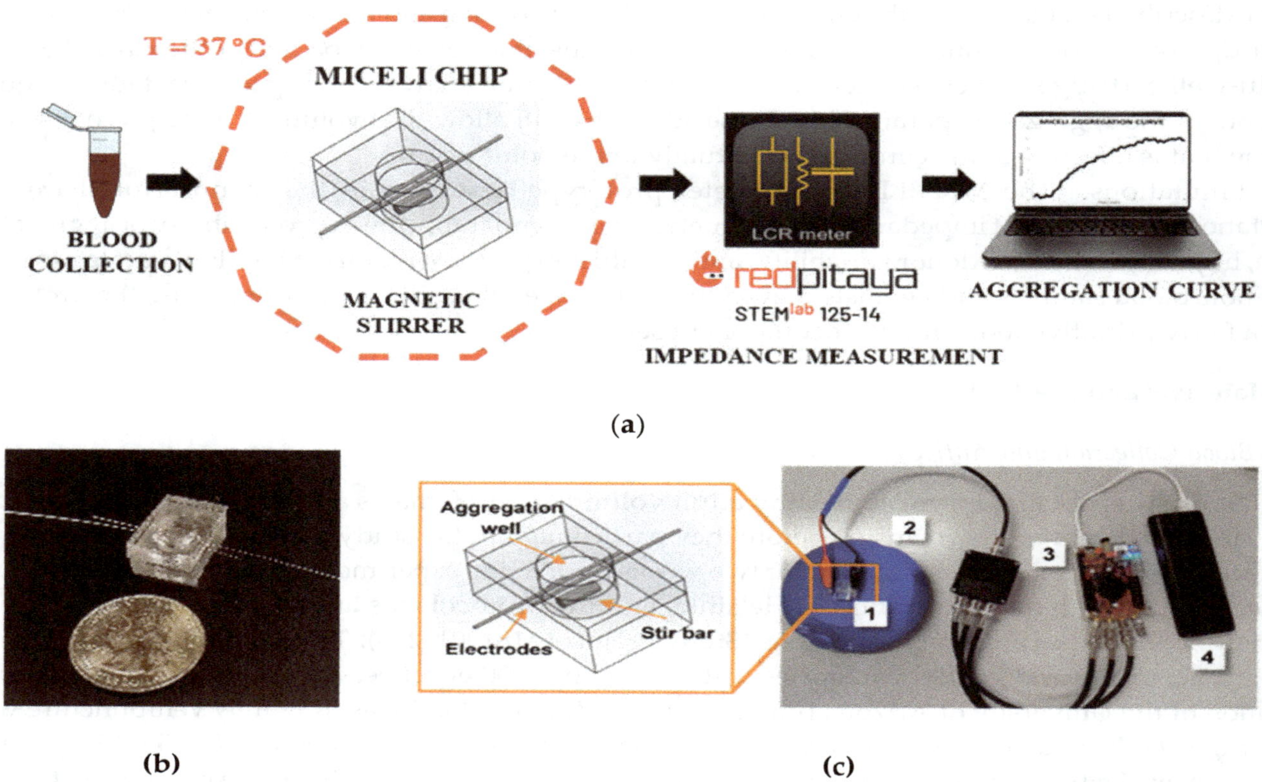

Figure 9. The MICELI system, a multicomponent prototype of the miniature impedance aggregometer: (**a**) workflow of the impedance aggregometry test using the MICELI system; (**b**) fully assembled MICELI cartridge; (**c**) the MICELI system components in a portable setup: 1—MICELI cartridge and its CAD drawing, 2—magnetic stirrer, 3—impedance analyzer, 4—power supply.

4.3. Impedance Aggregometry Using the MICELI System

Prior to an experiment, the thermostatic chamber was preheated to 37 °C. Then, 250 µL of platelet-containing sample (whole blood or PRP) were placed in the MICELI cartridge and impedance data acquisition was started. When whole blood sample was anticoagulated with ACD-A or sodium citrate, recalcification of the sample with 1 mM $CaCl_2$ was performed to reimburse physiological calcium concentration. Following a 3 min incubation, an aliquot of agonist solution was added and platelet aggregation as the sample impedance increased was recorded for 6 min. The increase of the sample impedance occurs as result of agonist-induced platelet aggregation on the surface of cartridge electrodes (Figure A3A,B). An aggregation curve, i.e., increase of impedance over time, was then recorded and the following aggregation parameters were calculated: area under the curve (AUC, Ohm*min), maximum aggregation (Amax, Ohm), and lag time (LT, s), i.e., a time gap between agonist introduction and aggregation start (Figure A3C).

The following biochemical agonists were applied to induce platelet aggregation: adenosine diphosphate (ADP, Sigma-Aldrich, St. Louis, MO, USA), thrombin receptor-activating peptide 6 (TRAP-6, AnaSpec Inc, Freemont, CA, USA), collagen (Helena Laboratories Corporation, Beaumont, TX, USA); epinephrine, and calcium ionophore A12387 (both from Sigma-Aldrich, St. Louis, MO). Agonists' final concentration and aliquot volume added to the reaction well are listed in Table 2:

Table 2. Final concentrations and aliquot volumes of biochemical agonists used to induce platelet aggregation in the MICELI system.

Agonist Type	Concentration	Volume, μL
ADP	1, 5, 10, 20 μM	5 [1]
TRAP-6	32 μM	6.4
Collagen	1, 5, 10 μg/mL	25 [1]
Epinephrine	10 μM	2.5
Calcium ionophore	5 μM	2.5

[1] The aliquot volume remained constant for all agonist concentrations tested.

4.4. *Impedance Aggregometry Using Multiplate® Analyzer*

The aggregometry test using the Multiplate® Analyzer impedance aggregometer (Roche Diagnostics, Milano, Italy) was performed following the manufacturer's protocol for hirudin-anticoagulated blood. Briefly, hirudin-anticoagulated blood was diluted with saline (1:1). Then, 600 μL of diluted blood was placed in a Multiplate® well and incubated for 3 min. Platelet aggregation was initiated by adding an aliquot of ADP (final concentration 1, 5, or 10 μM). Aggregation curve was recorded for 6 min and aggregation parameters, i.e., area under the curve (AUC, AU*min) and maximum aggregation (Amax, AU), were calculated by the device software. The instrument allows duplicate measurement of each well and automatically provides the mean value of the two aggregation readings.

4.5. *Statistical Analysis*

Aggregation tests using the MICELI aggregometer or the Multiplate® Analyzer, as well as hematocrit assay, were performed in duplicates for every condition tested. The arithmetic mean and SD were then calculated. Statistical analysis of the numerical data was performed using GraphPad Prism 8 software (GraphPad Software, Inc., CA, USA). Normal distribution of evaluated parameters was tested with the Shapiro–Wilk normality test. The one-way analysis of variance (ANOVA) test was used when normality hypothesis was satisfied for all the groups tested. Conversely, nonparametric Kruskal–Wallis one-way ANOVA test was performed. Statistical significance was assumed for p-values at least lower than 0.05.

5. Conclusions

We designed and fabricated the MICELI, a miniature, easy-to-use, accurate impedance aggregometer, capable of measuring platelet aggregation in a small volume of whole blood or PRP. Protocol optimization has demonstrated that maximum aggregation values could be obtained in 250 μL of blood sample, hirudin as a blood anticoagulant should be used, and the blood sample must be tested within two hours of blood sampling. The MICELI validation has proven the device capability to detect platelet aggregation stimulated by all conventional agonists, i.e., ADP, TRAP-6, collagen, and epinephrine. The aggregation values recorded by the MICELI strongly correlate with platelet count and are not significantly affected by hematocrit in hirudinized whole blood. Interestingly, the baseline impedance of the blood sample recorded by the MICELI prior to agonist introduction strongly correlates with the hematocrit, potentially allowing the electronic detection of the hematocrit using the MICELI hardware. The operational performance of the MICELI as compared to the Multiplate® has further confirmed the reliability of our prototype. Aggregation data by the MICELI strongly correlate with the Multiplate® values and show similar interdonor distribution. The MICELI, as a miniature, accurate, and easy-to-use impedance aggregometer, could be readily further translated to a commercial POC diagnostic device for real-time monitoring of platelet function to guide pharmacological hemostatic management and transfusion outcomes. Keeping in mind that the device is still a prototype, the results are encouraging and pave the way for future development as an automated, low-cost, and easy-to-use POC platelet function test.

6. Patents

International Application # PCT/US18/38955. Systems and Methods for Analyzing of Platelet Function. Filed by Arizona Board of Regents on behalf of the University of Arizona on 22 June 2018. Inventors: Alberto Redaelli and Marvin J. Slepian.

Author Contributions: Joint first authors: Y.R.-M. and S.B. Joint last authors: A.R. and M.J.S. Conceptualization, A.R. and M.J.S.; methodology, Y.R.-M., M.R., and F.C.; software, G.M. and C.F.; validation, G.M., C.F., A.D., M.R., M.S., and A.S.; formal analysis, Y.R.-M., G.M., and C.F.; investigation, Y.R.-M., G.M., and C.F.; resources, A.R. and M.J.S.; data curation, Y.R.-M. and S.B.; writing—Y.R.-M., S.B., G.M., and C.F.; writing—review and editing, Y.R.-M., A.R. and M.J.S.; visualization, G.M., C.F., A.D., and A.S.; supervision, M.C., A.R., and M.J.S. All authors have read and agree to the published version of the manuscript.

Acknowledgments: The authors of the manuscript would like to express their gratitude to Daniel E. Palomares, Annalisa Dimasi, and Adriana Ivich for provided technical support, and to John J. Jackson Sr. for administrative support of the project.

Abbreviations

POC	Point-of-care
LTA	Electrical impedance aggregometry
EIA	Light transmission aggregometry
PRP	Platelet-rich plasma
PPP	Platelet-poor plasma
ADP	Adenosine diphosphate
TRAP-6	Thrombin receptor-activating peptide 6
AUC	Area under the curve
AU	Arbitrary unit
ACD	Acid citrate dextrose
Amax	Maximal aggregation
LT	Lag time
VR	Volume ratio

Appendix A

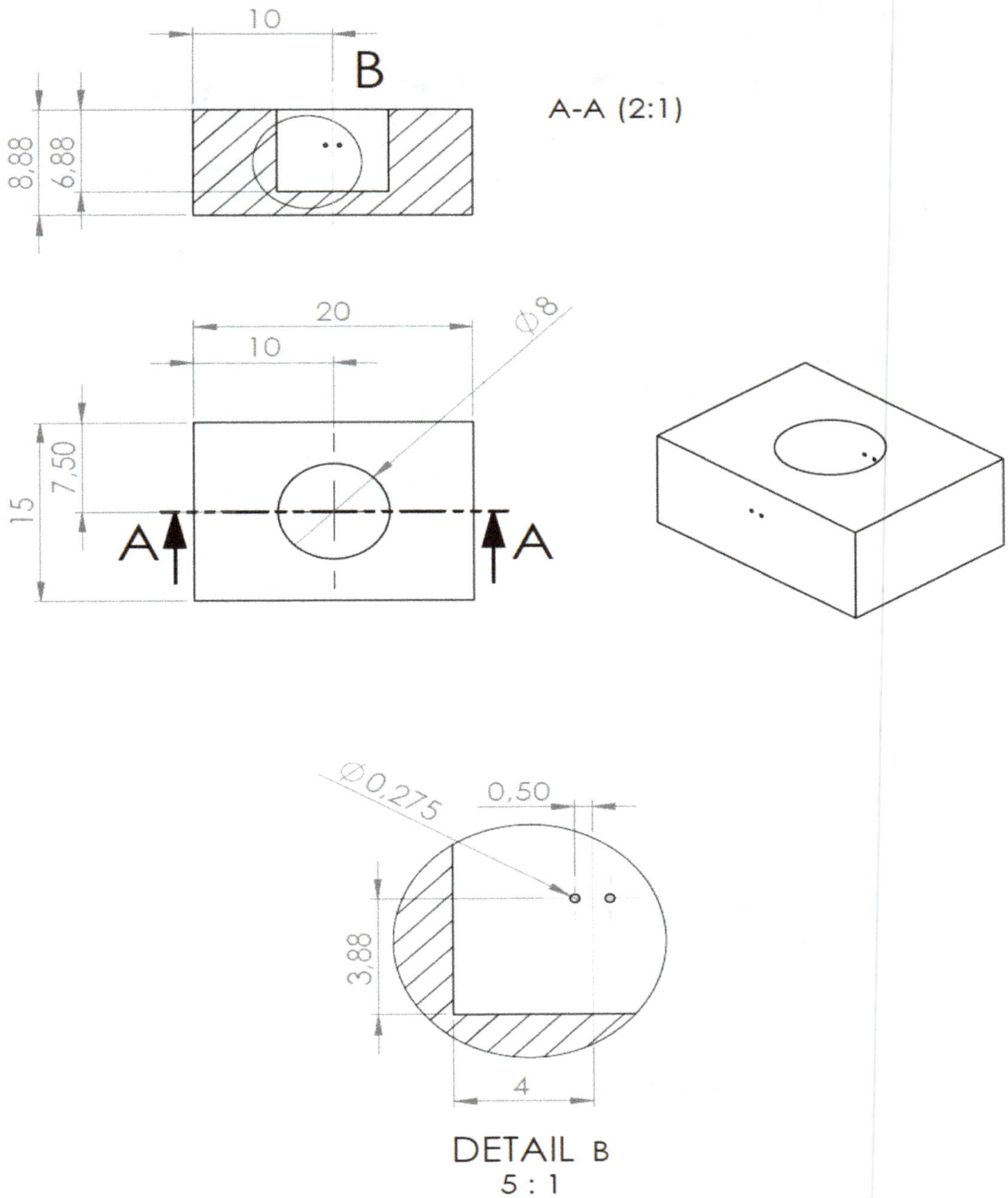

Figure A1. The drawings of the fully assembled cartridge utilized by the MICELI system with geometric dimensions of the cartridge, reaction well, and electrode channels.

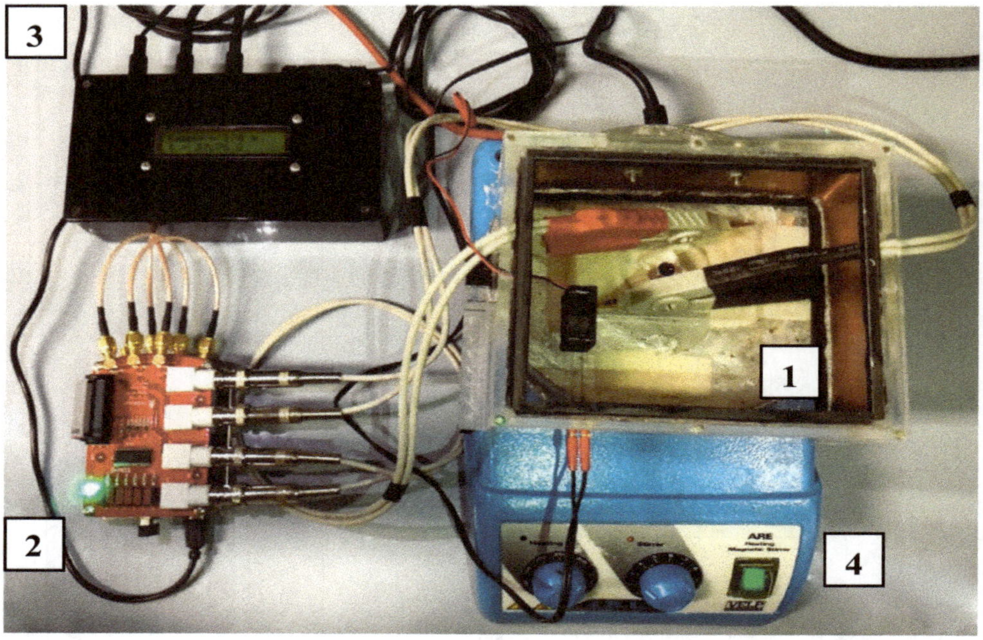

Figure A2. The MICELI aggregometer bench top setup with thermostatic chamber: 1) in-house manufactured thermostatic chamber; 2) impedance analyzer; 3) external unit controlling the thermostatic chamber; and 4) magnetic stirrer.

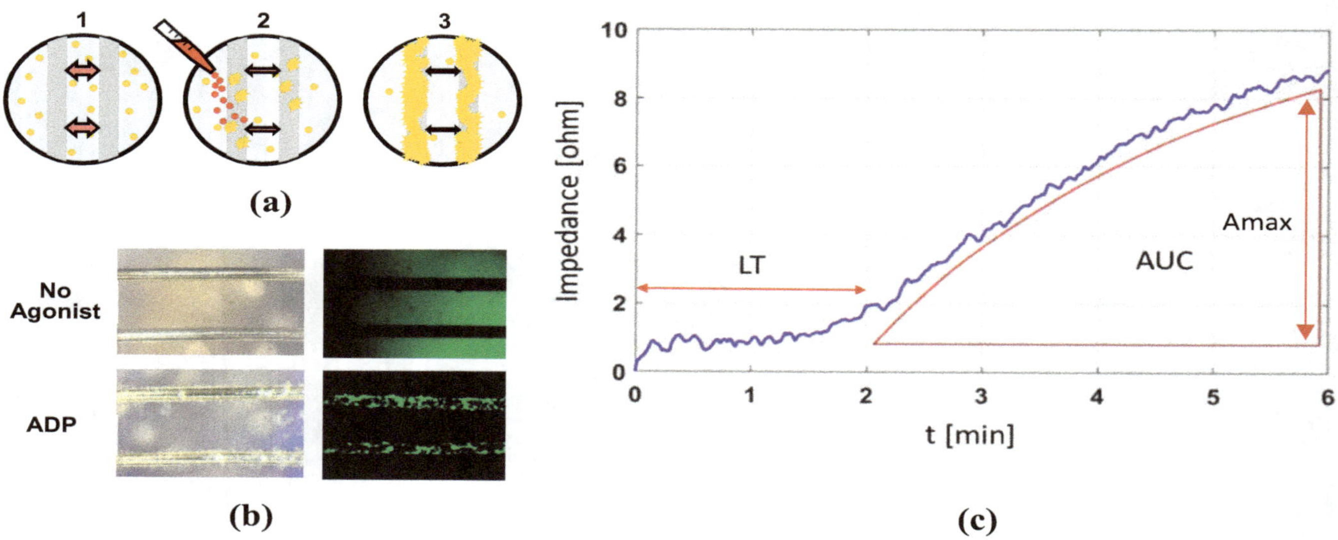

Figure A3. The impedance aggregometry: (**a**) basic principle of the electrical impedance aggregometry: 1—platelet-contained sample is placed in the cartridge with electrodes (basal level of sample impedance is recorded), 2—platelet agonist is added to initiate platelet aggregation on electrodes (platelets start to aggregate on electrode surface causing the increase of sample impedance), 3—platelets have coated the electrodes (the maximum impedance value is registered); (**b**) microscope images of the MICELI cartridge electrodes: left panel—platelet-rich plasma (bright field, 4x magnification); right panel—whole blood (fluorescence microscopy) (**c**) typical aggregation curve recorded by the MICELI: AUC—area under the curve, Amax—maximum aggregation, LT—lag time.

Table A1. The range of interdonor distribution of AUC values for aggregation, recorded by the MICELI and the Multiplate® Analyzer.

Statistical Value	MICELI (5 μM ADP)	Multiplate® (5 μM ADP)	Multiplate® User Manual (6.4 μM ADP)
n	6	6	64
Median	8.65	500	775
5° Percentile	5.83	410	474
95° Percentile	10.77	630	1076
Reference Range AUC	8.65–10.67	410–630	474–1076
5° and 50° Percentile Difference	−33%	−18%	−39%
50° and 95° Percentile Difference	26%	24%	39%

References

1. Holinstat, M. Normal platelet function. *Cancer Metastasis Rev.* **2017**, *36*, 195–198. [PubMed]
2. Koupenova, M.; Kehrel, B.E.; Corkrey, H.A.; Freedman, J.E. Thrombosis and platelets: An update. *Eur. Heart J.* **2016**, *38*, ehw550.
3. Cattaneo, M.; Hayward, C.P.M.; Moffat, K.A.; Pugliano, M.T.; Liu, Y.; Michelson, A.D. Results of a worldwide survey on the assessment of platelet function by light transmission aggregometry: A report from the platelet physiology subcommittee of the SSC of the ISTH. *J. Thromb. Haemost.* **2009**, *7*, 1029. [PubMed]
4. Antithrombotic Trialists' Collaboration. Collaborative meta-analysis of randomised trials of antiplatelet therapy for prevention of death, myocardial infarction, and stroke in high risk patients. *BMJ* **2002**, *324*, 71–86.
5. Grove, E.L.; Hossain, R.; Storey, R.F. Platelet function testing and prediction of procedural bleeding risk. *Thromb. Haemost.* **2013**, *109*, 817–824.
6. Cayla, G.; Cuisset, T.; Silvain, J.; Leclercq, F.; Manzo-Silberman, S.; Saint-Etienne, C.; Delarche, N.; Bellemain-Appaix, A.; Range, G.; El Mahmoud, R.; et al. Platelet function monitoring to adjust antiplatelet therapy in elderly patients stented for an acute coronary syndrome (ANTARCTIC): An open-label, blinded-endpoint, randomised controlled superiority trial. *Lancet* **2016**, *388*, 2015–2022.
7. Yeung, L.Y.Y.; Sarani, B.; Weinberg, J.A.; McBeth, P.B.; May, A.K. Surgeon's guide to anticoagulant and antiplatelet medications part two: Antiplatelet agents and perioperative management of long-term anticoagulation. *Trauma Surg. Acute Care Open* **2016**, *1*, e000022.
8. Vilahur, G.; Choi, B.G.; Zafar, M.U.; Viles-Gonzalez, J.F.; Vorchheimer, D.A.; Fuster, V.; Badimon, J.J. Normalization of platelet reactivity in clopidogrel-treated subjects. *J. Thromb. Haemost.* **2007**, *5*, 82–90.
9. Triulzi, D.J.; Assmann, S.F.; Strauss, R.G.; Ness, P.M.; Hess, J.R.; Kaufman, R.M.; Granger, S.; Slichter, S.J. The impact of platelet transfusion characteristics on posttransfusion platelet increments and clinical bleeding in patients with hypoproliferative thrombocytopenia. *Blood* **2012**, *119*, 5553–5562.
10. Kerkhoffs, J.-L.H.; Eikenboom, J.C.J.; van de Watering, L.M.G.; van Wordragen-Vlaswinkel, R.J.; Wijermans, P.W.; Brand, A. The clinical impact of platelet refractoriness: Correlation with bleeding and survival. *Transfusion* **2008**, *48*, 1959–1965.
11. Soliman, M.; Hartmann, M. Impedance Aggregometry Reveals Increased Platelet Aggregation during Liver Transplantation. *J. Clin. Med.* **2019**, *8*, 1803. [CrossRef] [PubMed]
12. Yorkgitis, B.K.; Ruggia-Check, C.; Dujon, J.E. Antiplatelet and anticoagulation medications and the surgical patient. *Am. J. Surg.* **2014**, *207*, 95–101. [CrossRef] [PubMed]
13. Velik-Salchner, C.; Maier, S.; Innerhofer, P.; Streif, W.; Klingler, A.; Kolbitsch, C.; Fries, D. Point-of-care whole blood impedance aggregometry versus classical light transmission aggregometry for detecting aspirin and clopidogrel: The results of a pilot study. *Anesth. Analg.* **2008**, *107*, 1798–1806. [CrossRef] [PubMed]
14. Koltai, K.; Kesmarky, G.; Feher, G.; Tibold, A.; Toth, K. Platelet aggregometry testing: Molecular mechanisms, techniques and clinical implications. *Int. J. Mol. Sci.* **2017**, *18*, 1–21. [CrossRef]
15. Born, G.V.R.; Cross, M.J. The aggregation of blood platelets. *J. Physiol.* **1963**, *168*, 178–195. [CrossRef]
16. Frontroth, J.P. *Haemostasis*; Springer Science Business Media: New York, NY, USA, 2013; Volume 992, pp. 227–240.
17. Hvas, A.; Favaloro, E.J. *Hemostasis and Thrombosis*; Springer: New York, NY, USA, 2017; Volume 1646, pp. 321–331.

18. Choi, J.-L.; Li, S.; Han, J.-Y. Platelet function tests: A review of progresses in clinical application. *Biomed Res. Int.* **2014**, *2014*, 456569. [CrossRef]
19. Cardinal, D.C.; Flower, R.J. The electronic aggregometer: A novel device for assessing platelet behavior in blood. *J. Pharmacol. Methods* **1980**, *3*, 135–158. [CrossRef]
20. Desconclois, C.; Valarche, V.; Boutekedjiret, T.; Raphael, M.; Dreyfus, M.; Proulle, V. Whole Blood Impedance Aggregometry: A New Tool for Severe Inherited Platelet Disorder Diagnosis? *Blood* **2011**, *118*, 5266. [CrossRef]
21. Mullins, E.S.; Stroop, D.; Ingala, D.; Tarango, C.; Luchtman-Jones, L.; Lane, A.; Gruppo, R.A. Comparison of Multiple Electrode Impedance-Based Platelet Aggregometry to Light Transmission Aggregometry for Diagnosis of Qualitative Platelet Defects. *Blood* **2017**, *130*, 1048.
22. 510(k) Number: K161329. Chrono-log Platelet Aggregometer, Model 490 4+4. Department of Health and Human Services Food and Drug Administration Indications for Use. Available online: https://www.accessdata.fda.gov/cdrh_docs/pdf16/K161329.pdf (accessed on 13 December 2019).
23. 510(k) Substantial Equivalence Determination Decision Summary: k103555. Multiplate 5.0 aggregometer. Platelet aggregometer assays: ADPtest (10 μM ADP) and ASPItest (0.5 mM Arachidonic Acid). Available online: https://www.accessdata.fda.gov/cdrh_docs/reviews/k103555.pdf (accessed on 13 December 2019).
24. Redaelli, A.; Slepian, M.J. Systems and Methods for Analyzing Platelet Function. Available online: http://hdl.handle.net/11311/1066190 (accessed on 22 June 2018).
25. Santoleri, A.; Dimasi, A.; Consolo, F.; Rasponi, M.; Fiore, G.B.; Slepian, M.J.; Redaelli, A. The MICELI (MICrofluidic, ELectical, Impedance): A New Portable Device for Bedside Platelet Aggregation Measurement. In Proceedings of the ISMCS 2017 25th Anniversary Scientific Congress "Limitations, Controversies, and Gaps in MCS: Pathway to Solutions", Controversies, Tucson, AZ, USA, 15–18 October 2017; p. 199.
26. US8617468B2—Platelet Aggregation Test and Device—Google Patents. Available online: https://patents.google.com/patent/US8617468B2/en (accessed on 13 December 2019).
27. Calatzis, A.; Kriger, B.; Wittwer, M. Cartridge Device For Blood Analysis. U.S. Patent No. 200,701,409,02, 21 June 2007.
28. Silver, W.P.; Keller, M.P.; Teel, R.; Silver, D. Effects of donor characteristics and in vitro time and temperature on platelet aggregometry platelet. *J. Vasc. Surg.* **1993**, *17*, 726–733. [CrossRef]
29. Zhou, L.; Schmaier, A.H. Platelet Aggregation Testing in Platelet-Rich Plasma Description of Procedures With the Aim to Develop Standards in the Field. *Am. J. Clin. Pathol.* **2005**, *123*, 172–183. [CrossRef] [PubMed]
30. Peerschke, E.I.B.; Castellone, D.D.; Stroobants, A.K.; Francis, J. Reference Range Determination for Whole-Blood Platelet Aggregation Using the Multiplate Analyzer. *Am. J. Clin. Pathol.* **2014**, *142*, 647–656. [CrossRef]
31. Kalb, M.L.; Potura, L.; Scharbert, G.; Kozek-Langenecker, S.A. The effect of ex vivo anticoagulants on whole blood platelet aggregation. *Platelets* **2009**, *20*, 7–11. [CrossRef] [PubMed]
32. Seyfert, U.T.; Haubelt, H.; Vogt, A.; Hellstern, P. Variables influencing Multiplate TM whole blood impedance platelet aggregometry and turbidimetric platelet aggregation in healthy individuals. *Platelets* **2007**, *18*, 199–206. [CrossRef]
33. Michelson, A.D.; Frelinger, A.L.; Furman, M.I. Current Options in Platelet Function Testing. *Am. J. Cardiol.* **2006**, *98*, S4–S10. [CrossRef]
34. Rasheed, H.; Tirmizi, A.H.; Salahuddin, F.; Rizvi, N.B.; Arshad, M.; Farooq, S.Z.; Saeed, S.A. Calcium Signaling in Human Platelet Aggregation Mediated by Platelet Activating Factor and Calcium Ionophore, A23187. *J. Biol. Sci.* **2004**, *4*, 117–121. [CrossRef]
35. Sun, P.; McMillan-Ward, E.; Mian, R.; Israels, S.J. Comparison of light transmission aggregometry and multiple electrode aggregometry for the evaluation of patients with mucocutaneous bleeding. *Int. J. Lab. Hematol.* **2019**, *41*, 133–140. [CrossRef]
36. Femia, E.A.; Scavone, M.; Lecchi, A.; Cattaneo, M. Effect of platelet count on platelet aggregation measured with impedance aggregometry (Multiplate^{TM} analyzer) and with light transmission aggregometry. *J. Thromb. Haemost.* **2013**, *11*, 2193–2196. [CrossRef]
37. Hanke, A.A.; Roberg, K.; Monaca, E.; Sellmann, T.; Weber, C.F.; Rahe-Meyer, N.; Görlinger, K. Impact of platelet count on results obtained from multiple electrode platelet aggregometry (multiplate^{TM}). *Eur. J. Med. Res.* **2010**, *15*, 214–219. [CrossRef]

38. Müller, M.R.; Salat, A.; Pulaki, S.; Stangl, P.; Ergun, E.; Schreiner, W.; Losert, U.; Wolner, E. Influence of hematocrit and platelet count on impedance and reactivity of whole blood for electrical aggregometry. *J. Pharmacol. Toxicol. Methods* **1995**, *34*, 17–22. [CrossRef]
39. Dézsi, D.A.; Merkely, B.; Skopál, J.; Barabás, E.; Várnai, K.; Faluközy, J.; Veress, G.; Alotti, N.; Aradi, D. Impact of Test Conditions on ADP-Induced Platelet Function Results With the Multiplate Assay: Is Further Standardization Required? *J. Cardiovasc. Pharmacol. Ther.* **2018**, *23*, 149–154. [CrossRef] [PubMed]
40. Jilma-Stohlawetz, P.; Ratzinger, F.; Schörgenhofer, C.; Jilma, B.; Quehenberger, P. Effect of preanalytical time-delay on platelet function as measured by multiplate, PFA-100 and VerifyNow. *Scand. J. Clin. Lab. Investig.* **2016**, *76*, 249–255. [CrossRef] [PubMed]
41. Johnston, L.R.; Larsen, P.D.; La Flamme, A.C.; Harding, S.A. Methodological considerations for the assessment of ADP induced platelet aggregation using the Multiplate® analyser. *Platelets* **2013**, *24*, 303–307. [CrossRef] [PubMed]
42. Wallén, N.H.; Ladjevardi, M.; Albert, J.; Bröijersén, A. Influence of different anticoagulants on platelet aggregation in whole blood; a comparison between citrate, low molecular mass heparin and hirudin. *Thromb. Res.* **1997**, *87*, 151–157. [CrossRef]
43. Bonello, L.; Tantry, U.S.; Marcucci, R.; Blindt, R.; Angiolillo, D.J.; Becker, R.; Bhatt, D.L.; Cattaneo, M.; Collet, J.P.; Cuisset, T.; et al. Consensus and future directions on the definition of high on-treatment platelet reactivity to adenosine diphosphate. *J. Am. Coll. Cardiol.* **2010**, *56*, 919–933. [CrossRef]
44. Mackie, I.J.; Jones, R.; Machin, S.J. Platelet impedance aggregation in whole blood and its inhibition by antiplatelet drugs. *J. Clin. Pathol.* **1984**, *37*, 874–878. [CrossRef]
45. Swart, S.; Pearson, D.; Wood, J.; Barnett, D. Effects of adrenaline and alpha adrenergic antagonists on platelet aggregation in whole blood: Evaluation of electrical impedance aggregometry. *Thromb. Res.* **1984**, *36*, 411–418. [CrossRef]
46. Beynon, C.; Scherer, M.; Jakobs, M.; Jung, C.; Sakowitz, O.W.; Unterberg, A.W. Initial experiences with Multiplate® for rapid assessment of antiplatelet agent activity in neurosurgical emergencies. *Clin. Neurol. Neurosurg.* **2013**, *115*, 2003–2008. [CrossRef]
47. Briggs, A.; Gates, J.D.; Kaufman, R.M.; Calahan, C.; Gormley, W.B.; Havens, J.M. Platelet dysfunction and platelet transfusion in traumatic brain injury. *J. Surg. Res.* **2015**, *193*, 802–806. [CrossRef]
48. Meißner, A.; Schlenke, P. Massive bleeding and massive transfusion. *Transfus. Med. Hemotherapy* **2012**, *39*, 73–84. [CrossRef] [PubMed]
49. Rahe-Meyer, N.; Winterhalter, M.; Boden, A.; Froemke, C.; Piepenbrock, S.; Calatzis, A.; Solomon, C. Platelet concentrates transfusion in cardiac surgery and platelet function assessment by multiple electrode aggregometry. *Acta Anaesthesiol. Scand.* **2009**, *53*, 168–175. [CrossRef] [PubMed]
50. U.S Census Bureau QuickFacts: Pima County, Arizona. Available online: https://www.census.gov/quickfacts/fact/table/pimacountyarizona/PST045218 (accessed on 12 December 2019).
51. CLSI. *Procedure for Determining Packed Cell Volume by the Microhematocrit Method*, 3rd ed.; CLSI document H07-A3; Clinical and Laboratory Standards Institute: Wayne, PA, USA, 2000.

7

Platelet Adhesion and Thrombus Formation in Microchannels: The Effect of Assay-Dependent Variables

Mariangela Scavone [1,†], Silvia Bozzi [2,†], Tatiana Mencarini [2], Gianmarco Podda [1], Marco Cattaneo [1] and Alberto Redaelli [2,*]

[1] Unità di Medicina 2, ASST Santi Paolo e Carlo, Dipartimento di Scienze della Salute, Università degli Studi di Milano, 20142 Milan, Italy; mariangela.scavone@guest.unimi.it (M.S.); gianmarco.podda@unimi.it (G.P.); marco.cattaneo@unimi.it (M.C.)

[2] Department of Electronics, Information and Bioengineering, Politecnico di Milano, 20133 Milan, Italy; silvia.bozzi@polimi.it (S.B.); tatiana.mencarini@mail.polimi.it (T.M.)

* Correspondence: alberto.redaelli@polimi.it

† These authors have equally contributed to this work.

Abstract: Microfluidic flow chambers (MFCs) allow the study of platelet adhesion and thrombus formation under flow, which may be influenced by several variables. We developed a new MFC, with which we tested the effects of different variables on the results of platelet deposition and thrombus formation on a collagen-coated surface. Methods: Whole blood was perfused in the MFC over collagen Type I for 4 min at different wall shear rates (WSR) and different concentrations of collagen-coating solutions, keeping blood samples at room temperature or 37 °C before starting the experiments. In addition, we tested the effects of the antiplatelet agent acetylsalicylic acid (ASA) (antagonist of cyclooxygenase-1, 100 μM) and cangrelor (antagonist of $P2Y_{12}$, 1 μM). Results: Platelet deposition on collagen (I) was not affected by the storage temperature of the blood before perfusion (room temperature vs. 37 °C); (II) was dependent on a shear rate in the range between 300/s and 1700/s; and (III) was influenced by the collagen concentration used to coat the microchannels up to a value of 10 μg/mL. ASA and cangrelor did not cause statistically significant inhibition of platelet accumulation, except for ASA at low collagen concentrations. Conclusions: Platelet deposition on collagen-coated surfaces is a shear-dependent process, not influenced by the collagen concentration beyond a value of 10 μg/mL. However, the inhibitory effect of antiplatelet drugs is better observed using low concentrations of collagen.

Keywords: platelets; thrombus formation; microfluidics; flow assays; platelet adhesion; platelet aggregation; shear rate; antiplatelet agents

1. Introduction

Platelets play a central pathogenic role in thrombosis; they aggregate at sites of atherosclerotic plaques, forming thrombi that can occlude the lumen of the artery [1–3]. In coronary arteries, this process causes acute coronary syndromes, which are a major cause of morbidity and mortality [4]. Atherosclerotic plaques favor the formation of thrombi by exposing thrombogenic surfaces, to which platelets adhere, and by causing stenosis of the arterial lumen, which contributes to platelet activation by increasing the shear rate of blood flow [5,6]. Patients with acute coronary syndrome are treated with antiplatelet drugs, inhibiting cyclooxygenase-1 (COX-1) (acetylsalicylic acid) and $P2Y_{12}$ receptors (clopidogrel, ticagrelor or prasugrel).

Microfluidic flow chambers (MFCs) designed to observe platelet adhesion and thrombus formation on surfaces coated with adhesive proteins, such as collagen, have become a valuable tool for research in hemostasis and thrombosis [7–9]. These devices can characterize platelet function under flow with low blood volume requirements and controlled conditions. They can mimic the anatomy of healthy and stenotic blood vessels [7,10], recreate a range of physiological and pathological shear stress conditions [11–13] and investigate platelet accumulation over different adhesive proteins [14,15]. Furthermore, these devices allow to study the inhibitory effect of antiplatelets agents on platelet adhesion and thrombus formation [16,17].

However, as a consequence of their high flexibility, MFCs are poorly standardized [18]. Limited studies have been performed to assess the variability of platelet accumulation in relation with the experimental variables. With regards to platelet deposition on collagen, most of the works have focused on the role of the mechanical stresses, reporting on the significant shear-rate dependency of the platelet adhesion and thrombus formation processes, e.g., [12,19,20]. The sensitivity to collagen concentration was investigated by Savage et al. [21] and Neeves and colleagues, who performed the most comprehensive study on the sources of variability in platelet accumulation on collagen, exploring also the effect of phenotypic and genetic factors [19]. The inhibitory effect of antiplatelet drugs on platelet accumulation has been studied by Diamond's research group in a number of works [16,17,22]. Despite these studies and some attempts to define common protocols [18,23], no real standardization has been achieved and the effect of a number of assay-dependent variables still remains to be fully elucidated.

Given this background, the current study aims at investigating a number of critical aspects in the study of platelet adhesion and thrombus formation with MFCs: The effect of the storage temperature of the blood samples before testing; the influences of wall shear rate and of the concentration of the adhesive protein collagen in the surface coating solution; and the inhibitory effect of some antiplatelet drugs.

2. Results

2.1. Effect of Blood Storage Temperature

We measured the effect of blood storage temperature at room temperature (RT) and 37 °C on platelet accumulation on collagen (200 µg/mL)-coated channels, at 300/s, 1100/s and 1700/s wall shear rates (Figure 1). In all conditions tested, no statistical difference was found between the results obtained at RT and 37 °C.

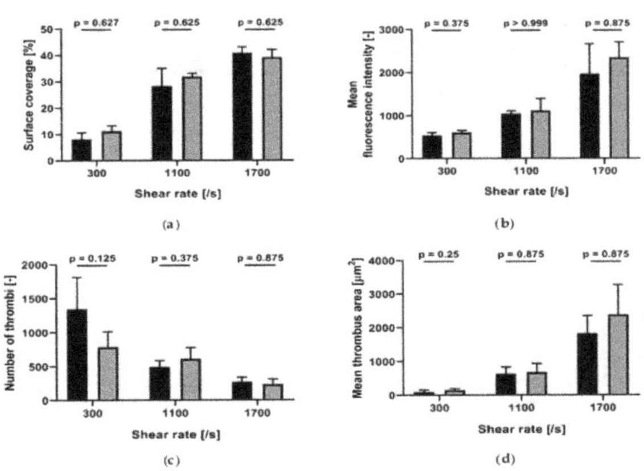

Figure 1. Effect of storage temperature (room temperature (RT) and 37 °C) on platelet accumulation on collagen (200 µg/mL)-coated channels at different wall shear rates (n = 4). Black: RT; grey: 37 °C. (**a**) Surface coverage; (**b**) mean fluorescence intensity; (**c**) number of thrombi; (**d**) mean thrombus area. Data are represented as column bar graphs, with means ± standard errors of the mean and analyzed by Wilcoxon matched pairs.

2.2. *Effect of Wall Shear Rate*

Platelet accumulation, as measured by surface coverage and fluorescence intensity, increased as a function of wall shear rate (Figure 2; Figure 3a,b). The number of thrombi significantly decreased while the mean thrombus area significantly increased as a function of wall shear rate (Figures 2 and 3c,d).

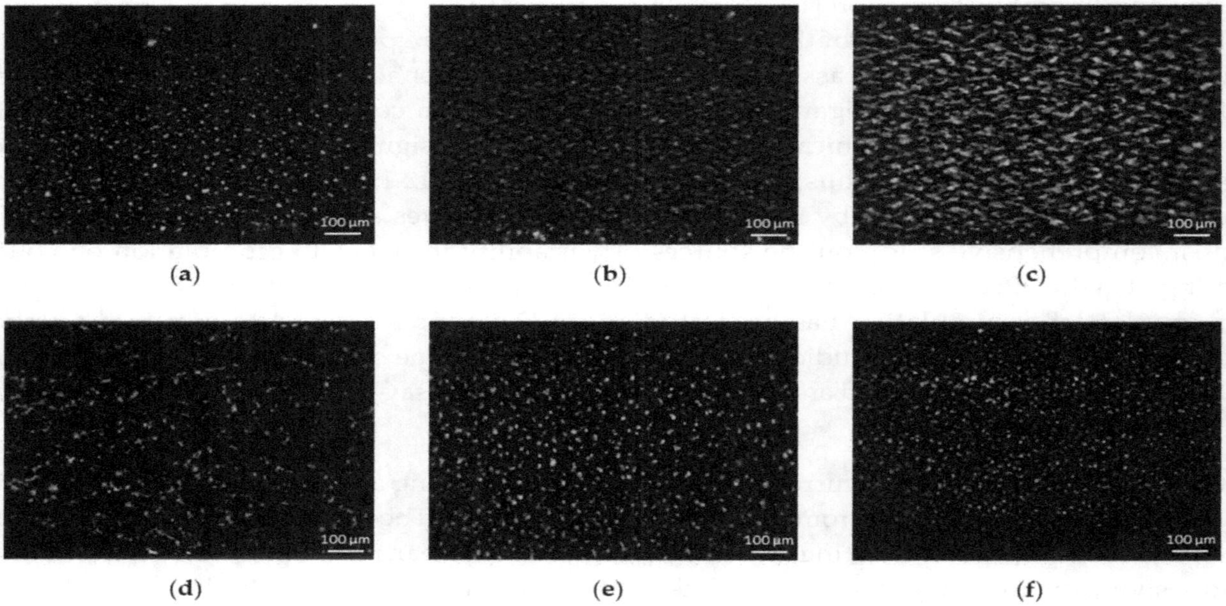

Figure 2. Representative images (6.3×) of platelet accumulation at different shear rates and collagen concentrations. Images are acquired after 4 min of perfusion. Flow is from left to right. The first row shows platelet accumulation over collagen (200 µg/mL)-coated perfusion chamber at (**a**) 300/s, (**b**) 1100/s and (**c**) 1700/s. The second row shows platelet accumulation at 300/s for collagen concentrations equal to (**d**) 10 µg/mL, (**e**) 50 µg/mL and (**f**) 100 µg/mL.

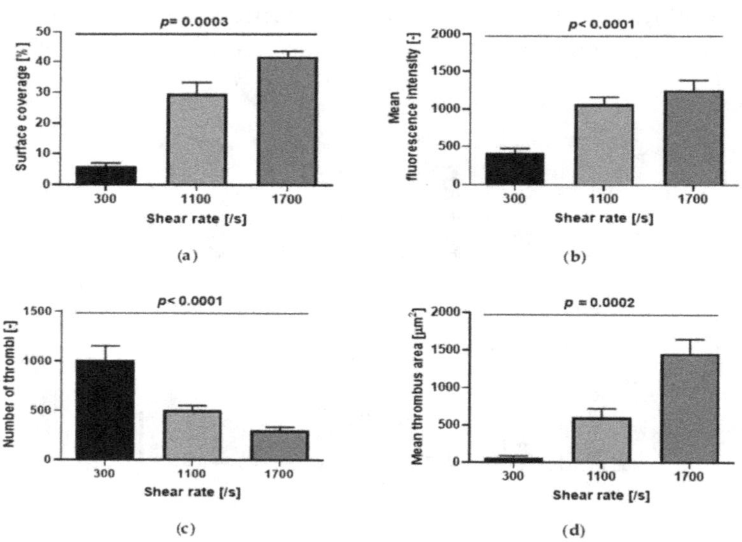

Figure 3. Effect of wall shear rate on platelet accumulation on collagen (200 µg/mL)-coated microchannels ($n = 7$). (**a**) Surface coverage; (**b**) mean fluorescence intensity; (**c**) number of thrombi; (**d**) mean thrombus area. Data are represented as column bar graphs, with means ± standard errors of the mean, and analyzed by Kruskal–Wallis tests or ANOVA tests as appropriate. Internal Contrasts: (**a**) 300/s vs. 1700/s, $p < 0.001$; (**b**) 300/s vs. 1100/s $p < 0.001$; 300/s vs. 1700/s, p < 0.001; (**c**) 300/s vs. 1100/s $p < 0.01$; 300/s vs. 1700/s, $p < 0.001$; (**d**) 300/s vs. 1700/s, $p < 0.001$.

2.3. *Effect of Collagen Concentration*

Surface coverage and fluorescence intensity at a shear rate of 300/s were affected by collagen concentration: They were lowest at 1 μg/mL and slowly increased with increasing collagen concentration up to 100 μg/mL; at 200 μg/mL the surface coverage tended to decrease (Figure 4). The internal contrast showed that there were no statistically significant differences over the range 5 to 200 μg/mL (Figures 2 and 4a,b). The number of thrombi significantly increased while their area significantly decreased as a function of collagen concentrations from 1 to 200 μg/mL (Figures 2 and 4c,d).

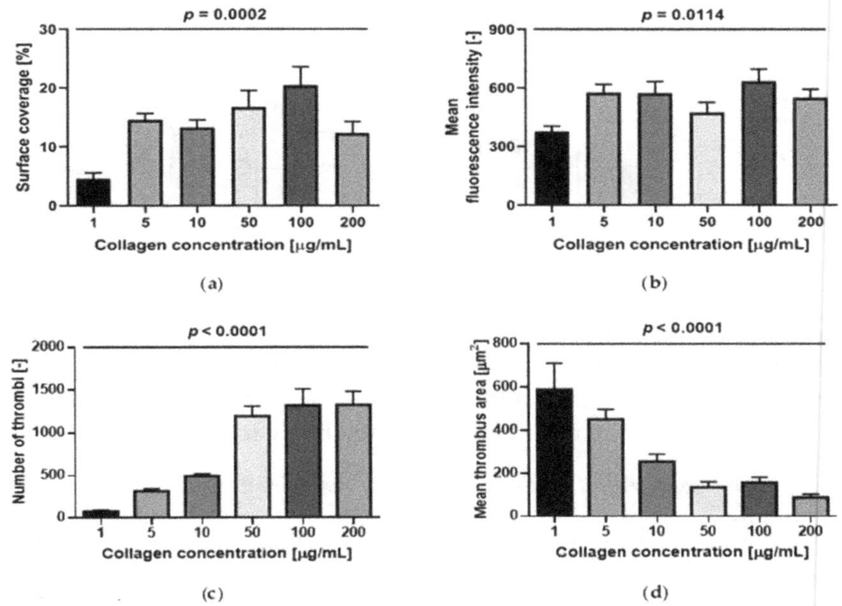

Figure 4. Effect of collagen concentration on platelet accumulation at 300/s ($n = 7$). (**a**) Surface coverage; (**b**) mean fluorescence intensity; (**c**) number of thrombi; (**d**) mean thrombus area. Data are represented as column bar graphs, with means ± standard errors of the mean, and analyzed by ANOVA tests plus Bonferroni's multiple comparison post-hoc tests. Internal Contrast: (**a**) 1 vs. 5 μg/mL of collagen, $p < 0.05$; 1 vs. 50 μg/mL of collagen, $p < 0.01$; 1 vs. 100 μg/mL of collagen, $p < 0.001$; (**b**) 1 vs. 100 μg/mL of collagen, $p < 0.05$; (**c**) 1 vs. 50, 100 and 200 μg/mL of collagen, $p < 0.001$; 5 vs. 50, 100 and 200 μg/mL, $p < 0.001$; 10 vs. 50 μg/mL of collagen, $p < 0.01$; 10 vs. 100 and 200 μg/mL of collagen, $p < 0.001$; (**d**) 1 vs. 10 μg/mL of collagen, $p < 0.01$; 1 vs. 50,100, 200 μg/mL, $p < 0.001$; 5 vs. 50, 200, $p < 0.01$; 5 vs. 100 μg/mL, $p < 0.05$.

2.4. *Effect of ASA on Platelet Accumulation*

- Collagen Concentration = 1 μg/mL. ASA (100 μM) induced a statistically significant decrease in surface coverage (4.4% vs. 1.6%; $p = 0.025$) and mean thrombus area, but did not significantly change mean fluorescence and number of thrombi (Figure 5).
- Collagen Concentration = 10 μg/mL. ASA (100 μM) did not induce a statistically significant change in surface coverage, tended to reduce mean fluorescence intensity and to increase the number of thrombi, but differences were borderline statistically significant and significantly decreased the mean thrombus area (Figure 5).
- Collagen Concentration = 100 μg/mL. ASA (100 μM) did not induce statistically significant modifications of surface coverage, fluorescence intensity, number of thrombi or mean thrombus area (Figure 5).

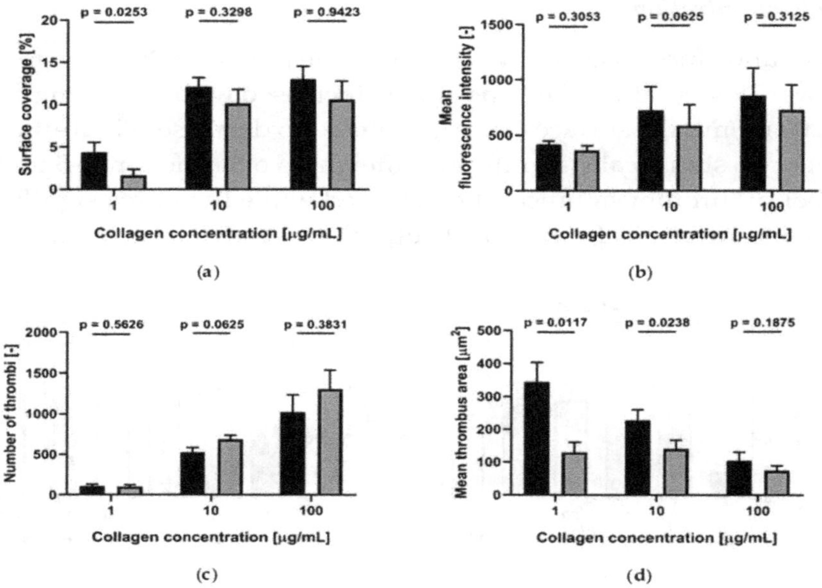

Figure 5. ASA inhibition of platelet accumulation at 300/s for different collagen concentrations (n = 5). Black: without ASA; grey: with ASA added in vitro. (**a**) Surface coverage; (**b**) mean fluorescence intensity: (**c**) number of thrombi; (**d**) mean thrombus area. Data are represented as column bar graphs, with means ± standard errors of the mean, and analyzed by Wilcoxon tests or t-tests as appropriate.

2.5. Cangrelor Inhibition of Platelet Accumulation

Collagen concentrations of 10 and 100 µg/mL. At shear rate of both 300/s and 1600/s, cangrelor (1 µM) did not induce a statistically significant reduction of surface coverage, mean fluorescence intensity, number of thrombi or mean thrombus area (Figures 6 and 7).

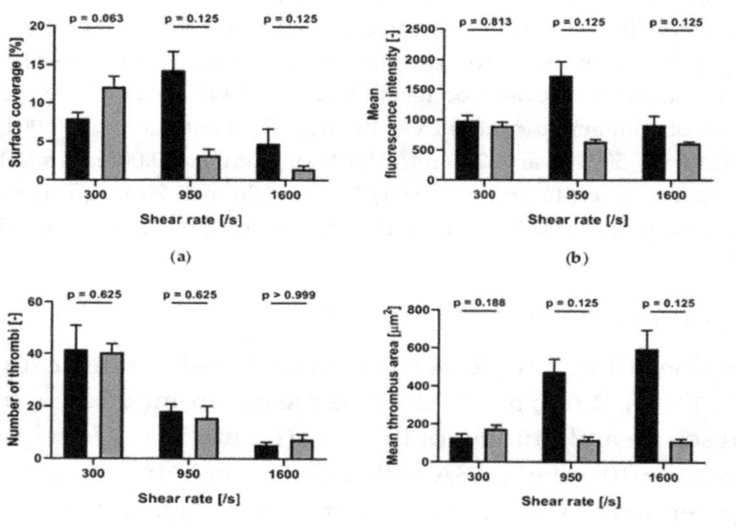

Figure 6. Effect of cangrelor at different shear rates on platelet accumulation on collagen (10 µg/mL)-coated perfusion chamber (n = 5). Black: without cangrelor; grey: with cangrelor added in vitro. (**a**) Surface coverage; (**b**) mean fluorescence intensity; (**c**) number of thrombi; (**d**) mean thrombus area. Data are represented as column bar graphs, with means ± standard errors of the mean, and analyzed by Wilcoxon matched pairs.

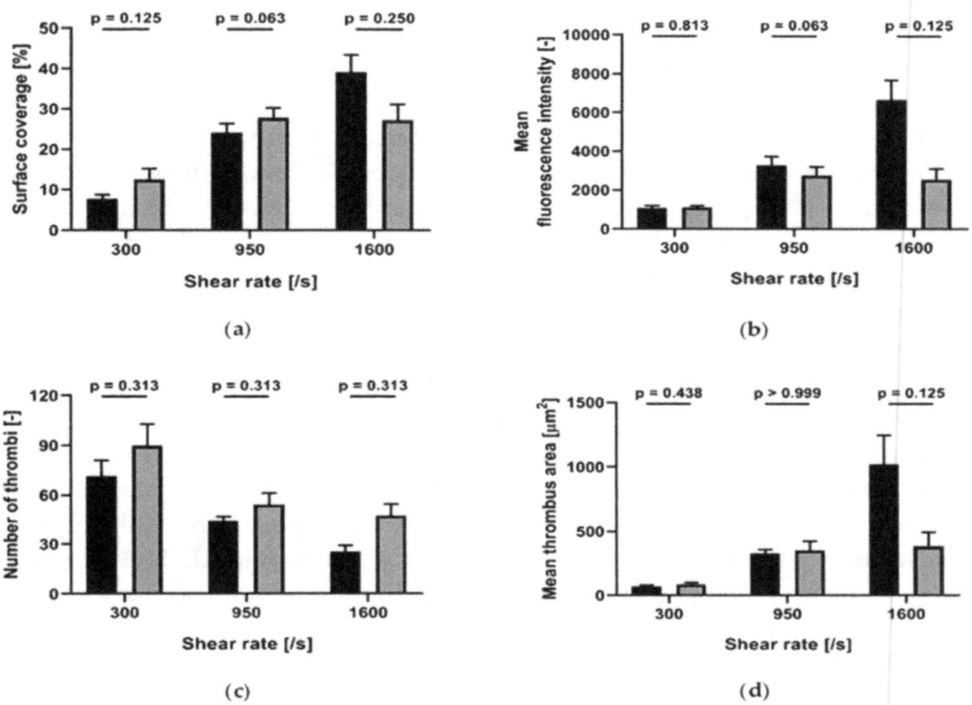

Figure 7. Effect of cangrelor at different shear rates on platelet accumulation on collagen (100 µg/mL)-coated m (n = 5). Black: without cangrelor; grey: with cangrelor added in vitro. (**a**) Surface coverage; (**b**) mean fluorescence intensity; (**c**) number of thrombi; (**d**) mean thrombus area. Data are represented as column bar graphs, with means ± standard errors of the mean, and analyzed by Wilcoxon matched pairs.

3. Discussion

Our study aimed at investigating different critical aspects of microfluidic platelet adhesion assays on collagen type I: (I) the storage temperature of blood before perfusion; (II) the wall shear rate; and (III) the concentration of collagen in the buffered solution used for coating. In addition, our study aimed at assessing the effect of antiplatelet agents targeting COX-1 (acetylsalicylic acid) and $P2Y_{12}$ (cangrelor) on platelet accumulation under the different experimental conditions considered in our study.

To this purpose, we developed a microfluidic device following the design recommendations given by [24], performed a number of experiments following the protocols suggested by [18] and evaluated the results by comprehensive image analysis, which included not only the typical outputs, such as surface coverage and mean fluorescence intensity, but also the number and area of thrombi, which, as outlined by [25], may change as a function of local flow conditions and platelet function.

Although it has been recommended to store blood samples for platelet function assays at room temperature [26], in our experiments no difference between the storage temperature of blood (at room temperature versus 37 °C) was found for all the WSR studied.

Platelet accumulation was a strongly shear-dependent process, as already observed by many authors, e.g., [19,25]. In our experiments, at 300/s several small and circular thrombi formed, whereas at higher shear rates (1100 and 1700/s), platelet aggregates elongated twofold in the direction of the flow. The same shear dependency of the morphology of platelet aggregates was found by Colace et al. [25], who calculated the width and length of growing thrombi on collagen in the range of shear rates from 100/s to 2000/s. Platelet accumulation increased with increasing shear rate, not only in the horizontal plane (i.e., the glass plane) as measured by surface coverage, but also vertically in the third dimension, as indicated by mean fluorescence intensity. All previous studies reported the same result despite each work observing the peak platelet deposition at a different shear rate: 300/s [19,20], 500/s [12] and 1500/s [21].

Collagen surface density influenced thrombus formation up to a threshold equal to 1 µg/mL with regard to the amount of platelet deposition (measured by SC and FI) and to 5 µg/mL for the variables quantifying the aggregate morphology (number of thrombi and mean thrombus area). At lower concentrations of collagen (1 and 5 µg/mL), a few dispersed, large thrombi have formed, while at higher concentrations (> 10 µg/mL) many small aggregates are present. Similar results were found by Savage et al. [21] who observed an increase in the total volume of thrombi from 0.01 µg/mL to 0.1 µg/mL and no further changes beyond 0.1 µg/mL, as well as by Neeves et al. [19] who observed a significant difference in surface coverage only up to a collagen concentration of 50 µg/mL.

Under our experimental conditions (shear rate equal to 300/s), ASA reduced surface coverage only on surfaces coated with the lowest concentration of collagen (1 µg/mL) but significantly decreased the mean thrombus area both at 1 µg/mL and 10 µg/mL. This is in agreement with Li and Diamond [22] who found that ex-vivo addition of ASA results in smaller platelet aggregates compared to responses measured with no drugs. In the same work, the authors also reported a decrease in total platelet accumulation (measured in terms of fluorescence intensity) over collagen at 200/s following high dose (500 µM) ASA addition, but the collagen concentration was not provided. The effect of ASA is known to be relevant only after some minutes of perfusion [16,17], when secondary aggregation mediated by thromboxane occurs, while initial platelet adhesion to collagen is not influenced by the presence of aspirin. This suggests that the perfusion time of our experiments should probably have been longer to observe the ASA inhibitory effect at high collagen concentrations. Cangrelor, an antagonist of the $P2Y_{12}$ receptor, did not cause any significant inhibition of platelet accumulation at any shear rate and concentration of collagen. However, a distinct trend was observed at the higher shear rates (950/s and 1600/s), indicating a marked decrease in surface coverage, fluorescence intensity and mean thrombus area due to the presence of cangrelor. This suggests that a statistical significance would probably be reached by increasing the number of experiments.

In summary, platelet deposition on collagen type I (I) is not affected by the storage temperature of the blood before perfusion (room temperature vs. 37 °C); (II) was a shear-dependent process in the range between 300/s and 1700/s; and (III) is influenced by the collagen concentration used to coat the microchannels up to a concentration of 5 µg/mL. Antiplatelet agents did not show statistically significant inhibitory effects on platelet interaction with collagen-coated surfaces, except for ASA at low concentrations of collagen.

By carefully examining the effect of several assay-dependent variables on platelet deposition on surfaces coated with collagen type I in a microfluidic device, we believe to have characterized different aspects of thrombus formation that should be taken into account before approaching these experiments.

4. Materials and Methods

4.1. Enrollment of Healthy Control Subjects

Healthy subjects (n = 15, 3 males; age range: 19–39 years) were recruited among workers of ASST Santi Paolo e Carlo and Università degli Studi di Milano. All studied subjects abstained from drugs known to affect platelet function for at least 10 days before enrolment. The study was approved by the institutional ethical committee of ASST Santi Paolo e Carlo; all subjects signed their informed consent.

4.2. Blood Sampling

Blood samples were collected in the morning from an antecubital vein using a 21-gauge butterfly needle with minimal stasis. The first 3 mL of blood was collected into K-EDTA tubes (Becton Dickinson vacutainer, North Ryde, NSW, Australia) for a complete blood count; the following 10–15 mL was collected in 250 µg/mL INN-desirudin (Canyon Pharmaceuticals, London, United Kingdom), gently mixed, and allowed "to rest" at room temperature for 15 min before use.

4.3. Fabrication of Microfluidic Devices

Microfluidic devices (Figure 8a) consisted of six independent channels (1000 μm wide, 100 μm high and 3.2 cm long). Channels were designed according to previous recommendations, which suggested an aspect ratio of 10:1 (width:height) in order to reduce wall effects and obtain homogenous platelet distribution across the width of the channel [12,19]. Microfluidic devices were fabricated in polydimethylsiloxane (PDMS, Sylgard™ 184, Dow Corning, Midland, MI, USA) from silicon masters using standard soft lithography techniques. PDMS was prepared by mixing pre-polymer and curing agent in a ratio 10:1 (w:w), degassed, poured over the master mold and cured at 80 °C for 3 h. Inlet and outlet fluidic ports were punched with a 1.5 mm diameter biopsy puncher. PDMS chips were permanently bonded to #0.6 microscope cover glasses via air plasma treatment.

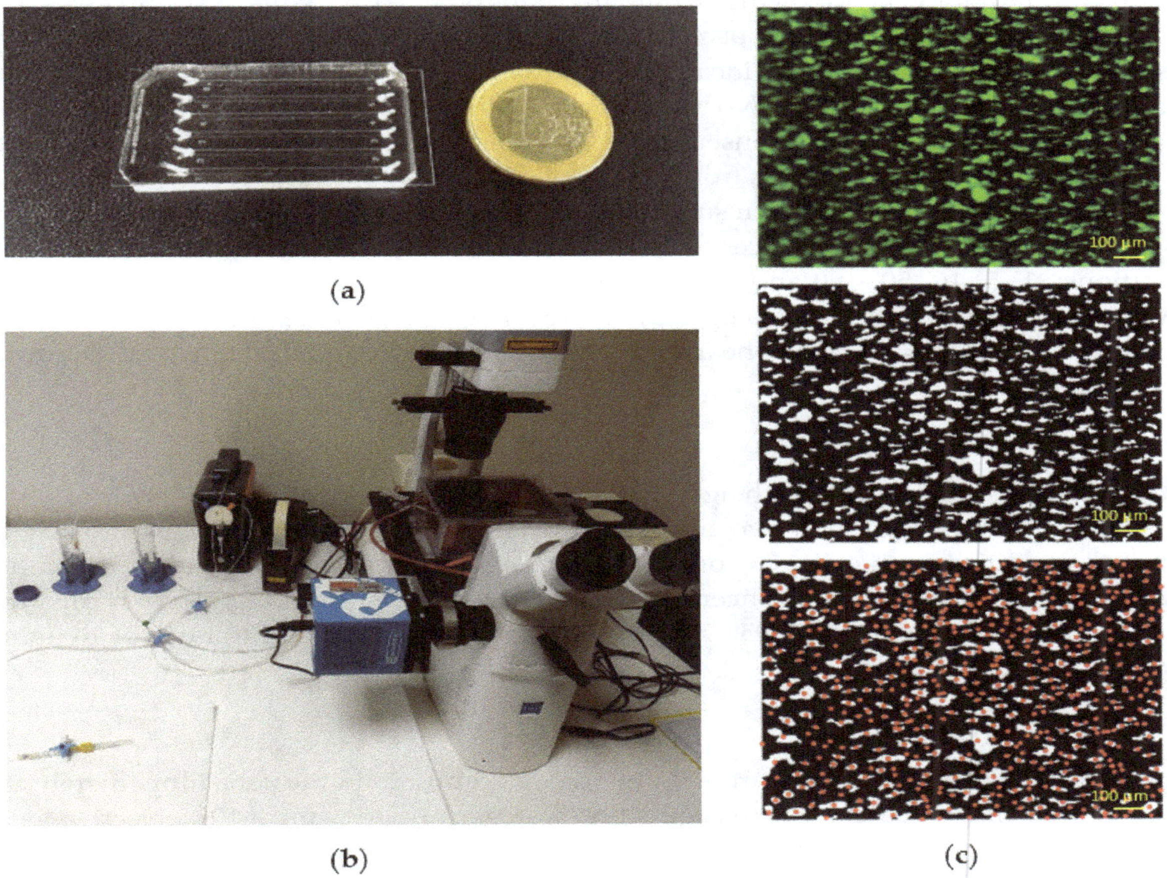

Figure 8. Photograph of the microfluidic chip (**a**) and of the experimental set-up (**b**). Results of the image-processing algorithm (**c**), from top to bottom: original image (6.3×), segmented binary image and identification of platelet aggregates (red dots represent individual thrombi).

4.4. Perfusion Experiments

Each microfluidic channel was incubated with Horm fibrillar collagen type I (Mascia Brunelli, Milano, Italy) at 4 °C overnight. An isotonic glucose solution (pH 2.7–2.9) was used to achieve the desired collagen concentration (1, 5, 10, 50, 100 or 200 μg/mL). Collagen pattering was obtained by pipetting 2.2 μL into the outlet port of the flow chamber. This allowed to fill only about 2/3 of the channel, leaving uncoated the first one-third. In this way, platelets were prevented from adhering in the proximity of the channel entry, where flow disturbances can induce altered platelet accumulation affecting downstream thrombus formation. Then, it was rinsed with filtered phosphate buffer saline (PBS). Before perfusion, it was incubated with 1% bovine serum albumin in PBS for 30 min at room temperature, to passivate the chamber surface that had not been coated by collagen. Subsequently, each channel was washed with PBS, connected with the flow system and checked for the absence of air bubbles. The microfluidic device was then placed on the stage of an inverted fluorescence microscope (Axiovert A1 FL, Zeiss, Milan, Italy) equipped with a 16-bit camera (Sony ICX-674 CCD Camera, Crisel Instruments, Rome, Italy) (Figure 8b). Fluorescence microscopy was performed with a mercury light source (HBO 50 AC L1, Zeiss, Milan, Italy) and a FITC filter set.

Blood samples were incubated with the green-fluorescent lipophilic dye 3,3′ Dihexyloxacarbocyanine Iodide (DiOC6, 1 μM; ThermoFisher Scientific, Milan, Italy), which tags platelets, at room temperature for 15 min. In most experiments, blood samples were kept at room temperature before testing; in some experiments, samples were divided in two equal parts, of which one was kept at RT and one at 37 °C for a maximum of 4h. In some experiments, 100 μM lysine-ASA (Flectadol, Sanofi-Aventis, Milan, Italy) or 1 μM cangrelor (The Medicines Company, Parsippany-Troy Hills, NJ, USA) was added to blood samples at RT for 10 or 3 min, respectively. Before the beginning of the experiments, whole blood samples that had been kept at RT were incubated at 37 °C for 3 min and then perfused for 4 min using a programmable syringe pump (Mirus Evo Nanopump, Cellix, Crisel Instruments, Rome, Italy) through Tygon tubes (0.5 mm inner diameter) inserted into the fluidic ports (Figure 8b). All experiments were performed at 37 °C within three hours from blood collection.

To investigate the effect of collagen surface density on platelet adhesion and thrombus formation, blood was perfused at 300/s over surfaces coated with buffered solutions containing 6 different collagen concentrations: 1, 5, 10, 50, 100 and 200 μg/mL ($n = 7$). To assess the effect of wall shear rate, blood was perfused over collagen (200 μg/mL) at three flow rates, corresponding to 300/s, 1100/s and 1700/s wall shear rates ($n = 7$). Wall shear rate γ was related to volumetric flow rate Q through the following equation:

$$Q = \frac{\gamma w h^2}{6} \tag{1}$$

where w and h refer to the width and height of the microfluidic channel, respectively.

Finally, the inhibitory effect of two antiplatelet drugs (ASA and cangrelor) was assessed under different flow conditions and collagen concentrations. ASA inhibition of platelet accumulation was investigated at 300/s and collagen concentrations equal to 1, 10 and 100 μg/mL ($n = 5$). The effect of cangrelor was evaluated at 300/s, 950/s and 1600/s with collagen concentrations of 10 μg/mL and 100 μg/mL.

4.5. Image Acquisition and Analysis

At the end of blood perfusion (4 min) an image was captured in the zone immediately after the beginning of the collagen patch. Image acquisition was performed using a 10× objective lens and a 0.64x video adapter, through the MicroManager software. The field view allowed to capture the full width of the channel. Real-time platelet accumulation was also monitored by acquiring an image every 2 s over the duration of the experiment.

Surface coverage, mean fluorescence intensity, number of thrombi and mean thrombus area were calculated using a custom MATLAB (MathWorks, Natick, MA, USA) script. To limit wall effects, images were cropped down to 80% of the channel width, where the wall shear rate is nearly uniform, within 5% of the value at the center [24]. Mean fluorescence intensity was estimated by averaging all the intensity values of the cropped image. Grayscale images were then converted into binary images by an automatic image thresholding method in order to eliminate operator bias. The Otsu's algorithm [27] was chosen as it provided a better performance than other global thresholding techniques (e.g., the triangle method). Otsu's algorithm is a statistical thresholding method, which partition the image intensity histogram into classes so that the between-class variance is maximized. After the thresholding operation, all the segmented images were checked one by one and when inconsistent results were found a background correction of the raw image was performed and then the image was thresholded again. After this step, pixel clusters smaller than the area of a single platelet were removed. The resulting images were used to calculate the surface coverage by dividing the number of pixels with values greater than zero by the total number of pixels. Finally, individual thrombi were identified, and the mean thrombus area was estimated. Figure 8c shows the results of the image analysis procedure: The central image represents the output of the thresholding operation, i.e., the segmented binary image from which surface coverage was calculated, while the image at the bottom shows the result of the object identification algorithm.

4.6. Statistical Analysis

Statistical analysis of the results was performed using GraphPad Prism 6 (GraphPad Software, Inc., CA, USA). Normal distribution of data was tested with the Shapiro–Wilk normality test. The One-way Analysis of Variance (ANOVA) test or t-tests were used when the normality hypothesis was satisfied. Conversely, non-parametric Kruskal–Wallis or Wilcoxon tests were performed. Statistical significance was assumed for p-values lower than 0.05.

Author Contributions: Conceptualization, M.S. and M.C.; formal analysis, M.S., S.B. and G.P.; investigation, M.S., S.B. and T.M.; methodology, M.S., S.B. and A.R.; software, S.B.; supervision, M.C. and A.R.; visualization, T.M.; writing—original draft, M.S., S.B. and M.C.; all authors read and approved the submitted version. All authors have read and agreed to the published version of the manuscript.

Acknowledgments: The authors are grateful to Elena Bossi and Raffaella Adami for their assistance in the laboratory analyses and to Marco Rasponi for his support in the lab-on-chip manufacturing process.

Abbreviations

MFC	microfluidic flow chambers
WSR	wall shear rates
ASA	acetylsalicylic acid
RT	room temperature
COX-1	cyclooxygenase-1
PDMS	polydimethylsiloxane
PBS	phosphate-buffered saline
K-EDTA	potassium ethylenediaminetetra-acetic acid

References

1. Kaplan, Z.S.; Jackson, S.P. The role of platelets in atherothrombosis. *Hematol. Am. Soc. Hematol. Educ. Program* **2011**, *2011*, 51–61. [CrossRef] [PubMed]
2. Badimon, L.; Padró, T.; Vilahur, G. Atherosclerosis, platelets and thrombosis in acute ischaemic heart disease. *Eur. Heart J. Acute Cardiovasc. Care* **2012**, *1*, 60–74. [CrossRef] [PubMed]
3. Corti, R.; Fuster, V.; Badimon, J. Pathogenetic concepts of acute coronary syndromes. *J. Am. Coll. Cardiol.* **2003**, *41*, 7S–14S. [CrossRef]
4. Sanchis-Gomar, F.; Perez-Quilis, C.; Leischik, R.; Lucia, A. Epidemiology of coronary heart disease and acute coronary syndrome. *Ann. Transl. Med.* **2016**, *4*, 256. [CrossRef] [PubMed]
5. Ruggeri, Z.M. Platelets in atherothrombosis. *Nat. Med.* **2002**, *8*, 1227–1234. [CrossRef] [PubMed]
6. Jennings, L.K. Role of platelets in atherothrombosis. *Am. J. Cardiol.* **2009**, *103*, 4A–10A. [CrossRef]
7. Westein, E.; de Witt, S.; Lamers, M.; Cosemans, J.M.; Heemskerk, J.W. Monitoring in vitro thrombus formation with novel microfluidic devices. *Platelets* **2012**, *23*, 501–509. [CrossRef]
8. Neeves, K.B.; Onasoga, A.A.; Wufsus, A.R. The use of microfluidics in hemostasis: Clinical diagnostics and biomimetic models of vascular injury. *Curr. Opin. Hematol.* **2013**, *20*, 417–423. [CrossRef]
9. Schoeman, R.M.; Marcus Lehmann, M.; Neeves, K.B. Flow chamber and microfluidic approaches for measuring thrombus formation in genetic bleeding disorders. *Platelets* **2017**, *28*, 463–471. [CrossRef]
10. Lui, M.; Gardiner, M.M.; Arthur, J.F.; Pinar, I.; Lee, W.M.; Ryan, K.; Carberry, J.; Andrews, R.K. Novel Stenotic Microchannels to Study Thrombus Formation in Shear Gradients: Influence of Shear Forces and Human Platelet-Related Factors. *Int. J. Mol. Sci.* **2019**, *20*, 2967. [CrossRef]

11. Gutierrez, E.; Petrich, B.G.; Shattil, S.J.; Ginsberg, M.H.; Groisman, A.; Kasirer-Friede, A. Microfluidic devices for studies of shear-dependent platelet adhesion. *Lab Chip* **2008**, *8*, 1486–1495. [CrossRef] [PubMed]
12. Neeves, K.B.; Maloney, S.F.; Fong, K.P.; Schmaier, A.A.; Kahn, M.L.; Brass, L.F.; Diamond, S.L. Microfluidic focal thrombosis model for measuring murine platelet deposition and stability: PAR4 signaling enhances shear-resistance of platelet aggregates. *J. Throm. Heamost.* **2008**, *6*. [CrossRef] [PubMed]
13. Hansen, R.R.; Wufsus, A.R.; Barton, S.T.; Onasoga, A.A.; Johnson-Paben, R.M.; Neeves, K.B. High Content Evaluation of Shear Dependent Platelet Function in a Microfluidic Flow Assay. *Ann. Biomed. Eng.* **2013**, *41*, 250–262. [CrossRef] [PubMed]
14. Maurer, E.; Schaff, M.; Receveur, N.; Bourdon, C.; Mercier, L.; Nieswandt, B.; Dubois, C.; Jandrot-Perrus, M.; Goetz, J.G.; Lanza, F.; et al. Fibrillar cellular fibronectin supports efficient platelet aggregation and procoagulant activity. *Thromb. Heamost.* **2015**, *114*, 1175–1188. [CrossRef] [PubMed]
15. Kent, N.J.; O'Brien, S.; Basabe-Desmonts, L.; Meade, G.R.; MacCraith, B.D.; Corcoran, B.G.; Kenny, D.; Ricco, A.J. Shear-Mediated Platelet Adhesion Analysis in Less Than 100 μL of Blood: Toward a POC Platelet Diagnostic. *IEEE Trans. Biomed. Eng.* **2011**, *58*, 826–830. [CrossRef]
16. Maloney, S.F.; Brass, L.; Diamond, S.L. $P2Y_{12}$ or $P2Y_1$ inhibitors reduce platelet deposition in a microfluidic model of thrombosis while apyrase lacks efficacy under flow conditions. *Integr. Biol.* **2010**, *2*, 183–192. [CrossRef]
17. Li, R.; Fries, S.; Li, X.; Grosser, T.; Diamond, S.L. Microfluidic assay of platelet deposition on collagen by perfusion of whole blood from healthy individuals taking ASA. *Clin. Chem.* **2013**, *59*, 1195–1204. [CrossRef]
18. Roest, M.; Reininger, A.; Zwaginga, J.J.; King, M.R.; Heemskerk, J.W. Biorheology Subcommittee of the SSC of the ISTH. Flow chamber-based assays to measure thrombus formation in vitro: Requirements for standardization. *J. Thromb. Haemost.* **2011**, *9*, 2322–2324. [CrossRef]
19. Neeves, K.B.; Onasoga, A.A.; Hansen, R.R.; Lilly, J.J.; Venckunaite, D.; Sumner, M.B.; Irish, A.T.; Brodsky, G. Sources of Variability in Platelet Accumulation on Type 1 Fibrillar Collagen in Microfluidic Flow Assays. *PLoS ONE* **2013**, *8*, e54680. [CrossRef]
20. Saelman, E.U.; Nieuwenhuis, H.K.; Hese, K.M.; de Groot, P.G.; Heijnen, H.F.; Sage, E.H.; Williams, S.; McKeown, L.; Gralnick, H.R.; Sixma, J.J. Platelet adhesion to collagen types I through VIII under conditions of stasis and flow is mediated by GPIa/IIa (alpha 2 beta 1-integrin). *Blood* **1994**, *83*, 1244–1250. [CrossRef]
21. Savage, B.; Almus-Jacobs, F.; Ruggeri, Z.M. Specific synergy of multiple substrate-receptor interactions in platelet thrombus formation under flow. *Cell* **1998**, *94*, 657–666. [CrossRef]
22. Li, R. Diamond, S.R. Detection of platelet sensitivity to inhibitors of COX-1, P2Y1, and $P2Y_{12}$ using a whole blood microfluidic flow assay. *Thromb. Res.* **2014**, *133*, 209–210. [CrossRef] [PubMed]
23. Van Kruchten, R.; Cosemans, J.M.; Heemskerk, J.W. Measurement of whole blood thrombus formation using parallel-plate flow chambers—A practical guide. *Platelets* **2012**, *23*, 229–242. [CrossRef] [PubMed]
24. Sarvepalli, D.P.; Schmidtke, D.W.; Nollert, M.U. Design considerations for a microfluidic device to quantify the platelet adhesion to collagen at physiological shear rates. *Ann. Biomed. Eng.* **2009**, *37*, 1331–1341. [CrossRef]
25. Colace, T.; Falls, E.; Zheng, X.L.; Diamond, S.L. Analysis of morphology of platelet aggregates formed on collagen under laminar blood flow. *Ann. Biomed. Eng.* **2011**, *39*, 922–929. [CrossRef]
26. McCarty, O.J.T.; Ku, D.; Sugimoto, M.; King, M.R.; Cosemans, M.E.M.; Neeves, K.B. Dimensional analysis and scaling relevant to flow models of thrombus formation: Communication from the SSC of the ISTH. *J. Thromb. Haemost.* **2016**, *14*, 619–622. [CrossRef]
27. Otsu, N. A threshold selection method from gray level histograms. *IEEE Trans. Syst. Man Cybern.* **1979**, *9*, 62–66. [CrossRef]

8

The "Janus Face" of Platelets in Cancer

Maria Valeria Catani *, Isabella Savini, Valentina Tullio and Valeria Gasperi *

Department of Experimental Medicine, Tor Vergata University of Rome, 00133 Rome, Italy; savini@uniroma2.it (I.S.); valentinatullio.nu@gmail.com (V.T.)

* Correspondence: catani@uniroma2.it (M.V.C.); gasperi@med.uniroma2.it (V.G.)

Abstract: Besides their vital role in hemostasis and thrombosis, platelets are also recognized to be involved in cancer, where they play an unexpected central role: They actively influence cancer cell behavior, but, on the other hand, platelet physiology and phenotype are impacted by tumor cells. The existence of this platelet-cancer loop is supported by a large number of experimental and human studies reporting an association between alterations in platelet number and functions and cancer, often in a way dependent on patient, cancer type and treatment. Herein, we shall report on an update on platelet-cancer relationships, with a particular emphasis on how platelets might exert either a protective or a deleterious action in all steps of cancer progression. To this end, we will describe the impact of (i) platelet count, (ii) bioactive molecules secreted upon platelet activation, and (iii) microvesicle-derived miRNAs on cancer behavior. Potential explanations of conflicting results are also reported: Both intrinsic (heterogeneity in platelet-derived bioactive molecules with either inhibitory or stimulatory properties; features of cancer cell types, such as aggressiveness and/or tumour stage) and extrinsic (heterogeneous characteristics of cancer patients, study design and sample preparation) factors, together with other confounding elements, contribute to "the Janus face" of platelets in cancer. Given the difficulty to establish the univocal role of platelets in a tumor, a better understanding of their exact contribution is warranted, in order to identify an efficient therapeutic strategy for cancer management, as well as for better prevention, screening and risk assessment protocols.

Keywords: microvesicles; miRNAs; paraneoplastic thrombocytosis and thrombocytopenia; platelet activation; platelet-derived bioactive molecules; platelet-tumor crosstalk

1. Introduction

Platelets were described for the first time in 1882, when the Italian pathologist Giulio Bizzozero identified in the blood vessels "very thin platelets, disc-shaped, with parallel surfaces or rarely lens-shaped structures, round or oval and with a diameter 2–3 times smaller than the diameter of the red cells ... " that "when they are circulating in the blood stream of a living animal a small injury to the vessel wall, or contact with a foreign body is sufficient for them to become viscous, to adhere to one another and so form a white thrombus" [1]. Platelets, indeed, are small, anucleated cytoplasmic fragments derived from large (30–100 μm) progenitor cells, the megakaryocytes, formed in bone marrow, lungs and blood [2–5], which are vitally involved in thrombosis and hemostasis [6]. Nevertheless, besides this primary function, platelets are also recognized to affect immune and inflammatory responses, thus participating in regulation of biological mechanisms underlying a broad range of human disorders. A large body of experimental and clinical evidences, indeed, shows that platelet activation and dysfunction are implicated in diabetes, cardiovascular disease, chronic back pain, sepsis, Alzheimer's disease, multiple sclerosis, psychiatric disorders and other central and peripheral pathological conditions [7–11]. In this context, the unexpected central role of platelets in cancer biology is noteworthy: They actively influence cancer cell behavior, but, on the other hand, platelet physiology and phenotype are impacted by tumor cells [12]. Indeed, a large number of experimental and human investigations support the hypothesis that tumor cells are able to modulate the RNA profile, number and activity of platelets that, once "educated", would regulate the tumor microenvironment and progression in a way dependent on the patient, cancer type and treatment. Nonetheless, the exact molecular mechanisms underlying this platelet-cancer loop are not yet well defined, often due to contradictory data.

Therefore, based on this background, the present review will focus on platelet-cancer crosstalk and their mutual impact, especially considering how platelets might exert either a protective or a deleterious role in all steps of cancer progression.

2. Paraneoplastic Thrombocytosis and Thrombocytopenia in Cancer

In healthy subjects, platelet count ranges from 150,000 to 450,000/μL, with age-, sex-, race- and genetic background-specific reference intervals [13,14]. A fine-tuned control of both platelet number and function exists, being ensured by the delicate balance among their (i) production, (ii) maintenance in the circulation (average life span of 8–10 days) and (iii) clearance of senescent cells (via hepatic and splenic macrophages, as well by apoptosis) [15–17].

In the light of the central roles played by platelets in a multitude of biological events, abnormalities in their number that often accompany various pathologies are clearly relevant, and this also applies to cancer.

Although, with some inter-individual variations, a platelet count of $\geq 450 \times 10^9$/L is a generally accepted value used to identify a clinically significant thrombocytosis [18], which has a multitude of potential etiologies. In particular, it can be classified as (i) primary thrombocytosis, when it occurs as the result of genetic or chronic myeloid disorders [19–21] or (ii) secondary or reactive thrombocytosis, when it occurs as a comorbidity of another underlying disease independent of a vascular event, including cancer. In the latter case, the pathology is called paraneoplastic thrombocytosis [22,23]. According to the clinical evidence that patients with a high platelet count have a higher risk to develop venous thromboembolism (VTE) [24], cancer patients frequently show activated coagulation pathways, resulting in a four-fold increase in thrombosis risk [25].

The first evidence of paraneoplastic thrombocytosis dates back to 1964, when Levin and Conley found that, among their hospitalized cancer patients, at least 40% had thrombocytosis [26]. Since then, an ever-growing body of studies has reported a significant association between thrombocytosis and solid tumors, with a range of thrombocytosis incidence at initial diagnosis of 4–55% [27–31]. This evidence may assume clinical implications, if we consider that a large amount of retrospective and meta-analysis studies point out to the correlation among higher platelet count and tumor progression, advanced-stage disease, vascular thromboembolic complications and poor survival in patients with different solid tumors, such as esophageal cancer, bladder cancer, inflammatory breast cancer and epithelial ovarian cancers (see Table 1) [28,31–38].

Conversely, such a correlation has not been found by other authors [29,39–41], or it appears strictly dependent on inflammatory components, as described by a recent retrospective study of 3654 patients with stage I–III breast cancer, of whom 6.5% had a diagnosis of Inflammatory Breast Cancer (IBC), the most aggressive form of breast tumors [37]. What emerged from this study is that thrombocytosis, more prevalent in IBC patients, correlated with poor overall survival in these subjects, but not in non-IBC individuals [37].

It must not be overlooked that some investigations reporting a correlation among platelet count, metastasis and shortened survival had some limitations, such as a low platelet count threshold ($<200 \times 10^9$/L) not clinically correct to define a patient as having real thrombocytosis [33,34,37,42]. In addition, heterogeneity in sample size, clinical stages, treatment and follow-up, smoking history and inclusion/exclusion criteria may make it difficult to establish a univocal association between thrombocytosis and poor prognosis in cancer patients.

Table 1. Main findings on the relationship between high platelet count [1] and cancer.

Cancer	Study	Platelet Cut-Off	Main Findings	Ref.
Oesophageal	- Retrospective - 584 adenocarcinoma patients with or without pre-operative neoadjuvant chemo-radiation therapy - 2.4% with high PLT	450×10^9/L	- Death rates: 50% with normal PLT (median survival time: 76.9 months), 86% with high PLT (median survival time: 23.2 months) - No differences in age, gender, tumor T or N stages - Median survival time in patients without neoadjuvant therapy: 35.8 months with high PLT, 112 months with normal PLT (HR = 3.02, p = 0.032) - Median survival time in patients with neoadjuvant therapy: 16.2 months with high PLT, 52.1 months with normal PLT (HR = 2.31, p = 0.021)	[43]
	- Retrospective - 374 squamous cell carcinoma patients with non-neoadjuvant therapy - 21.1% with high PLT	293×10^9/L	- PLT increased in patients with large and deep tumors, nodal involvement, and distant metastasis - CRP levels increased in patients with high PLT (p = 0.001) - Worse survival in patients with high PLT, especially in advanced tumor stage patients	[33]
	- Retrospective - 425 squamous cell carcinoma patients subjected to esophagectomy - 48.4% with high PLT	205×10^9/L	- Overall 5-year survival: 60.7% with PLT below cut off value, 31.6% with PLT above cut off value (p < 0.001) - 5-year survival with no involvement of nodes: similar rates independent of PLT - 5-year survival with involvement of nodes: 32.0% with PLT below cut off value, 12.7% with PLT above cut off value (p = 0.004)	[34]
	- Retrospective - 119 squamous cell carcinoma patients subjected to esophagectomy - 20.2% with high PLT	300×10^9/L	- No association between high PLT and disease-free (HR = 0.918, 95% CI = 0.524 – 1.608, p = 0.765) or overall (HR = 1.072, 95% CI = 0.618 – 1.891, p = 0.809) survival	[41]
	- Retrospective - 112 patients subjected to esophagectomy - 4% with high PLT	400×10^9/L	- No correlation between PLT and patient survival (p < 0.644)	[39]
	- Retrospective - 381 patients (93% squamous cell carcinoma and 7% adenocarcinoma) subjected to esophagectomy - 3.4% with high PLT	400×10^9/L	- Higher PLT in patients with adenocarcinoma (p = 0.003) - No correlation between PLT and prognostic factors - No correlation among PLT, site and degree of tumor penetration, lymph node involvement, distant metastasis, degree of differentiation, vascular, lymphatic and perineural invasion, presence of multiple cancers	[40]

Table 1. *Cont.*

Cancer	Study	Platelet Cut-Off	Main Findings	Ref.
Cervical	- Meta-analysis - 6521 patients	400×10^9/L (seven studies) 300×10^9/L (six studies) 200×10^9/L (five studies) Not specified (one study)	- High PLT before surgery and/or chemotherapy associated with poor overall (HR = 1.50, 95% CI = 1.19 – 1.88, $p = 0.001$), progression-free (HR = 1.33, 95% CI = 1.07 – 1.64, $p = 0.010$) and recurrence-free (HR = 1.66, 95% CI = 1.20 – 2.28, $p = 0.002$) survival	[32]
Epithelial ovarian	- Retrospective - 619 patients - 31% with high PLT	450×10^9/L	- High PLT associated with advanced-stage disease, vascular thromboembolic complications, higher preoperative levels of cancer antigen 125 and shortened survival	[28]
Lung	- Retrospective - 234 patients with Stage I non-small cell lung cancer	300×10^9/L	- Correlation among PLT, disease progression (HR = 5.314, 95% CI = 2.750 – 10.269, $p < 0.05$) and death (HR = 3.139, 95% CI = 1.227 – 8.034, $p < 0.05$)	[31]
	- Meta-analysis - 5884 patients - 6.9–58.5% with high PLT	400×10^9/L (6 studies) 300×10^9/L (5 studies) 214.5×10^9/L (1 study)	- High PLT associated with overall survival (HR = 1.74, 95% CI = 1.39-2.19, $p < 0.001$), advanced TNM stage (OR = 2.65, 95% CI = 1.77 – 3.97, $p = 0.367$), smoking history (OR = 2.70, 95% CI = 1.79 – 4.08, $p = 0.373$) - No correlation between high PLT associated and squamous cell carcinoma (OR = 1.54, 95% CI = 0.77 – 3.07, $p = 0.017$)	[29]

CI: confidence interval; CRP: C-reactive protein; HR: hazard ratio; OR: odds ratio; PLT: platelet count. [1] With respect to the cut-off values set by the authors.

However, available literature data suggest that thrombocytosis is a paraneoplastic event not depending on elongation of platelet half-life survival [44], but on increased thrombopoietin (TPO)-dependent thrombopoiesis, together with the action of inflammatory cancer-derived cytokines. TPO is normally produced and secreted by the liver, kidney and bone marrow at a fixed rate and it promotes megakaryocyte growth and platelet generation, by binding to its receptor c-MPL and triggering activation of the Janus kinase (JAK)/signal transducer and activator of the transcription (STAT) pathway [45]. It is well documented that increased circulating TPO levels might be one of the mechanisms accounting for cancer-related thrombocytosis, as demonstrated by elevated TPO levels in plasma of cancer patients with a high platelet count [46–48]. Two different and complementary mechanisms have been proposed, both encompassing tumors that represent a TPO source per se and, moreover, secrete factors targeting hepatic TPO synthesis. In particular, it has been reported that certain cancer cells, besides expressing TPO receptors on their surface [49], are also able to produce and release functional TPO [47,50], thus contributing to the rise in blood TPO levels. Additionally, cancer cells release a plethora of humoral factors and cytokines, and some of them have been shown to upregulate hepatic TPO biosynthesis; this is the case of the pleiotropic cytokine interleukin (IL)-6, a major mediator of inflammation and activator of STAT3 [51], whose deregulated overexpression has been associated with tumor progression [28,52–55]. Both IL-6 and its receptors (IL-6R and sIL-6R) are, indeed, upregulated in tumors [56–61] and their increased content in plasma of cancer patients correlates with a poor diagnosis [28,56,61,62], thus indicating clinical utility of IL-6 as a biomarker or therapeutic target in cancer management. An elegant model proposed by several authors suggests that IL-6 plays a crucial role in inducing cancer-related thrombocytosis, via up-regulation of hepatic TPO transcription [28,53–55]. In particular, this molecular model hypothesizes that cancer cells release large amounts of IL-6 that, in turn, determines complex chains of events (i.e., an increase in platelet count, tumor growth and metastasis) reinforcing themselves through a feed-forward loop. This hypothesis has been confirmed by the study of Stone and co-workers [28], who analyzed 619 patients with epithelial ovarian cancer and of whom 30% had thrombocytosis at the time of initial diagnosis. The researchers found that TPO and IL-6 levels were high in patients who had thrombocytosis, as compared with those who did not, and that an increase in IL-6 levels positively correlated with plasma TPO levels and thrombocytosis, while negatively correlating with patient survival. Further proofs of the crucial role of IL-6 in paraneoplastic thrombocytosis have also been provided by molecular/genetic and pharmacological experiments: Silencing of *Il-6* and *tpo* genes fully abrogated thrombocytosis in murine ovarian cancer, and siltuximab (humanized anti-IL-6 antibody) significantly reduced tumor growth and platelet count, both in murine and human ovarian cancers [28].

Other circulating factors released by cancer cells and known to stimulate thrombopoiesis and megakaryopoiesis are granulocyte colony-stimulating factor (G-CSF) and granulocyte-macrophage colony-stimulating factor (GM-CSF), whose blood levels are increased in cancer patients with thrombocytosis [63].

A more in-depth analysis of basal cytokine profile in 81 newly diagnosed IBC patients revealed that patients with thrombocytosis, although not differing in IL-6 levels with respect to IBC subjects without thrombocytosis, showed a positive correlation between serum levels of Growth-Regulated Oncogene (GRO) and Transforming Growth Factor (TGF)-β and IBC-related thrombocytosis [37]. In this context, it should be underlined that both cancer cells and activated platelets are able to release GRO and TGF-β [64–66], thus suggesting that the observed increase in their content might be a consequence rather than a cause of thrombocytosis. In addition, the study has several limitations, above all the lowering of the thrombocytosis threshold from 450 to 300 $\times 10^9$/L. Therefore, more studies are needed to establish a real relationship between these two cytokines and platelets in the context of tumor biology.

While thrombocytosis is more frequently reported to be associated with increased mortality, some findings also suggest the presence of cancer-related thrombocytopenia. For example, a strong trend toward increased mortality has been found in thrombocytopenic patients (hazard ratio (HR) = 1.50, but without reaching statistical significance) [43], although it is conceivable that thrombocytopenia might be a surrogate for general debility and/or other clinical factors, such as possible sepsis and hematological abnormalities that could contribute to overall mortality.

Thrombocytopenia is a frequent complication in solid tumors [67]. The degree and incidence of this disease depends on the type of malignancy, tumor stage and treatment approach [68]. It has also been described as a complement of local cancer recurrence and may be considered a paraneoplastic syndrome [69] Some tumors can alter the platelet count below 100×10^9/L, leading to thrombocytopenia, and therefore cancer patients have a high risk of hemorrhagic complications [68]. The first evidence of low platelet count and bleeding episodes in patients with malignancies came from Gaydos in 1962: He demonstrated that bleeding episodes in patients with leukemia were frequently associated with a decreased platelet count [70]. Since then, other studies reported similar bleeding events in solid tumor patients [71,72].

Single nucleotide polymorphisms (SNPs) and mutations in genes encoding for cytokines and transcription factors are both two major causes of thrombocytopenia in solid tumors, including lung, breast, ovary and colorectal cancers [72]. Just as an example, the -31 T > C SNP of the *il-1β* gene was up-regulated in solid tumors associated with thrombocytopenia [73,74]. It is unclear how IL-1β can induce thrombocytopenia in solid tumors, but it is known that -31 T > C SNP can increase susceptibility to thrombocytopenia in these malignancies [72]. A strong association between IL-1β -31 T > C SNP and *Helicobacter pylori* infection has also been reported: The two phenomena collaborate themselves to increase the risk of gastric cancer with hemorrhagic complications [75,76]. IL-6 is also involved in thrombocytopenic mechanisms, as well as in paraneoplastic thrombocytosis. The IL-6 -174 G/C SNP has been reported in many malignancies, including adenocarcinoma, lung, colorectal, gastric and ovarian cancers [73,77,78]. This polymorphism also has been correlated with a poor prognosis, because it can induce antibody production against platelets, increasing the risk of thrombocytopenia [72].

As mentioned before, several transcription factors can be associated with thrombocytopenia, due to their involvement in platelet production. For example, overexpression of GATA3, a member of the GATA family of transcription factors that control maturation of hematopoietic stem cells, can stimulate platelet clearance, worsening the prognosis in breast cancer [79]. Another example is represented by Homeobox (HOA) genes that control proliferation and maturation of hematopoietic stem cells: Hypermethylation of *hoxa11*, for example, increased the incidence of thrombocytopenia and risk of poor prognosis in lung, gastric and breast cancers [80,81]. Although few studies have evaluated genetic changes in the incidence of thrombocytopenia, it seems that investigations on the mechanisms accounting for this phenomenon may be useful for the prevention of bleeding in solid tumors and for choosing the appropriate treatment [72].

Clinical observations of the impact of platelet count on cancer biology have been supported by studies employing platelet-depleted or transgenic mice [82]. A significant reduction in neovascularization has been observed in different transgenic mice-rendered thrombocytopenia [83], with GPIbα/IL4R transgenic mice (lacking the receptor for the von Willebrand factor as well as other adhesive and pro-coagulant proteins) showing the most severe phenotype [83,84]. Furthermore, in a mammary carcinoma murine model, platelet depletion increased the efficacy of the chemotherapy, by favoring drug delivery and tumoricidal action [85]. Thus, currently available animal models of platelet dysfunction may provide a framework for better understanding the molecular mechanisms through which thrombocytosis or thrombocytopenia impact cancer progression.

3. Platelet-Derived Bioactive Compounds

It is now recognized that tumor cells and platelets strictly influence each other, thus establishing the so-called "platelet-tumor loop". The need for recruiting platelets is supported by several evidences: (i) tumor cells sequester platelets in order to escape themselves from immune system surveillance [86,87]; (ii) platelets associate with cancer emboli, thus prolonging survival of circulating tumor cells and promoting their arrest and adhesion to endothelium for transmigrating to metastatic sites [86,88]; and (iii) platelets secrete a plethora of tumor, angiogenic, growth and permeability factors, which can regulate tumor growth, epithelial to mesenchymal transition and metastasis (see below). Noteworthy, circulating hyperactivated platelets, as well as exhausted platelets (i.e., with totally or partially depleted granules, as a consequence of previous activation), are commonly found in subjects with different tumor types, concomitant with the high incidence of VTE [89].

In order to grow and develop metastasis, tumor cells concurrently influence platelet behavior by up-regulating synthesis and/or release of several compounds able to promote platelet activation and aggregation [90–92]. For example, when compared to benign tumors, malignant cells show higher generation of thrombin, one of the most potent platelet activators with strong pro-coagulating properties [93], but also able (when bound to thrombomodulin expressed on endothelium) to attenuate the thrombotic cascade [94,95]. Accordingly, blood thrombin concentrations negatively predict both success to surgery/chemotherapy and survival of patients with gynecological tumors [96]. A cohort study enrolling 112 patients with different cancers revealed that, although without association with disease state, rise in thrombin levels were dependent on tumor site, with lung cancers having more significant increases compared to brain and pancreas cancers [97]. Detrimental effects of increased thrombin generation seem to rely on its ability in promoting, in concert with its targets (among them, protease-activated receptor-1 and -4), cancer adhesion to platelets or endothelium via up-regulation of pro-inflammatory cytokines, adhesion molecules, angiogenic factors, and matrix-degrading proteases, thereby dramatically increasing tumor growth, angiogenesis, invasion, and metastasis [98]. Proofs of the "tumor cell-induced platelet aggregation" also come from quantitative analysis of circulating neutrophil elastase (which proteolytically activates integrin αIIbβ3) and serglycin (a pro-apoptotic and half-life cytokine regulating protein) in melanoma patients; while levels of the former were found up-regulated in cancer subjects, content of the latter was low [99], thus indicating the role of activated platelets in promoting cancer progression.

Although data on the molecular mechanisms underlying platelet hyperactivation are still not well-defined and often contradictory, nonetheless they suggest the involvement of some platelet bioactive molecules. Among them, those contained in α-granules play a pivotal role during platelet-cancer crosstalk (Table 2).

Table 2. Main platelet-derived proteins involved in cancer.

Molecule	Main Findings	Role in Cancer	Ref.
P-selectin	↑ tumor cell extravasation by promoting cancer cell interaction with platelets and endothelium ↑ platelet activation by increasing thrombin generation	**NEGATIVE**	[98,100–103]
TF	↑ monocyte TF exposure Clotting cascade activation ↑ cancer cell survival, ↑ angiogenesis, ↑ tumor growth, ↑ metastasis		[100]
VEGF	↑cancer growth and angiogenesis ↑ MK maturation		[101]
EGF	↑ mesenchymal and epithelial cell proliferation ↑ pro-angiogenic effect of other cytokines		[102]
Ang-1	↑ vessel development and maturation		[103]
PDGF-BB	↑ cancer cell proliferation, survival and invasion ↑ tumor stroma changes ↑ blood vessel maturation		[104,105]
	↓ tumor cell growth and dissemination ↓ metastasis	**POSITIVE**	[106]
Endostatin, TSP-1, angiostatin	↓ angiogenesis		[107,108]

Ang-1: angiopoietin-1; EGF: endothelial growth factor; MK: megakaryocytes; NK: natural kill cells; PAI-1: plasminogen activator inhibitor-1; PDGF-BB: Platelet-derived growth factor BB; TF: tumor necrosis factor; TSP-1: thrombospondin-1; VEGF Vascular endothelial growth factor.

Besides P-selectin and other clotting proteins (such as thrombospondin, fibrinogen, fibronectin integrin αIIβ3, integrin αVβ3, factor V, and the von Willebrand factor), α-granules contain growth and pro-angiogenic factors (including Platelet-Derived Growth Factor (PDGF), Vascular Endothelial Growth Factor (VEGF), TGF-β, Epidermal Growth Factor (EGF) and Angiopoetin-1 (Ang-1)), which are secreted following platelet activation and, either directly or indirectly, promote tumorigenesis. For example, besides directly promoting cancer progression and tumor cell extravasation (by facilitating interaction of cancer cells with platelets and endothelium) [109,110], P-selectin may indirectly exacerbate cancer evolution by triggering thrombin generation [111] and rapid monocyte exposure of tissue factor (TF) [100]. The latter protein, primarily involved in activation of the clotting cascade, positively affects tumor growth and metastasis: TF (produced by cancer cells) present in the tumor microenvironment may increase cell survival and/or angiogenesis, while TF present in the bloodstream (deriving from both monocytes and circulating cancer cells) has been shown to enhance thrombosis, tumor growth and metastasis. Several studies showed that P-selectin (either soluble or membrane-bound) changes in cancer patients, although controversial data have been documented. Some clinical studies found high P-selectin levels in cancer patients, with or without correlation to tumor clinical advancement [97,112]; conversely, recent observational and longitudinal studies enrolling patients with heterogeneous cancers found a decrease in platelet surface expression of this protein, which, together with diminished integrin αIIbβ3 exposure, thrombin and collagen receptor responsiveness and monocyte–platelet aggregate formation, correlates with risk of mortality and VTE [113,114]. Nevertheless, a lack of data on platelet activation changes in relation to cancer evolution makes it difficult to clarify whether decreased platelet reactivity is a consequence of continuous pre-activation in patients with poor prognosis or, rather, it represents a cancer-independent event, or even it is the result of a lack of protective effects exerted by activated platelets.

PDGF also has been implicated in tumorigenesis: It acts on tumor cells, thereby favoring proliferation, survival and invasion [104,105], and, in the meanwhile, creates a favorable microenvironment for tumor cells by inducing changes into tumor stroma and promoting blood vessel maturation (especially in advanced stages of angiogenesis) [115]. Accordingly, anti-PDGF drugs significantly inhibit tumor growth and metastasis, although downregulation of PDGF-BB (one of the five isoforms) signaling is associated with tumor cell dissemination and metastasis [106]. In the latter case, a protective role of PDGF-BB has been suggested, since its overexpression, by increasing tumor pericyte content, decreases colorectal and pancreatic cancer growth [116]. However, data on PDGF-BB content in cancer patients are controversial: Concentrations of secreted PDGF-BB, for example, have been found to be significantly high in the serum of colorectal carcinoma patients [117], but low in liver cancer patients with recurrence [118].

Among platelet-derived growth factors/cytokines, VEGF, EGF and Ang-1 play a crucial role in cancer angiogenesis [101–103]. Concerning the pro-angiogenic effect, the mutual influence of platelets and tumor cells also is not fully determined. A recent study performed on twenty-four women with active breast cancer and ten healthy controls showed that breast cancer and its chemotherapeutic treatment influence platelet phenotype, by increasing VEGF release and modulating the response to antiplatelet therapy [119]. Noticeably, an autocrine–paracrine loop, occurring in the bone marrow microenvironment and involving VEGFR-1-dependent megakaryocyte maturation has been documented [120]. Such evidence, together with the finding that tumor-derived IL-6 leads to enhanced megakaryocyte VEGF expression and a higher platelet VEGF load (concomitantly associated with fast tumor growth kinetics and poor diagnosis), strongly suggest a cooperation between platelets and cancer in promoting angiogenesis [121]. Conversely, low EGF levels have been found in cancer subjects with recurrence, and an inverse correlation between its concentrations and survival [118] has been documented in these subjects. Low serum Ang-1 levels, related to poor diagnosis, were also found in subjects with certain types of cancer [118]. On the contrary, patients with lung and ovarian cancer show high Ang-1 concentrations not related to patient survival [122].

In proteomic studies, α-granules also have been shown to contain a plethora of angiogenesis inhibitors, including endostatin, platelet factor-4, thrombospondin-1, α_2-macroglobulin, plasminogen activator inhibitor-1 and angiostatin. Therefore, activated platelets can organize regulatory proteins in the α-granules in order to selectively address the release of pro- or anti-angiogenic factors, depending on which different sets of α-granules are segregated [107,108]. Although cancer cells might have the ability to provoke preferential release of pro-angiogenic mediators from platelet α-granules, in order to create a dynamic microenvironment favorable for their growth and survival [107], several experimental and clinical data point out that platelet secretion in the tumorigenic microenvironment might also be oriented towards an anti-angiogenic effect. Such a hypothesis is supported by several findings: (i) significantly higher endostatin levels have been found in hepatocellular carcinoma patients, as well as in gastric cancer subjects [123,124]; (ii) increases in circulating thrombospondin-1 have been found to positively correlate with survival of patients with gynecological and non-small cell lung cancer [125,126]; and (iii) higher levels of angiostatin have been found in serum of prostate cancer patients [127], as well as in urine of patients with epithelial ovarian cancer [128].

Platelets also secrete other factors, which can interfere with all steps of cancer development and metastasis. Serotonin, a monoamine synthesized by enterochromaffin cells in the intestinal mucosa, is largely (about 95%) sequestered by platelets in dense granules, from which it is released in response to various stimuli [12]. A large body of experimental data support both stimulatory and inhibitory properties of serotonin on tumor onset and progression [129–134]. Such different behavior is much likely due to the ability of serotonin to act in a concentration-dependent manner, as well as in its capability to activate distinct signaling pathways, depending on the receptor subtype present at various tumor stages. Coherently, high serotonin levels, correlating with the tumor stage, distant metastases and a poor prognosis, have been found in the serum of patients with different solid tumors [131,135–137], whereas low concentrations have been found in hepatocellular carcinoma patients, who showed recurrence after partial hepatectomy [118]; in untreated breast adenocarcinoma or malignant melanoma patients, where the phenotype (also associated with a high ATP/ADP ratio and index of delta storage pool deficiency) was more marked in regionally spread malignant tumors [138].

Platelets are also important producers of eicosanoids, lipids derived from polyunsaturated fatty acids (PUFAs), through the catalytic action of cyclooxygenase (COX) and lipoxygenase (LOX). Eicosanoids are crucially involved in several pathophysiological conditions [139], and the intersection between changes in certain platelet-derived eicosanoids and cancer appears particularly intriguing. For example, a multi-omics analysis of serum from metastatic melanoma patients revealed a rise in the concentration of 12-hydroxyeicosatraenoic acid (HETE) and 15-HETE eicosanoids, which are respectively produced by platelet 12-LOX and COX-1 [99]. These two platelet-derived eicosanoids have also been shown to exert pro-malignant effects in several cancer types, by activating mitogen and angiogenic pathways [139–142]. Coherently, blockage of COX-1 by aspirin causes the loss of platelet ability to transform human colon carcinoma cells into mesenchymal-like cells [139,143] and the long-term use of low-dose aspirin is associated with a reduction in risk of various cancers [144].

Finally, activated platelets release lysophosphatidic acid (LPA), a bioactive lipid growth factor, which has been shown to promote cell proliferation, survival, migration, tumor cell invasion and reversal of differentiation, through multiple G protein-coupled receptor (LPA1-6) cascades [145]. Several studies have found a relationship between plasma LPA levels and ovarian carcinoma; for example, a meta-analysis, comparing LPA levels in the serum of 980 ovarian cancer patients, 872 benign controls and 668 healthy controls, showed higher LPA plasma levels in the cancer group with respect to the benign and healthy control samples [146], thus suggesting that the raised detection of plasma LPA might be a potential diagnostic biomarker. Nonetheless, this finding is not supported by a recent lipidomic study that did not find any change in the content of this lipid in serum of ovarian cancer patients [147]. Discrepancies in the results can be explained taking into the account that, due to its susceptibility to sample processing procedure (e.g., plasma storage time at room temperature and anticoagulant used for blood drawing), LPA can artificially increase. In the light of this finding, therefore, it appears important to consider these confounding factors, in order to reduce to the minimum potential errors in measuring plasma LPA.

4. Platelet Microvesicle-Derived miRNAs

Upon activation, platelets release microvesicles (MVs), which are vesicular fragments with a diameter ranging from 0.5 to 1 μm, that express parental antigens (such as P-selectin and integrin αIIbβ3) and contain a plethora of mediators (growth factors, cytokines, inflammatory molecules, mRNAs and miRNAs) able to exert biological effects. These platelet MVs, accounting for 70–90% of all MVs circulating in the bloodstream, contribute to regulation of the tumor microenvironment and cancer-cell interactions [148–150]. Accordingly, cancer patients usually show increased levels of circulating MVs (in a way, that is different depending on tumor type, but it is directly proportional to tumor stage), which may be prognostic for monitoring tumor progression and response to specific therapeutics [151].

Although the role of MVs in cancer progression is multi-faceted and not fully understood, nonetheless it is becoming clear that MVs represent one of crucial determinants in tumor biology. Firstly, surface expression of platelet antigens leads to shedding of MVs displaying pro-coagulant and pro-thrombotic features. Therefore, MVs, together with activated platelets, enhance coagulation (which is further exacerbated by cancer-triggered activation of more platelets), thus playing an additional role in cancer progression [152]. Secondly, MVs are able to enhance angiogenesis: They stimulate the expression of pro-angiogenic molecules [including matrix metalloproteinase (MMP)-9, VEGF, IL-8 and hepatocyte growth factor (HGF)] in tumor cells [153] and drive capillary tube formation by stimulating endothelial cells [27,82,154].

Noteworthy, a bidirectional effect occurs as cancer cells can induce platelet activation and MV release; subsequently, a paracrine positive feedback mechanism is established, since MVs, taken up by cancer cells, potentiate the invasive phenotype through stimulation of migration [148,155]. Interestingly, although different cancer cells are able to induce platelet-derived MV release, only the most aggressive ones are responsive to MV action and, furthermore, only some subsets of MVs can positively feed back to cancer cells [155]. These findings suggest that i) cancer/platelet interplay is complex and strongly dependent on features of tumor cell type and ii) composition of MVs may differ depending on the stimulus given to platelets [155]. These findings may clarify discrepancies observed by authors in studies using different platelet preparations, agonists and cancer cell types. Just as an example, our recent work [156] showed that MVs, once internalized by cancer cells, inhibit migration rather than enhancing invasive properties; this may be explained considering that MV release was induced not by cancer cells but, instead, by a different stimulus (namely, arachidonic acid) that led to MV enrichment of specific bioactive molecules. Therefore, depending on MV composition, the effects on tumors may be completely different. A further point of discrimination may reside in MV internalization, since we found that, albeit taken up, the bioactive molecules delivered to cancer cells had different stability and, thus, exert their action in a time-and concentration-dependent fashion [156].

Among the bioactive molecules contained inside MVs, microRNAs (miRNAs) deserve a particular mention. Human platelets, indeed, contain an abundant repertoire of miRNAs that are released through MVs; depending on the nature of the agonists or stimuli activating platelets, the miRNA content of MVs can vary, but it always mirrors the content found in the platelets from which MVs derive [157]. It is exactly this heterogeneity of composition that may account for the observed differences in terms of molecular targets, mechanisms of action and effects on cancer cells (Table 3) [152]. If, on one side, cancer-promoting effects of MVs have been described (especially related to their content in growth factors, inflammatory cytokines and angiogenic factors), it is also true, on the other hand, that the ability to deliver miRNAs to recipient cells (including cancer cells) suggests a potential tumor-suppressive role.

Table 3. Main findings on the cancer-related role of miRNAs, potentially delivered by platelet MVs.

miRNA	Experimental Settings	Main Findings	Targets	Ref.
miR-223	Platelet MV delivey to lung A549 cancer cells	↑ cell invasion	EPB41L3	[158]
	Transfection of breast MCF-7 and prostate PC-3 cancer cells	↓ vitality, ↑ effects of the anti-tumor celastrol	NF-κB	[159]
	Transfection of breast MDA-MB-231 and MCF-7 cancer cells Incubation of MDA-MB-231 cells with CM derived from stable transduced MEFs or HEK293 cells	↓ migration, ↑ anoikis cell death, ↑ sensitivity to chemotherapy	STAT5A	[160]
	Transfection of breast MCF-7, SKBR3, MDA-MB-231 and MDA-MB-435 cancer cells	↑ sensitivity to TRAIL-induced apoptosis	HAX-1	[161]
	Transient transfection of primary endothelial cells	↓ formation of new blood vessels	endothelial β1 integrin	[162]
miR-939	Platelet MV delivey to ovarian SKOV3 cancer cells	↑ epithelial to mesenchymal transition	E-cadherin and vimentin	[163]
miR-24-3p	Transfection of small-cell lung H446 cancer cells	resistance to etoposide plus cisplatin therapy	ATG4A	[164]
miR-24	Platelet MV delivey to Lewis lung and colon MC-38 carcinoma cells	↓ tumor growth, ↑ apoptosis	mt-Nd2 and Snora75	[165]
miR-130a	miRNA microarray in drug-resistant ovarian A2780 carcinoma cells Transfection of cervix HeLa carcinoma cells	drug resistance	M-CSF	[166]
miR-27a, miR-451	Transfection of MDR ovarian A2780 and cervical KB-V1 carcinoma cells		MDR1	[167]
miR-let-7a, miR-27b	Platelet MV delivey to primary endothelial cells	↑ endothelial tube formation	thrombospondin-1	[168,169]
miR-126	Transfection of breast MDAMB231 and MCF7 cancer cells	↓ cancer progression	ADAM9	[170]
	Transfection of breast BT-549 cancer cells Platelet MV delivey to breast BT-549, MDA-MB-468, BT-20 and MCF-7 cancer cells	cell cycle arrest, ↓ migration, ↑ sensitivity to cisplatin	ND	[156]
	Transfection of lung A549, Y-90 and SPC-A1 carcinoma cells	↑ proliferation	VEGF	[171]

ND: not determined. ADAM9: ADAM Metallopeptidase Domain 9; ATG4A: autophagy-associated gene 4A; CM: conditioned medium; EPB41L3: Erythrocyte Membrane Protein Band 4.1 Like 3; HAX-1: HS-1-associated protein X-1; β; M-CSF: macrophage colony-stimulating factor; MDR1: multidrug resistance gene; MEF: mouse embryonic fibroblasts; mt-Nd2: Mitochondrial NADH dehydrogenase 2; MV: microvesicle; NF-κB: nuclear factor-κB; Snora75: Small Nucleolar RNA, H/ACA Box 75; STAT5A: signal transducer and activator of transcription 5A; TRAIL: TNF-related apoptosis-inducing ligand; VEGF: vascular endothelial growth factor.

Platelet MVs can, indeed, be viewed as intercellular carriers that transfer inside cells specific molecules able to negatively modulate gene expression, with both positive and negative consequences. In this context, it should be underlined that most of the studies reporting differential effects of specific miRNAs in cancer employed transfection experiments, where miRNA expression was artificially increased and, to the best of our knowledge, only few of them checked the effects in more physiological (delivery of platelet-derived MVs) conditions.

Transfer of miRNAs to target cells has been shown to promote tumor progression [157]: miR-939 delivered by platelet MVs induces, in ovarian cancer cells, epithelial to mesenchymal transition, by down-regulating E-cadherin and up-regulating vimentin expression [163], while miR-223 has been shown to stimulate lung cancer cell invasion, by targeting tumor suppressor EPB41L3 [158]. Besides targeting tumor suppressor genes and oncogenes, several miRNAs enriched in MVs (miR-223, miR-24, miR-27a, miR-155, miR-195, let-7a/b) may also be implicated in therapy resistance. In small-cell lung cancer, miR-24-3p contributes to resistance to combination therapy (etoposide plus cisplatin), by targeting the autophagy-associated gene 4A [164]; other miRNAs that may be involved in drug resistance include miR-130a (which targets the pro-metastatic macrophage colony-stimulating factor (M-CSF)), and miR-27a and miR-451 (which target the multi drug resistance transporter 1) [166,167,172,173].

MVs also deliver angiogenic signals [27]: Transfer of miRNA let-7a or miR-27b in endothelial cells down-regulates the expression of the anti-angiogenic modulator thrombospondin-1, thus enhancing platelet-dependent endothelial tube formation [168,169].

However, what the available literature data suggest is that platelet-derived MVs may support cancer progression and metastatic dissemination at late stages, while it seems likely that they exert tumor suppressive roles at earlier stages. Michael's group found that circulating MVs directly infiltrating lung and colon cancer cells deliver miR-24 that suppress tumor growth; this miRNA localizes to mitochondria where it inhibits mt-Nd2 and Snora75, resulting in mitochondrial dysfunction and induction of apoptotic cell death [165]. Similarly, miR-223 inhibits migration, stimulates anoikis cell death and enhances chemo-sensitivity in different cancer cell types [159–161]. We and others have demonstrated that also platelet-specific miR-126 exert tumor suppressive roles; this miRNA may be a predictor for tumor relapse in postmenopausal breast cancer patients treated with tamoxifen [174], impairs cancer progression through direct repression of MMP-9 [170], and MV-mediated delivery into breast cancer cells induces cell cycle arrest, inhibition of migration and sensitivity to cisplatin [156]. Finally, miR-126 and miR-223 exert antagonistic effects on angiogenesis: miR-126 stimulates VEGF-induced proliferation in endothelial cells [171], while miR-223 exerts an inhibitory effect on formation of new blood vessels, by targeting endothelial β1 integrin [162].

The ability of MVs to acquire distinct roles, depending on their repertoire of proteins and miRNAs, suggest that they may be used as biomarkers with diagnostic and therapeutic implications [175]. For example, plasma levels of platelet MVs, together with VEGF, IL-6 and RANTES, have been found to be increased in patients with stage IV gastric cancer [176]. Elevated amounts of endothelial and platelet MVs (that significantly decreased after chemotherapy) have been found in the plasma of non-small cell lung cancer patients, thus suggesting a predictive role for prognostic clinical outcome [177].

5. Conclusions

Although some symptoms of cancer, such as breast lumps, are classic "alarm" symptoms, others are ambiguous and more likely caused by other conditions. Accordingly, different studies demonstrated that alterations in platelet number or/and activity often occur in cancer patients. This finding, together with the evidence that platelets may basically affect all steps of tumor development, prompts researchers to carry out more studies for fully understanding the mechanisms underlying cancer-related platelet dysfunction.

Environmental cancer-related stimuli encountered by platelets are intricate, as are the intracellular signaling pathways regulating platelet responses to the stimuli themselves. Moreover, due to heterogeneity in the cargo of growth factors, cytokines, microRNAs and other bioactive molecules and platelets may potentially release either stimulators or inhibitors in all cancer steps (Figure 1).

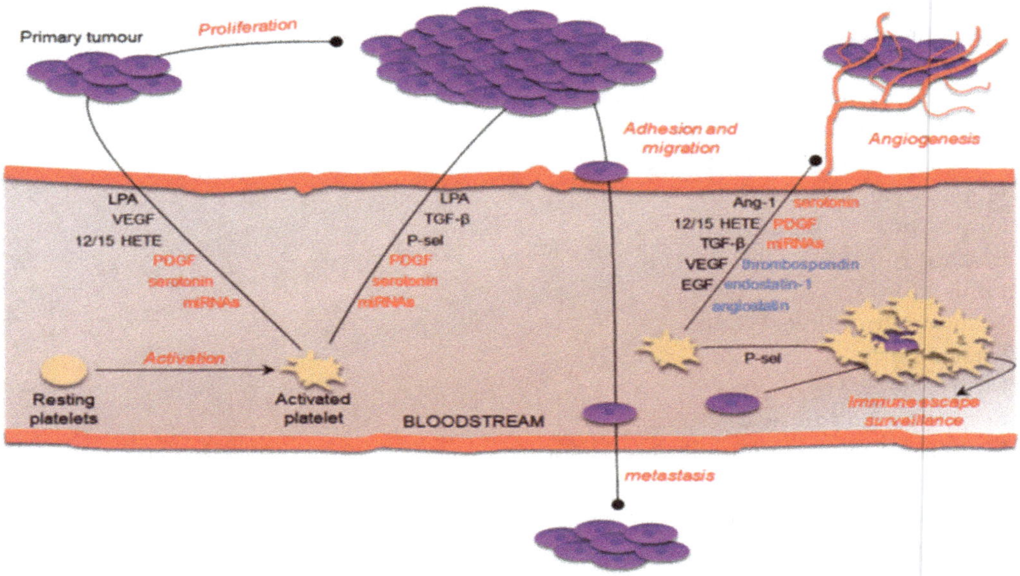

Figure 1. Schematic representation of the main platelet effects on tumor biology. See text for details. In black: platelet-derived bioactive molecules with positive effects. In blue: platelet-derived compounds with negative effects. In red: platelet-derived compounds with both positive and negative effects. Lines with dot indicate either stimulation or inhibition, depending on the platelet-derived bioactive molecule. 12/15 HETEs: 12 and 15 hydroxyeicosatraenoic acid; Ang-1: Angiopoietin; LPA: lysophosphatidic acid; EGF: Endothelial Growth Factor; P-sel: P-selectin; PDGF: Platelet-Derived Growth Factor; TGF-β: tumor growth factor-β; VEGF: vascular endothelial growth factor.

Accordingly, on one hand, there are a great deal of proofs of deadly interaction between platelets and cancer cells, but, on the other hand, some experimental and clinical data also indicate a protective role. Besides to complexity of platelet signaling in cancer, the scenario is further complicated by other confounding factors extrinsically related to platelets: (i) most of results come from retrospective studies, analyzing a wide range of patients with heterogeneous characteristics, such as age, sex, race, cancer type and stage, as well as treatments not always related to cancer; (ii) some studies have been carried out in cancer patients post-diagnosis and changes in platelets have not been monitored with respect to cancer progression over time, so that it is difficult to establish the exact contribution of platelets and if their changes are the cause or rather a consequence of tumor; and (iii) some of the bioactive compounds, whose plasma concentrations have been shown to be correlated with cancer progression, are also released by other cells, including cancer cells themselves; therefore, the effect on cancer, whether positive or negative, is not necessarily due to platelets.

Other questions still need to be answered: Why do platelet changes occur in some types of cancer and not in others? How does the platelet–cancer relationship change with age, sex and cancer progression over time? Although the heterogeneity and adaptive potential of the tumor features make a "one-size-fits-all" approach for targeting platelet–cancer interactions difficult, a better understanding of the interplay might provide efficient tools for cancer prevention, screening, risk assessment and management.

Author Contributions: M.V.C. and V.G. conceptualized the work and did the original draft preparation; V.T. provided the resources for this work and contributed to draft preparation; I.S. reviewed the text; V.G. drew the table and figures. All authors have read and agreed to the published version of the manuscript.

Abbreviations

Ang-1	Angiopoetin-1
COX	Cyclooxygenase
EGF	Epidermal growth factor
G-CSF	Granulocyte colony-stimulating factor
GM-CSF	Granulocyte-macrophage colony-stimulating factor
GRO	Growth-regulated oncogene
HETE	Hydroxyeicosatraenoic acid
HGF	Hepatocyte growth factor
HOA	Homeobox
HR	Hazard ratio
JAK	Janus kinase
IBC	Inflammatory breast cancer
IL	Interleukin
LOX	Lipoxygenase
LPA	Lysophosphatidic acid
M-CSF	Macrophage colony-stimulating factor
miRNA	MicroRNA
MMP	Matrix metalloproteinase
MV	Microvesicle
PDGF	Platelet-derived growth factor
PUFA	Polyunsaturated fatty acid
SNP	Single nucleotide polymorphisms
STAT	Signal transducer and activator of transcription
TF	Tissue factor
TGF	Transforming growth factor
TPO	Thrombopoietin
VEGF	Vascular endothelial growth factor
VTE	Venous thromboembolism

References

1. Bizzozero, J. Ueber einen neuen Formbestandtheil des Blutes und dessen Rolle bei der Thrombose und der Blutgerinnung—Untersuchungen. *Arch. Pathol. Anat. Physiol. Klin. Med.* **1882**, *90*, 261–332. [CrossRef]
2. Lefrançais, E.; Ortiz-Muñoz, G.; Caudrillier, A.; Mallavia, B.; Liu, F.; Sayah, D.M.; Thornton, E.E.; Headley, M.B.; David, T.; Coughlin, S.R.; et al. The lung is a site of platelet biogenesis and a reservoir for haematopoietic progenitors. *Nature* **2017**, *544*, 105–109. [CrossRef] [PubMed]
3. Schulze, H.; Stegner, D. Imaging platelet biogenesis in vivo. *Res. Pract. Thromb. Haemost.* **2018**, *2*, 461–468. [CrossRef] [PubMed]
4. Humphrey, J.H. Origin of Blood Platelets. *Nature* **1955**, *176*, 38. [CrossRef] [PubMed]
5. Wright, J.H. The histogenesis of the blood platelets. *J. Morphol.* **1910**, *21*, 263–278. [CrossRef]
6. Vinholt, P.J. The role of platelets in bleeding in patients with thrombocytopenia and hematological disease. *Clin. Chem. Lab. Med.* **2019**, *57*, 1808–1817. [CrossRef]
7. Mezger, M.; Nording, H.; Sauter, R.; Graf, T.; Heim, C.; von Bubnoff, N.; Ensminger, S.M.; Langer, H.F. Platelets and Immune Responses During Thromboinflammation. *Front. Immunol.* **2019**, *10*, 1731. [CrossRef]
8. Assinger, A.; Schrottmaier, W.C.; Salzmann, M.; Rayes, J. Platelets in Sepsis: An Update on Experimental Models and Clinical Data. *Front. Immunol.* **2019**, *10*, 1687. [CrossRef]
9. Mohammed, S.; Yu, J. Platelet-rich plasma injections: An emerging therapy for chronic discogenic low back pain. *J. Spine Surg.* **2018**, *4*, 115–122. [CrossRef]
10. Nieswandt, B.; Kleinschnitz, C.; Stoll, G. Ischaemic stroke: A thrombo-inflammatory disease? *J. Physiol.* **2011**, *589*, 4115–4123. [CrossRef]
11. Saluk-Bijak, J.; Dziedzic, A.; Bijak, M. Pro-Thrombotic Activity of Blood Platelets in Multiple Sclerosis. *Cells* **2019**, *8*, 110. [CrossRef] [PubMed]

12. Ghoshal, K.; Bhattacharyya, M. Overview of Platelet Physiology: Its Hemostatic and Nonhemostatic Role in Disease Pathogenesis. *Sci. World J.* **2014**, *2014*, 781857. [CrossRef] [PubMed]
13. Biino, G.; Santimone, I.; Minelli, C.; Sorice, R.; Frongia, B.; Traglia, M.; Ulivi, S.; Di Castelnuovo, A.; Gögele, M.; Nutile, T.; et al. Age- And Sex-Related Variations in Platelet Count in Italy: A Proposal of Reference Ranges Based on 40987 Subjects' Data. *PLoS ONE* **2013**, *8*, e54289. [CrossRef] [PubMed]
14. Eicher, J.D.; Lettre, G.; Johnson, A.D. The genetics of platelet count and volume in humans. *Platelets* **2018**, *29*, 125–130. [CrossRef]
15. Catani, M.V.; Gasperi, V.; Evangelista, D.; Finazzi Agrò, A.; Avigliano, L.; Maccarrone, M. Anandamide extends platelets survival through CB1-dependent Akt signaling. *Cell. Mol. Life Sci.* **2010**, *67*, 601–610. [CrossRef]
16. Grozovsky, R.; Giannini, S.; Hoffmeister, K.M. Novel mechanisms of platelet clearance and thrombopoietin regulation. *Curr. Opin. Hematol.* **2017**, *22*, 445–451. [CrossRef]
17. Grozovsky, R.; Giannini, S.; Falet, H.; Hoffmeister, K.M. Novel mechanisms of platelet clearance and thrombopoietin regulation. *Curr. Opin. Hematol.* **2015**, *22*, 445–451. [CrossRef]
18. Tefferi, A.; Thiele, J.; Orazi, A.; Kvasnicka, H.M.; Barbui, T.; Hanson, C.A.; Barosi, G.; Verstovsek, S.; Birgegard, G.; Mesa, R.; et al. Proposals and rationale for revision of the World Health Organization diagnostic criteria for polycythemia vera, essential thrombocythemia, and primary myelofibrosis: Recommendations from an ad hoc international expert panel. *Blood* **2007**, *110*, 1092–1097. [CrossRef]
19. Barbui, T.; Thiele, J.; Gisslinger, H.; Finazzi, G.; Vannucchi, A.M.; Tefferi, A. The 2016 revision of WHO classification of myeloproliferative neoplasms: Clinical and molecular advances. *Blood Rev.* **2016**, *30*, 453–459. [CrossRef]
20. Song, J.; Hussaini, M.; Zhang, H.; Shao, H.; Qin, D.; Zhang, X.; Ma, Z.; Hussnain Naqvi, S.M.; Zhang, L.; Moscinski, L.C. Comparison of the Mutational Profiles of Primary Myelofibrosis, Polycythemia Vera, and Essential Thrombocytosis. *Am. J. Clin. Pathol.* **2017**, *147*, 444–452. [CrossRef]
21. Chia, T.L.; Chesney, T.R.; Isa, D.; Mnatzakanian, G.; Colak, E.; Belmont, C.; Hirpara, D.; Veigas, P.V.; Acuna, S.A.; Rizoli, S.; et al. Thrombocytosis in splenic trauma: In-hospital course and association with venous thromboembolism. *Injury* **2017**, *48*, 142–147. [CrossRef] [PubMed]
22. Griesshammer, M.; Bangerter, M.; Sauer, T.; Wennauer, R.; Bergmann, L.; Heimpel, H. Aetiology and clinical significance of thrombocytosis: Analysis of 732 patients with an elevated platelet count. *J. Intern. Med.* **1999**, *245*, 295–300. [CrossRef] [PubMed]
23. Santhosh-Kumar, C.R.; Yohannan, M.D.; Higgy, K.E.; Al-Mashhadani, S.A. Thrombocytosis in adults: Analysis of 777 patients. *J. Intern. Med.* **1991**, *229*, 493–495. [CrossRef] [PubMed]
24. Khorana, A.A.; Francis, C.W.; Culakova, E.; Lyman, G.H. Risk factors for chemotherapy-associated venous thromboembolism in a prospective observational study. *Cancer* **2005**, *104*, 2822–2829. [CrossRef] [PubMed]
25. Wang, S.; Li, Z.; Xu, R. Human cancer and platelet interaction, a potential therapeutic target. *Int. J. Mol. Sci.* **2018**, *19*, 1246. [CrossRef] [PubMed]
26. Levin, J. Thrombocytosis Associated With Malignant Disease. *Arch. Intern. Med.* **1964**, *114*, 497. [CrossRef]
27. Wojtukiewicz, M.Z.; Sierko, E.; Hempel, D.; Tucker, S.C.; Honn, K.V. Platelets and cancer angiogenesis nexus. *Cancer Metastasis Rev.* **2017**, *36*, 249–262. [CrossRef]
28. Stone, R.L.; Nick, A.M.; McNeish, I.A.; Balkwill, F.; Han, H.D.; Bottsford-Miller, J.; Rupaimoole, R.; Armaiz-Pena, G.N.; Pecot, C.V.; Coward, J.; et al. Paraneoplastic Thrombocytosis in Ovarian Cancer. *N. Engl. J. Med.* **2012**, *366*, 610–618. [CrossRef]
29. Zhang, X.; Ran, Y. Prognostic role of elevated platelet count in patients with lung cancer: A systematic review and meta-analysis. *Int. J. Clin. Exp. Med.* **2015**, *8*, 5379–5387.
30. Gu, D.; Szallasi, A. Thrombocytosis portends adverse prognosis in colorectal cancer: A meta-analysis of 5,619 patients in 16 individual studies. *Anticancer Res.* **2017**, *37*, 4717–4726.
31. Ji, Y.; Sheng, L.; Du, X.; Qiu, G.; Su, D. Elevated platelet count is a strong predictor of poor prognosis in stage i non-small cell lung cancer patients. *Platelets* **2015**, *26*, 138–142. [CrossRef] [PubMed]
32. Cao, W.; Yao, X.; Cen, D.; Zhi, Y.; Zhu, N.; Xu, L. Prognostic role of pretreatment thrombocytosis on survival in patients with cervical cancer: A systematic review and meta-analysis. *World J. Surg. Oncol.* **2019**, *17*, 132. [CrossRef] [PubMed]

33. Shimada, H.; Oohira, G.; Okazumi, S.; Matsubara, H.; Nabeya, Y.; Hayashi, H.; Takeda, A.; Gunji, Y.; Ochiai, T. Thrombocytosis associated with poor prognosis in patients with esophageal carcinoma1 1No competing interests declared. *J. Am. Coll. Surg.* **2004**, *198*, 737–741. [CrossRef] [PubMed]
34. Feng, J.F.; Huang, Y.; Lu, W.S.; Chen, Q.X. Preoperative platelet count in esophageal squamous cell carcinoma: Is it a prognostic factor? *Langenbeck's Arch. Surg.* **2013**, *398*, 1115–1122. [CrossRef] [PubMed]
35. Wan, S.; Lai, Y.; Myers, R.E.; Li, B.; Hyslop, T.; London, J.; Chatterjee, D.; Palazzo, J.P.; Burkart, A.L.; Zhang, K.; et al. Preoperative platelet count associates with survival and distant metastasis in surgically resected colorectal cancer patients. *J. Gastrointest. Cancer* **2013**, *44*, 293–304. [CrossRef]
36. Lee, M.; Kim, S.W.; Nam, E.J.; Yim, G.W.; Kim, S.; Kim, Y.T. The impact of pretreatment thrombocytosis and persistent thrombocytosis after adjuvant chemotherapy in patients with advanced epithelial ovarian cancer. *Gynecol. Oncol.* **2011**, *122*, 238–241. [CrossRef]
37. Harano, K.; Kogawa, T.; Wu, J.; Yuan, Y.; Cohen, E.N.; Lim, B.; Reuben, J.M.; Ueno, N.T. Thrombocytosis as a prognostic factor in inflammatory breast cancer. *Breast Cancer Res. Treat.* **2017**, *166*, 819–832. [CrossRef]
38. Moschini, M.; Suardi, N.; Pellucchi, F.; Rocchini, L.; La Croce, G.; Capitanio, U.; Briganti, A.; Damiano, R.; Montorsi, F.; Colombo, R. Impact of preoperative thrombocytosis on pathological outcomes and survival in patients treated with radical cystectomy for bladder carcinoma. *Anticancer Res.* **2014**, *34*, 3225–3230.
39. Dutta, S.; Crumley, A.B.C.; Fullarton, G.M.; Horgan, P.G.; McMillan, D.C. Comparison of the prognostic value of tumour and patient related factors in patients undergoing potentially curative resection of gastric cancer. *Am. J. Surg.* **2012**, *204*, 294–299. [CrossRef]
40. Aminian, A.; Karimian, F.; Mirsharifi, R.; Alibakhshi, A.; Dashti, H.; Jahangiri, Y.; Safari, S.; Ghaderi, H.; Noaparast, M.; Hasani, S.; et al. Significance of platelet count in esophageal carcinomas. *Saudi J. Gastroenterol.* **2011**, *17*, 134. [CrossRef]
41. Wang, J.; Liu, H.; Shao, N.; Tan, B.; Song, Q.; Jia, Y.; Cheng, Y. The clinical significance of preoperative plasma fibrinogen level and platelet count in resectable esophageal squamous cell carcinoma. *World J. Surg. Oncol.* **2015**, *13*, 157. [CrossRef] [PubMed]
42. Liu, H.B.; Gu, X.L.; Ma, X.Q.; Lv, T.F.; Wu, Y.; Xiao, Y.Y.; Yuan, D.M.; Li, Y.F.; Song, Y. Preoperative platelet count in predicting lymph node metastasis and prognosis in patients with non-small cell lung cancer. *Neoplasma* **2012**, *60*, 203–208. [CrossRef] [PubMed]
43. Agoston, A.T.; Srivastava, A.; Zheng, Y.; Bueno, R.; Odze, R.D.; Szallasi, Z. Paraneoplastic thrombocytosis is associated with increased mortality and increased rate of lymph node metastasis in oesophageal adenocarcinoma. *Pathology* **2017**, *49*, 471–475. [CrossRef] [PubMed]
44. Tranum, B.L.; Haut, A. Thrombocytosis: Platelet kinetics in neoplasia. *J. Lab. Clin. Med.* **1974**, *84*, 615–619.
45. Kuter, D.J. The biology of thrombopoietin and thrombopoietin receptor agonists. *Int. J. Hematol.* **2013**, *98*, 10–23. [CrossRef]
46. Tsukishiro, S.; Suzumori, N.; Nishikawa, H.; Arakawa, A.; Suzumori, K. Preoperative serum thrombopoietin levels are higher in patients with ovarian cancer than with benign cysts. *Eur. J. Obstet. Gynecol. Reprod. Biol.* **2008**, *140*, 67–70. [CrossRef]
47. Furuhashi, M.; Miyabe, Y.; Oda, H. A Case of Thrombopoietin-Producing Ovarian Carcinoma Confirmed by Immunohistochemistry. *Gynecol. Oncol.* **1999**, *74*, 278–281. [CrossRef]
48. Weryńska, B.; Ramlau, R.; Podolak-Dawidziak, M.; Jankowska, R.; Prajs, I.; Usnarska-Zubkiewicz, L.; Kuliczkowski, K. Serum thrombopoietin levels in patients with reactive thrombocytosis due to lung cancer and in patients with essential thrombocythemia. *Neoplasma* **2003**, *50*, 447–451.
49. Columbyova, L.; Loda, M.; Scadden, D.T. Thrombopoietin Receptor Expression in Human Cancer Cell Lines and Primary Tissues. *Cancer Res.* **1995**, *55*, 3509–3512.
50. Besbes, S.; Shah, S.; Al-Dybiat, I.; Mirshahi, S.; Helfer, H.; Najah, H.; Fourgeaud, C.; Pocard, M.; Ghedira, I.; Soria, J.; et al. Thrombopoietin Secretion by Human Ovarian Cancer Cells. *Int. J. Cell Biol.* **2017**, *2017*, 1873834. [CrossRef]
51. Hodge, D.R.; Hurt, E.M.; Farrar, W.L. The role of IL-6 and STAT3 in inflammation and cancer. *Eur. J. Cancer* **2005**, *41*, 2502–2512. [CrossRef] [PubMed]
52. Guo, Y.; Xu, F.; Lu, T.; Duan, Z.; Zhang, Z. Interleukin-6 signaling pathway in targeted therapy for cancer. *Cancer Treat. Rev.* **2012**, *38*, 904–910. [CrossRef] [PubMed]

53. Kaser, A.; Brandacher, G.; Steurer, W.; Kaser, S.; Offner, F.A.; Zoller, H.; Theurl, I.; Widder, W.; Molnar, C.; Ludwiczek, O.; et al. Interleukin-6 stimulates thrombopoiesis through thrombopoietin: Role in inflammatory thrombocytosis. *Blood* **2001**, *98*, 2720–2725. [CrossRef] [PubMed]
54. Wolber, E.-M.; Fandrey, J.; Frackowski, U.; Jelkmann, W. Hepatic Thrombopoietin mRNA Is Increased in Acute Inflammation. *Thromb. Haemost.* **2001**, *86*, 1421–1424. [PubMed]
55. Wolber, E.M.; Jelkmann, W. Interleukin-6 increases thrombopoietin production in human hepatoma cells HepG2 and Hep3B. *J. Interferon Cytokine Res.* **2000**, *20*, 499–506. [CrossRef]
56. Kumari, N.; Dwarakanath, B.S.; Das, A.; Bhatt, A.N. Role of interleukin-6 in cancer progression and therapeutic resistance. *Tumor Biol.* **2016**, *37*, 11553–11572. [CrossRef] [PubMed]
57. Conze, D.; Weiss, L.; Regen, P.S.; Rincón, M.; Weaver, D.; Bhushan, A.; Johnson, P. Autocrine production of interleukin 6 causes multidrug resistance in breast cancer cells. *Cancer Res.* **2001**, *61*, 8851–8858.
58. Nagasaki, T.; Hara, M.; Nakanishi, H.; Takahashi, H.; Sato, M.; Takeyama, H. Interleukin-6 released by colon cancer-associated fibroblasts is critical for tumour angiogenesis: Anti-interleukin-6 receptor antibody suppressed angiogenesis and inhibited tumour-stroma interaction. *Br. J. Cancer* **2014**, *110*, 469–478. [CrossRef]
59. Higashihara, M.; Sunaga, S.; Tange, T.; Oohashi, H.; Kurokawa, K. Increased secretion of lnterleukin-6 in malignant mesothelioma cells from a patient with marked thrombocytosis. *Cancer* **1992**, *70*, 2105–2108. [CrossRef]
60. Shinriki, S.; Jono, H.; Ota, K.; Ueda, M.; Kudo, M.; Ota, T.; Oike, Y.; Endo, M.; Ibusuki, M.; Hiraki, A.; et al. Humanized anti-interleukin-6 receptor antibody suppresses tumor angiogenesis and in vivo growth of human oral squamous cell carcinoma. *Clin. Cancer Res.* **2009**, *15*, 5426–5434. [CrossRef]
61. Chen, F.; Teachey, D.T.; Pequignot, E.; Frey, N.; Porter, D.; Maude, S.L.; Grupp, S.A.; June, C.H.; Melenhorst, J.J.; Lacey, S.F. Measuring IL-6 and sIL-6R in serum from patients treated with tocilizumab and/or siltuximab following CAR T cell therapy. *J. Immunol. Methods* **2016**, *434*, 1–8. [CrossRef] [PubMed]
62. Vainer, N.; Dehlendorff, C.; Johansen, J.S. Systematic literature review of IL-6 as a biomarker or treatment target in patients with gastric, bile duct, pancreatic and colorectal cancer. *Oncotarget* **2018**, *9*, 29820–29841. [CrossRef] [PubMed]
63. Suzuki, A.; Takahashi, T.; Nakamura, K.; Tsuyuoka, R.; Okuno, Y.; Enomoto, T.; Fukumoto, M.; Imura, H. Thrombocytosis in patients with tumors producing colony-stimulating factor. *Blood* **1992**, *80*, 2052–2059. [CrossRef] [PubMed]
64. Assoian, R.K.; Komoriya, A.; Meyers, C.A.; Miller, D.M.; Sporn, M.B. Transforming growth factor-β in human platelets. Identification of a major storage site, purification, and characterization. *J. Biol. Chem.* **1983**, *258*, 7155–7160. [PubMed]
65. Lian, S.; Zhai, X.; Wang, X.; Zhu, H.; Zhang, S.; Wang, W.; Wang, Z.; Huang, J. Elevated expression of growth-regulated oncogene-alpha in tumor and stromal cells predicts unfavorable prognosis in pancreatic cancer. *Medicine* **2016**, *95*, e4328. [CrossRef] [PubMed]
66. Yung, M.M.H.; Tang, H.W.M.; Cai, P.C.H.; Leung, T.H.Y.; Ngu, S.F.; Chan, K.K.L.; Xu, D.; Yang, H.; Ngan, H.Y.S.; Chan, D.W. GRO-α and IL-8 enhance ovarian cancer metastatic potential via the CXCR2-mediated TAK1/NFκB signaling cascade. *Theranostics* **2018**, *8*, 1270–1285. [CrossRef]
67. Phakathi, B.P.; Mannell, A.; Nietz, S. Early stage breast cancer with concomittant primary hyperparathyroidism and autoimmune thrombocytopenia: A case report. *S. Afr. J. Surg.* **2018**, *56*, 64–65. [CrossRef]
68. Liebman, H.A. Thrombocytopenia in cancer patients. *Thromb. Res.* **2014**, *133*, S63–S69. [CrossRef]
69. Khasraw, M.; Faraj, H.; Sheikha, A. Thrombocytopenia in solid tumors. *Eur. J. Clin. Med. Oncol.* **2010**, *2*, 89–92.
70. Gaydos, L.A.; Freireich, E.J.; Mantel, N. The quantitative relation between platelet count and hemorrhage in patients with acute leukemia. *N. Engl. J. Med.* **1962**, *266*, 905–909. [CrossRef]
71. Avvisati, G.; Tirindelli, M.C.; Annibali, O. Thrombocytopenia and hemorrhagic risk in cancer patients. *Crit. Rev. Oncol. Hematol.* **2003**, *48*, 13–16. [CrossRef] [PubMed]
72. Ghanavat, M.; Ebrahimi, M.; Rafieemehr, H.; Maniati, M.; Behzad, M.M.; Shahrabi, S. Thrombocytopenia in solid tumors: Prognostic significance. *Oncol. Rev.* **2019**, *13*, 43–48. [CrossRef] [PubMed]
73. Pooja, S.; Chaudhary, P.; Nayak, L.V.; Rajender, S.; Saini, K.S.; Deol, D.; Kumar, S.; Bid, H.K.; Konwar, R. Polymorphic variations in IL-1β, IL-6 and IL-10 genes, their circulating serum levels and breast cancer risk in Indian women. *Cytokine* **2012**, *60*, 122–128. [CrossRef]

74. Rodríguez-Berriguete, G.; Sánchez-Espiridión, B.; Cansino, J.R.; Olmedilla, G.; Martínez-Onsurbe, P.; Sánchez-Chapado, M.; Paniagua, R.; Fraile, B.; Royuela, M. Clinical significance of both tumor and stromal expression of components of the IL-1 and TNF-α signaling pathways in prostate cancer. *Cytokine* **2013**, *64*, 555–563. [CrossRef] [PubMed]
75. Xue, H.; Lin, B.; Ni, P.; Xu, H.; Huang, G. Interleukin-1B and interleukin-1 RN polymorphisms and gastric carcinoma risk: A meta-analysis. *J. Gastroenterol. Hepatol.* **2010**, *25*, 1604–1617. [CrossRef] [PubMed]
76. Ying, H.Y.; Yu, B.W.; Yang, Z.; Yang, S.S.; Bo, L.H.; Shan, X.Y.; Wang, H.J.; Zhu, Y.J.; Wu, X.S. Interleukin-1B 31 C>T polymorphism combined with Helicobacter pylori-modified gastric cancer susceptibility: Evidence from 37 studies. *J. Cell. Mol. Med.* **2016**, *20*, 526–536. [CrossRef] [PubMed]
77. Hefler, L.A.; Grimm, C.; Ackermann, S.; Malur, S.; Radjabi-Rahat, A.R.; Leodolter, S.; Beckmann, M.W.; Zeillinger, R.; Koelbl, H.; Tempfer, C.B. An interleukin-6 gene promoter polymorphism influences the biological phenotype of ovarian cancer. *Cancer Res.* **2003**, *63*, 3066–3068.
78. Talar-Wojnarowska, R.; Gasiorowska, A.; Smolarz, B.; Romanowicz-Makowska, H.; Kulig, A.; Malecka-Panas, E. Clinical significance of interleukin-6 (Il-6) gene polymorphism and Il-6 serum level in pancreatic adenocarcinoma and chronic pancreatitis. *Dig. Dis. Sci.* **2009**, *54*, 683–689. [CrossRef]
79. Takaku, M.; Grimm, S.A.; Wade, P.A. GATA3 in breast cancer: Tumor suppressor or oncogene? *Gene Expr.* **2015**, *16*, 163–168. [CrossRef]
80. Li, Q.; Chen, C.; Ren, X.; Sun, W. DNA methylation profiling identifies the HOXA11 gene as an early diagnostic and prognostic molecular marker in human lung adenocarcinoma. *Oncotarget* **2017**, *8*, 33100–33109. [CrossRef]
81. Xia, B.; Shan, M.; Wang, J.; Zhong, Z.; Geng, J.; He, X.; Vu, T.; Zhang, D.; Pang, D. Homeobox A11 hypermethylation indicates unfavorable prognosis in breast cancer. *Oncotarget* **2017**, *8*, 9794–9805. [CrossRef] [PubMed]
82. Franco, A.T.; Ware, J. Pathophysiology 2: The Role of Platelets in Cancer Biology. In *Cancer Treatment and Research*; Springer: Cham, Switzerland, 2019; pp. 37–54.
83. Kisucka, J.; Butterfield, C.E.; Duda, D.G.; Eichenberger, S.C.; Saffaripour, S.; Ware, J.; Ruggeri, Z.M.; Jain, R.K.; Folkman, J.; Wagner, D.D. Platelets and platelet adhesion support angiogenesis while preventing excessive hemorrhage. *Proc. Natl. Acad. Sci. USA* **2006**, *103*, 855–860. [CrossRef] [PubMed]
84. Jain, S.; Zuka, M.; Liu, J.; Russell, S.; Dent, J.; Guerrero, J.A.; Forsyth, J.; Maruszak, B.; Gartner, T.K.; Felding-Habermann, B.; et al. Platelet glycoprotein Ibα supports experimental lung metastasis. *Proc. Natl. Acad. Sci. USA* **2007**, *104*, 9024–9028. [CrossRef] [PubMed]
85. Demers, M.; Ho-Tin-Noé, B.; Schatzberg, D.; Yang, J.J.; Wagner, D.D. Increased Efficacy of Breast Cancer Chemotherapy in Thrombocytopenic Mice. *Cancer Res.* **2011**, *71*, 1540–1549. [CrossRef] [PubMed]
86. Seth, R.; Tai, L.H.; Falls, T.; De Souza, C.T.; Bell, J.C.; Carrier, M.; Atkins, H.; Boushey, R.; Auer, R.A. Surgical stress promotes the development of cancer metastases by a coagulation-dependent mechanism involving natural killer cells in a murine model. *Ann. Surg.* **2013**, *258*, 158–168. [CrossRef]
87. Clar, K.L.; Hinterleitner, C.; Schneider, P.; Salih, H.R.; Maurer, S. Inhibition of NK reactivity against solid tumors by platelet-derived RANKL. *Cancers* **2019**, *11*, 277. [CrossRef]
88. Chaffer, C.L.; Weinberg, R.A. A perspective on cancer cell metastasis. *Science* **2011**, *331*, 1559–1564. [CrossRef]
89. Farge, D.; Bounameaux, H.; Brenner, B.; Cajfinger, F.; Debourdeau, P.; Khorana, A.A.; Pabinger, I.; Solymoss, S.; Douketis, J.; Kakkar, A. International clinical practice guidelines including guidance for direct oral anticoagulants in the treatment and prophylaxis of venous thromboembolism in patients with cancer. *Lancet Oncol.* **2016**, *17*, e452–e466. [CrossRef]
90. Zarà, M.; Canobbio, I.; Visconte, C.; Canino, J.; Torti, M.; Guidetti, G.F. Molecular mechanisms of platelet activation and aggregation induced by breast cancer cells. *Cell. Signal.* **2018**, *48*, 45–53. [CrossRef]
91. Reddel, C.; Tan, C.; Chen, V. Thrombin Generation and Cancer: Contributors and Consequences. *Cancers* **2019**, *11*, 100. [CrossRef]
92. Chang, J.; Jiang, L.; Wang, Y.; Yao, B.; Yang, S.; Zhang, B.; Zhang, M.-Z. 12/15 lipoxygenase regulation of colorectal tumorigenesis is determined by the relative tumor levels of its metabolite 12-HETE and 13-HODE in animal models. *Oncotarget* **2015**, *6*, 2879. [CrossRef] [PubMed]
93. Duvernay, M.T.; Temple, K.J.; Maeng, J.G.; Blobaum, A.L.; Stauffer, S.R.; Lindsley, C.W.; Hamm, H.E. Contributions of Protease-Activated Receptors PAR1 and PAR4 to Thrombin-Induced GPIIbIIIa Activation in Human Platelets. *Mol. Pharmacol.* **2017**, *91*, 39–47. [CrossRef] [PubMed]

94. Dahlbäck, B. Pro- and anticoagulant properties of factor V in pathogenesis of thrombosis and bleeding disorders. *Int. J. Lab. Hematol.* **2016**, *38*, 4–11. [CrossRef] [PubMed]
95. Xiao, J.; Melvin, R.L.; Salsbury, F.R. Mechanistic insights into thrombin's switch between "slow" and "fast" forms. *Phys. Chem. Chem. Phys.* **2017**, *19*, 24522–24533. [CrossRef] [PubMed]
96. Abu Saadeh, F.; Langhe, R.; Galvin, D.M.; Toole, S.O.; O'Donnell, D.M.; Gleeson, N.; Norris, L.A. Procoagulant activity in gynaecological cancer patients; The effect of surgery and chemotherapy. *Thromb. Res.* **2016**, *139*, 135–141. [CrossRef] [PubMed]
97. Reitter, E.M.; Kaider, A.; Ay, C.; Quehenberger, P.; Marosi, C.; Zielinski, C.; Pabinger, I. Longitudinal analysis of hemostasis biomarkers in cancer patients during antitumor treatment. *J. Thromb. Haemost.* **2016**, *14*, 294–305. [CrossRef]
98. Adams, G.N.; Rosenfeldt, L.; Frederick, M.; Miller, W.; Waltz, D.; Kombrinck, K.; McElhinney, K.E.; Flick, M.J.; Monia, B.P.; Revenko, A.S.; et al. Colon cancer growth and dissemination relies upon thrombin, Stromal PAR-1, and fibrinogen. *Cancer Res.* **2015**, *75*, 4235–4243. [CrossRef]
99. Muqaku, B.; Eisinger, M.; Meier, S.M.; Tahir, A.; Pukrop, T.; Haferkamp, S.; Slany, A.; Reichle, A.; Gerner, C. Multi-omics analysis of serum samples demonstrates reprogramming of organ functions via systemic calcium mobilization and platelet activation in metastatic melanoma. *Mol. Cell. Proteom.* **2017**, *16*, 86–99. [CrossRef]
100. McCarty, O.J.T.; Mousa, S.A.; Bray, P.F.; Konstantopoulos, K. Immobilized platelets support human colon carcinoma cell tethering, rolling, and firm adhesion under dynamic flow conditions. *Blood* **2000**, *96*, 1789–1797. [CrossRef]
101. Li, S.S.; Ip, C.K.M.; Tang, M.Y.H.; Tang, M.K.S.; Tong, Y.; Zhang, J.; Hassan, A.A.; Mak, A.S.C.; Yung, S.; Chan, T.M.; et al. Sialyl Lewisx-P-selectin cascade mediates tumor–mesothelial adhesion in ascitic fluid shear flow. *Nat. Commun.* **2019**, *10*, 2406. [CrossRef]
102. Del Conde, I.; Nabi, F.; Tonda, R.; Thiagarajan, P.; López, J.A.; Kleiman, N.S. Effect of P-selectin on phosphatidylserine exposure and surface-dependent thrombin generation on monocytes. *Arterioscler. Thromb. Vasc. Biol.* **2005**, *25*, 1065–1070. [CrossRef] [PubMed]
103. Ivanov, I.I.; Apta, B.H.R.; Bonna, A.M.; Harper, M.T. Platelet P-selectin triggers rapid surface exposure of tissue factor in monocytes. *Sci. Rep.* **2019**, *9*, 13397. [CrossRef] [PubMed]
104. Jiang, L.; Luan, Y.; Miao, X.; Sun, C.; Li, K.; Huang, Z.; Xu, D.; Zhang, M.; Kong, F.; Li, N. Platelet releasate promotes breast cancer growth and angiogenesis via VEGF-integrin cooperative signalling. *Br. J. Cancer* **2017**, *117*, 695–703. [CrossRef] [PubMed]
105. Yang, J.G.; Wang, L.L.; Ma, D.C. Effects of vascular endothelial growth factors and their receptors on megakaryocytes and platelets and related diseases. *Br. J. Haematol.* **2018**, *180*, 321–334. [CrossRef] [PubMed]
106. Sigismund, S.; Avanzato, D.; Lanzetti, L. Emerging functions of the EGFR in cancer. *Mol. Oncol.* **2018**, *12*, 3–20. [CrossRef] [PubMed]
107. Vega, S.; Kondo, A.; Suzuki, M.; Arai, H.; Jiapaer, S.; Sabit, H.; Nakada, M.; Ikeuchi, T.; Ishijima, M.; Arikawa-Hirasawa, E.; et al. Fibulin-7 is overexpressed in glioblastomas and modulates glioblastoma neovascularization through interaction with angiopoietin-1. *Int. J. Cancer* **2019**, *145*, 2157–2169. [CrossRef] [PubMed]
108. Nissen, L.J.; Cao, R.; Hedlund, E.M.; Wang, Z.; Zhao, X.; Wetterskog, D.; Funa, K.; Bråkenhielm, E.; Cao, Y. Angiogenic factors FGF2 and PDGF-BB synergistically promote murine tumor neovascularization and metastasis. *J. Clin. Investig.* **2007**, *117*, 2766–2777. [CrossRef]
109. Kumar, S.; Lu, B.; Davra, V.; Hornbeck, P.; Machida, K.; Birge, R.B. Crk tyrosine phosphorylation regulates PDGF-BB-inducible Src activation and breast tumorigenicity and metastasis. *Mol. Cancer Res.* **2018**, *16*, 173–183. [CrossRef]
110. Pietras, K.; Pahler, J.; Bergers, G.; Hanahan, D. Functions of paracrine PDGF signaling in the proangiogenic tumor stroma revealed by pharmacological targeting. *PLoS Med.* **2008**, *5*, 0123–0138. [CrossRef]
111. Hosaka, K.; Yang, Y.; Seki, T.; Nakamura, M.; Andersson, P.; Rouhi, P.; Yang, X.; Jensen, L.; Lim, S.; Feng, N.; et al. Tumour PDGF-BB expression levels determine dual effects of anti-PDGF drugs on vascular remodelling and metastasis. *Nat. Commun.* **2013**, *4*, 2129. [CrossRef]
112. Battinelli, E.M.; Markens, B.A.; Italiano, J.E. Release of angiogenesis regulatory proteins from platelet alpha granules: Modulation of physiologic and pathologic angiogenesis. *Blood* **2011**, *118*, 1359–1369. [CrossRef] [PubMed]

113. Italiano, J.E.; Richardson, J.L.; Patel-Hett, S.; Battinelli, E.; Zaslavsky, A.; Short, S.; Ryeom, S.; Folkman, J.; Klement, G.L. Angiogenesis is regulated by a novel mechanism: Pro- and antiangiogenic proteins are organized into separate platelet α granules and differentially released. *Blood* **2008**, *111*, 1227–1233. [CrossRef] [PubMed]
114. Dymicka-Piekarska, V.; Matowicka-Karna, J.; Osada, J.; Kemona, H.; Butkiewicz, A.M. Changes in platelet CD 62P expression and soluble P-selectin concentration in surgically treated colorectal carcinoma. *Adv. Med. Sci.* **2006**, *51*, 304–308. [PubMed]
115. Riedl, J.; Kaider, A.; Marosi, C.; Prager, G.W.; Eichelberger, B.; Assinger, A.; Pabinger, I.; Panzer, S.; Ay, C. Decreased platelet reactivity in patients with cancer is associated with high risk of venous thromboembolism and poor prognosis. *Thromb. Haemost.* **2017**, *117*, 90–98. [CrossRef] [PubMed]
116. Riedl, J.; Kaider, A.; Marosi, C.; Prager, G.; Eichelberger, B.; Koder, S.; Panzer, S.; Pabinger, I.; Ay, C. PO-63—Exhausted platelets in cancer patients with high risk of venous thromboembolism and poor prognosis. *Thromb. Res.* **2016**, *140*, S199–S200. [CrossRef]
117. McCarty, M.F.; Somcio, R.J.; Stoeltzing, O.; Wey, J.; Fan, F.; Liu, W.; Bucana, C.; Ellis, L.M. Overexpression of PDGF-BB decreases colorectal and pancreatic cancer growth by increasing tumor pericyte content. *J. Clin. Investig.* **2007**, *117*, 2114–2122. [CrossRef]
118. Üçüncü, M.; Serilmez, M.; Sarı, M.; Bademler, S.; Karabulut, S. The Diagnostic Significance of PDGF, EphA7, CCR5, and CCL5 Levels in Colorectal Cancer. *Biomolecules* **2019**, *9*, 464. [CrossRef]
119. Aryal, B.; Shimizu, T.; Kadono, J.; Furoi, A.; Komokata, T.; Kitazono, I.; Koriyama, C.; Yamakuchi, M.; Hashiguchi, T.; Imoto, Y. Post-resection exhaustion of intra-platelet serotonin: Also an indicator of early hepatocellular carcinoma recurrence? *J. Cancer* **2017**, *8*, 3984–3991. [CrossRef]
120. Holmes, C.E.; Levis, J.E.; Schneider, D.J.; Bambace, N.M.; Sharma, D.; Lal, I.; Wood, M.E.; Muss, H.B. Platelet phenotype changes associated with breast cancer and its treatment. *Platelets* **2016**, *27*, 703–711. [CrossRef]
121. Dirix, L.Y.; Salgado, R.; Weytjens, R.; Colpaert, C.; Benoy, I.; Huget, P.; Van Dam, P.; Prové, A.; Lemmens, J.; Vermeulen, P. Plasma fibrin D-dimer levels correlate with tumour volume, progression rate and survival in patients with metastatic breast cancer. *Br. J. Cancer* **2002**, *86*, 389–395. [CrossRef]
122. Sallinen, H.; Heikura, T.; Koponen, J.; Kosma, V.M.; Heinonen, S.; Ylä-Herttuala, S.; Anttila, M. Serum angiopoietin-2 and soluble VEGFR-2 levels predict malignancy of ovarian neoplasm and poor prognosis in epithelial ovarian cancer. *BMC Cancer* **2014**, *14*, 696. [CrossRef] [PubMed]
123. Faried, A.; Mobarak, L.; El Gohary, K.K.; El-deeb, H.H.; El-feky, S.; Ahmed, A.; Zaki, N.A.; Alkhalegy, A.A. Serum levels of Arginase Isoenzyme Activity, Alpha- Fetoprotein-L3 and Endostatin as Biomarkers for Hepatocellular Carcinoma in Egyptian Patients. *Donn. J. Biomed. Res.* **2016**, *3*, 1–5.
124. Wang, Z.H.; Zhu, Z.T.; Xiao, X.Y.; Sun, J. Correlation of serum levels of endostatin with tumor stage in gastric cancer: A systematic review and meta-analysis. *BioMed Res. Int.* **2015**, *2015*, 623939. [CrossRef] [PubMed]
125. Cymbaluk-Ploska, A.; Chudecka-Glaz, A.; Pius-Sadowska, E.; Machalilski, B.; Menkiszak, J. Thrombospondin-I concentrations behavior in plasma of patients with ovarian cancer. *Cancer Biomark.* **2017**, *20*, 31–39. [CrossRef]
126. Rouanne, M.; Adam, J.; Goubar, A.; Robin, A.; Ohana, C.; Louvet, E.; Cormier, J.; Mercier, O.; Dorfmüller, P.; Fattal, S.; et al. Osteopontin and thrombospondin-1 play opposite roles in promoting tumor aggressiveness of primary resected non-small cell lung cancer. *BMC Cancer* **2016**, *16*, 483. [CrossRef]
127. Wiśniewski, T.; Zyromska, A.; Makarewicz, R.; Zekanowska, E. Osteopontin and angiogenic factors as new biomarkers of prostate cancer. *Urol. J.* **2019**, *16*, 134–140.
128. Drenberg, C.D.; Saunders, B.O.; Wilbanks, G.D.; Chen, R.; Nicosia, R.F.; Kruk, P.A.; Nicosia, S.V. Urinary angiostatin levels are elevated in patients with epithelial ovarian cancer. *Gynecol. Oncol.* **2010**, *117*, 117–124. [CrossRef]
129. Nocito, A.; Dahm, F.; Jochum, W.; Jae, H.J.; Georgiev, P.; Bader, M.; Graf, R.; Clavien, P.A. Serotonin regulates macrophage-mediated angiogenesis in a mouse model of colon cancer allografts. *Cancer Res.* **2008**, *68*, 5152–5158. [CrossRef]
130. Kelly, C.M.; Juurlink, D.N.; Gomes, T.; Duong-Hua, M.; Pritchard, K.I.; Austin, P.C.; Paszat, L.F. Selective serotonin reuptake inhibitors and breast cancer mortality in women receiving tamoxifen: A population based cohort study. *BMJ* **2010**, *340*, 355. [CrossRef]
131. Alpini, G.; Invernizzi, P.; Gaudio, E.; Venter, J.; Kopriva, S.; Bernuzzi, F.; Onori, P.; Franchitto, A.; Coufal, M.; Frampton, G.; et al. Serotonin metabolism is dysregulated in cholangiocarcinoma, which has implications for tumor growth. *Cancer Res.* **2008**, *68*, 9184–9193. [CrossRef]

132. Svejda, B.; Kidd, M.; Timberlake, A.; Harry, K.; Kazberouk, A.; Schimmack, S.; Lawrence, B.; Pfragner, R.; Modlin, I.M. Serotonin and the 5-HT7 receptor: The link between hepatocytes, IGF-1 and small intestinal neuroendocrine tumors. *Cancer Sci.* **2013**, *104*, 844–855. [CrossRef]
133. Müller, K.; Gilbertz, K.P.; Meineke, V. Serotonin and ionizing radiation synergistically affect proliferation and adhesion molecule expression of malignant melanoma cells. *J. Dermatol. Sci.* **2012**, *68*, 89–98. [CrossRef] [PubMed]
134. Lübbe, A.S.; Huhnt, W. Microvessel diameters of human colon adenocarcinoma during acute treatment with serotonin. *Int. J. Microcirc. Exp.* **1994**, *14*, 218–225. [CrossRef] [PubMed]
135. Dowling, P.; Hughes, D.J.; Larkin, A.M.; Meiller, J.; Henry, M.; Meleady, P.; Lynch, V.; Pardini, B.; Naccarati, A.; Levy, M.; et al. Elevated levels of 14-3-3 proteins, serotonin, gamma enolase and pyruvate kinase identified in clinical samples from patients diagnosed with colorectal cancer. *Clin. Chim. Acta* **2015**, *441*, 133–141. [CrossRef] [PubMed]
136. Abdel-Razik, A.; Elhelaly, R.; Elzehery, R.; El-Diasty, A.; Abed, S.; Elhammady, D.; Tawfik, A. Could serotonin be a potential marker for hepatocellular carcinoma? A prospective single-center observational study. *Eur. J. Gastroenterol. Hepatol.* **2016**, *28*, 599–605. [CrossRef] [PubMed]
137. Jungwirth, N.; Haeberle, L.; Schrott, K.M.; Wullich, B.; Krause, F.S. Serotonin used as prognostic marker of urological tumors. *World J. Urol.* **2008**, *26*, 499–504. [CrossRef] [PubMed]
138. Mannucci, P.M.; Cattaneo, M.; Teresa Canciani, M.; Maniezzo, M.; Vaglini, M.; Cascinelli, N. Early presence of activated ('exhausted') platelets in malignant tumors (breast adenocarcinoma and malignant melanoma). *Eur. J. Cancer Clin. Oncol.* **1989**, *25*, 1413–1417. [CrossRef]
139. Crescente, M.; Menke, L.; Chan, M.V.; Armstrong, P.C.; Warner, T.D. Eicosanoids in platelets and the effect of their modulation by aspirin in the cardiovascular system (and beyond). *Br. J. Pharmacol.* **2019**, *176*, 988–999. [CrossRef]
140. Guo, Y.; Zhang, W.; Giroux, C.; Cai, Y.; Ekambaram, P.; Dilly, A.K.; Hsu, A.; Zhou, S.; Maddipati, K.R.; Liu, J.; et al. Identification of the orphan G protein-coupled receptor GPR31 as a receptor for 12-(S)-hydroxyeicosatetraenoic acid. *J. Biol. Chem.* **2011**, *286*, 33832–33840. [CrossRef]
141. Porro, B.; Songia, P.; Squellerio, I.; Tremoli, E.; Cavalca, V. Analysis, physiological and clinical significance of 12-HETE: A neglected platelet-derived 12-lipoxygenase product. *J. Chromatogr. B Anal. Technol. Biomed. Life Sci.* **2014**, *964*, 26–40. [CrossRef]
142. Rauzi, F.; Kirkby, N.S.; Edin, M.L.; Whiteford, J.; Zeldin, D.C.; Mitchell, J.A.; Warner, T.D. Aspirin inhibits the production of proangiogenic 15(S)-HETE by platelet cyclooxygenase-1. *FASEB J.* **2016**, *30*, 4256–4266. [CrossRef] [PubMed]
143. Kanikarla-Marie, P.; Kopetz, S.; Hawk, E.T.; Millward, S.W.; Sood, A.K.; Gresele, P.; Overman, M.; Honn, K.; Menter, D.G. Bioactive lipid metabolism in platelet "first responder" and cancer biology. *Cancer Metastasis Rev.* **2018**, *37*, 439–454. [CrossRef] [PubMed]
144. Tsoi, K.K.F.; Ho, J.M.W.; Chan, F.C.H.; Sung, J.J.Y. Long-term use of low-dose aspirin for cancer prevention: A 10-year population cohort study in Hong Kong. *Int. J. Cancer* **2019**, *145*, 267–273. [CrossRef] [PubMed]
145. Lin, Y.C.; Chen, C.C.; Chen, W.M.; Lu, K.Y.; Shen, T.L.; Jou, Y.C.; Shen, C.H.; Ohbayashi, N.; Kanaho, Y.; Huang, Y.L.; et al. LPA1/3 signaling mediates tumor lymphangiogenesis through promoting CRT expression in prostate cancer. *Biochim. Biophys. Acta-Mol. Cell Biol. Lipids* **2018**, *1863*, 1305–1315. [CrossRef] [PubMed]
146. Li, Y.Y.; Zhang, W.C.; Zhang, J.L.; Zheng, C.J.; Zhu, H.; Yu, H.M.; Fan, L.M. Plasma levels of lysophosphatidic acid in ovarian cancer versus controls: A meta-analysis. *Lipids Health Dis.* **2015**, *14*, 72. [CrossRef] [PubMed]
147. Yagi, T.; Shoaib, M.; Kuschner, C.; Nishikimi, M.; Becker, L.B.; Lee, A.T.; Kim, J. Challenges and inconsistencies in using lysophosphatidic acid as a biomarker for ovarian cancer. *Cancers* **2019**, *11*, 520. [CrossRef]
148. Mezouar, S.; Mege, D.; Darbousset, R.; Farge, D.; Debourdeau, P.; Dignat-George, F.; Panicot-Dubois, L.; Dubois, C. Involvement of platelet-derived microparticles in tumor progression and thrombosis. *Semin. Oncol.* **2014**, *41*, 346–358. [CrossRef]
149. Goubran, H.A.; Kotb, R.R.; Stakiw, J.; Emara, M.E.; Burnouf, T. Regulation of Tumor Growth and Metastasis: The Role of Tumor Microenvironment. *Cancer Growth Metastasis* **2014**, *7*, CGM-S11285. [CrossRef]
150. Goubran, H.A.; Stakiw, J.; Radosevic, M.; Burnouf, T. Platelet-cancer interactions. *Semin. Thromb. Hemost.* **2014**, *40*, 296–305.
151. Rak, J. Microparticles in cancer. *Semin. Thromb. Hemost.* **2010**, *36*, 888–906. [CrossRef]

152. Lazar, S.; Goldfinger, L.E. Platelet Microparticles and miRNA Transfer in Cancer Progression: Many Targets, Modes of Action, and Effects Across Cancer Stages. *Front. Cardiovasc. Med.* **2018**, *5*, 13. [CrossRef] [PubMed]
153. Janowska-Wieczorek, A.; Wysoczynski, M.; Kijowski, J.; Marquez-Curtis, L.; Machalinski, B.; Ratajczak, J.; Ratajczak, M.Z. Microvesicles derived from activated platelets induce metastasis and angiogenesis in lung cancer. *Int. J. Cancer* **2005**, *113*, 752–760. [CrossRef] [PubMed]
154. Prokopi, M.; Pula, G.; Mayr, U.; Devue, C.; Gallagher, J.; Xiao, Q.; Boulanger, C.M.; Westwood, N.; Urbich, C.; Willeit, J.; et al. Proteomic analysis reveals presence of platelet microparticles in endothelial progenitor cell cultures. *Blood* **2009**, *114*, 723–732. [CrossRef]
155. Zarà, M.; Guidetti, G.; Boselli, D.; Villa, C.; Canobbio, I.; Seppi, C.; Visconte, C.; Canino, J.; Torti, M. Release of Prometastatic Platelet-Derived Microparticles Induced by Breast Cancer Cells: A Novel Positive Feedback Mechanism for Metastasis. *TH Open* **2017**, *1*, e155–e163. [CrossRef] [PubMed]
156. Gasperi, V.; Vangapandu, C.; Savini, I.; Ventimiglia, G.; Adorno, G.; Catani, M.V. Polyunsaturated fatty acids modulate the delivery of platelet microvesicle-derived microRNAs into human breast cancer cell lines. *J. Nutr. Biochem.* **2019**, *74*, 108242. [CrossRef] [PubMed]
157. Provost, P. The clinical significance of platelet microparticle-associated microRNAs. *Clin. Chem. Lab. Med.* **2017**, *55*, 657–666. [CrossRef]
158. Liang, H.; Yan, X.; Pan, Y.; Wang, Y.; Wang, N.; Li, L.; Liu, Y.; Chen, X.; Zhang, C.Y.; Gu, H.; et al. MicroRNA-223 delivered by platelet-derived microvesicles promotes lung cancer cell invasion via targeting tumor suppressor EPB41L3. *Mol. Cancer* **2015**, *14*, 58. [CrossRef]
159. Cao, L.; Zhang, X.; Cao, F.; Wang, Y.; Shen, Y.; Yang, C.; Uzan, G.; Peng, B.; Zhang, D. Inhibiting inducible miR-223 further reduces viable cells in human cancer cell lines MCF-7 and PC3 treated by celastrol. *BMC Cancer* **2015**, *15*, 873. [CrossRef]
160. Pinatel, E.M.; Orso, F.; Penna, E.; Cimino, D.; Elia, A.R.; Circosta, P.; Dentelli, P.; Brizzi, M.F.; Provero, P.; Taverna, D. miR-223 is a coordinator of breast cancer progression as revealed by bioinformatics predictions. *PLoS ONE* **2014**, *9*, e84859. [CrossRef]
161. Sun, X.; Li, Y.; Zheng, M.; Zuo, W.; Zheng, W. MicroRNA-223 increases the sensitivity of triple-negative breast cancer stem cells to TRAIL-Induced apoptosis by targeting HAX-1. *PLoS ONE* **2016**, *11*, e0162754. [CrossRef]
162. Shi, L.; Fisslthaler, B.; Zippel, N.; Frömel, T.; Hu, J.; Elgheznawy, A.; Heide, H.; Popp, R.; Fleming, I. MicroRNA-223 antagonizes angiogenesis by targeting β1 integrin and preventing growth factor signaling in endothelial cells. *Circ. Res.* **2013**, *113*, 1320–1330. [CrossRef] [PubMed]
163. Tang, M.; Jiang, L.; Lin, Y.; Wu, X.; Wang, K.; He, Q.; Wang, X.; Li, W. Platelet microparticle-mediated transfer of miR-939 to epithelial ovarian cancer cells promotes epithelial to mesenchymal transition. *Oncotarget* **2017**, *8*, 97464–97475. [CrossRef] [PubMed]
164. Pan, B.; Chen, Y.; Song, H.; Xu, Y.; Wang, R.; Chen, L. Mir-24-3p downregulation contributes to VP16-DDP resistance in small-cell lung cancer by targeting ATG4A. *Oncotarget* **2015**, *6*, 317–331. [CrossRef] [PubMed]
165. Michael, J.V.; Wurtzel, J.G.T.; Mao, G.F.; Rao, A.K.; Kolpakov, M.A.; Sabri, A.; Hoffman, N.E.; Rajan, S.; Tomar, D.; Madesh, M.; et al. Platelet microparticles infiltrating solid tumors transfer miRNAs that suppress tumor growth. *Blood* **2017**, *130*, 567–580. [CrossRef] [PubMed]
166. Sorrentino, A.; Liu, C.G.; Addario, A.; Peschle, C.; Scambia, G.; Ferlini, C. Role of microRNAs in drug-resistant ovarian cancer cells. *Gynecol. Oncol.* **2008**, *111*, 478–486. [CrossRef] [PubMed]
167. Zhu, H.; Wu, H.; Liu, X.; Evans, B.R.; Medina, D.J.; Liu, C.G.; Yang, J.M. Role of MicroRNA miR-27a and miR-451 in the regulation of MDR1/P-glycoprotein expression in human cancer cells. *Biochem. Pharmacol.* **2008**, *76*, 582–588. [CrossRef]
168. Anene, C.; Graham, A.M.; Boyne, J.; Roberts, W. Platelet microparticle delivered microRNA-Let-7a promotes the angiogenic switch. *Biochim. Biophys. Acta-Mol. Basis Dis.* **2018**, *1864*, 2633–2643. [CrossRef]
169. Miao, X.; Rahman, M.F.U.; Jiang, L.; Min, Y.; Tan, S.; Xie, H.; Lee, L.; Wang, M.; Malmström, R.E.; Lui, W.O.; et al. Thrombin-reduced miR-27b attenuates platelet angiogenic activities in vitro via enhancing platelet synthesis of anti-angiogenic thrombospondin-1. *J. Thromb. Haemost.* **2018**, *16*, 791–801. [CrossRef]
170. Wang, C.Z.; Yuan, P.; Li, Y. miR-126 regulated breast cancer cell invasion by targeting ADAM9. *Int. J. Clin. Exp. Pathol.* **2015**, *8*, 6547–6553. [PubMed]
171. Liu, B.; Peng, X.C.; Zheng, X.L.; Wang, J.; Qin, Y.W. MiR-126 restoration down-regulate VEGF and inhibit the growth of lung cancer cell lines in vitro and in vivo. *Lung Cancer* **2009**, *66*, 169–175. [CrossRef]

172. Penson, R.T.; Oliva, E.; Skates, S.J.; Glyptis, T.; Fuller, A.F.; Goodman, A.; Seiden, M.V. Expression of multidrug resistance-1 protein inversely correlates with paclitaxel response and survival in ovarian cancer patients: A study in serial samples. *Gynecol. Oncol.* **2004**, *93*, 98–106. [CrossRef] [PubMed]
173. Nam, E.J.; Yoon, H.; Kim, S.W.; Kim, H.; Kim, Y.T.; Kim, J.H.; Kim, J.W.; Kim, S. MicroRNA expression profiles in serous ovarian carcinoma. *Clin. Cancer Res.* **2008**, *14*, 2690–2695. [CrossRef] [PubMed]
174. Hoppe, R.; Achinger-Kawecka, J.; Winter, S.; Fritz, P.; Lo, W.Y.; Schroth, W.; Brauch, H. Increased expression of miR-126 and miR-10a predict prolonged relapse-free time of primary oestrogen receptor-positive breast cancer following tamoxifen treatment. *Eur. J. Cancer* **2013**, *49*, 3598–3608. [CrossRef] [PubMed]
175. Dovizio, M.; Bruno, A.; Contursi, A.; Grande, R.; Patrignani, P. Platelets and extracellular vesicles in cancer: Diagnostic and therapeutic implications. *Cancer Metastasis Rev.* **2018**, *37*, 455–467. [CrossRef]
176. Kim, H.K.; Song, K.S.; Park, Y.S.; Kang, Y.H.; Lee, Y.J.; Lee, K.R.; Kim, H.K.; Ryu, K.W.; Bae, J.M.; Kim, S. Elevated levels of circulating platelet microparticles, VEGF, IL-6 and RANTES in patients with gastric cancer: Possible role of a metastasis predictor. *Eur. J. Cancer* **2003**, *39*, 184–191. [CrossRef]
177. Wang, C.C.; Tseng, C.C.; Chang, H.C.; Huang, K.T.; Fang, W.F.; Chen, Y.M.; Yang, C.T.; Hsiao, C.C.; Lin, M.C.; Ho, C.K.; et al. Circulating microparticles are prognostic biomarkers in advanced non-small cell lung cancer patients. *Oncotarget* **2017**, *8*, 75952–75967. [CrossRef]

9

Heat-Shock Protein 27 (HSPB1) is Upregulated and Phosphorylated in Human Platelets during ST-Elevation Myocardial Infarction

Bjoern F. Kraemer [1], Hanna Mannell [2,3], Tobias Lamkemeyer [4], Mirita Franz-Wachtel [5] and Stephan Lindemann [6,7,8,*]

1 Medizinische Klinik und Poliklinik I, Klinikum der Universität München, Marchioninistrasse 15, 81377 Munich, Germany; bjoern.kraemer@klinik-ebe.de
2 Walter Brendel Centre of Experimental Medicine, University Hospital, Ludwig-Maximilians-University, Marchioninistr. 27, 81377 Munich, Germany; Hanna.Mannell@med.uni-muenchen.de
3 Biomedical Center, Ludwig-Maximilians-University, Großhaderner Str. 9, 82152 Planegg, Germany
4 Cluster of Excellence Cologne (CEDAD), Mass Spectrometry Facility at the Institute for Genetics, University of Köln, Josef-Stelzmann-Str. 26, 50931 Köln, Germany; tobias_lamkemeyer@yahoo.de
5 Proteasome Center Tuebingen, University of Tuebingen, Auf der Morgenstelle 15, 72076 Tübingen, Germany; mirita.franz@uni-tuebingen.de
6 FB 20–Medizin, Philipps Universität Marburg, Baldingerstraße, 35032 Marburg, Germany
7 Medizinische Klinik II, Klinikum Warburg, Hüffertstr. 50, 34414 Warburg, Germany
8 Medizinische Klinik und Poliklinik III, Universitätsklinikum Tübingen, Otfried-Muller-Str. 10, 72076 Tübingen, Germany
* Correspondence: lindemanns@aol.com

Abstract: Heat-shock proteins are a family of proteins which are upregulated in response to stress stimuli including inflammation, oxidative stress, or ischemia. Protective functions of heat-shock proteins have been studied in vascular disease models, and malfunction of heat-shock proteins is associated with vascular disease development. Heat-shock proteins however have not been investigated in human platelets during acute myocardial infarction ex vivo. Using two-dimensional electrophoresis and immunoblotting, we observed that heat-shock protein 27 (HSPB1) levels and phosphorylation are significantly increased in platelets of twelve patients with myocardial infarction compared to patients with nonischemic chest pain (6.4 ± 1.0-fold versus 1.0 ± 0.9-fold and 5.9 ± 1.8-fold versus 1.0 ± 0.8-fold; $p < 0.05$). HSP27 (HSPB1) showed a distinct and characteristic intracellular translocation from the cytoskeletal fraction into the membrane fraction of platelets during acute myocardial infarction that did not occur in the control group. In this study, we could demonstrate for the first time that HSP27 (HSPB1) is upregulated and phosphorylated in human platelets during myocardial infarction on a cellular level ex vivo with a characteristic intracellular translocation pattern. This HSP27 (HSPB1) phenotype in platelets could thus represent a measurable stress response in myocardial infarction and potentially other acute ischemic events.

Keywords: platelets; myocardial infarction; HSP27; heat-shock proteins

---

1. Introduction

Heat-shock proteins (HSPs) are expressed in many cells of the cardiovascular system such as endothelial cells, cardiac muscle cells, monocytes, and platelets. They are upregulated in response to inflammation, oxidative stress, or ischemia [1] and protect cells against extracellular stress factors. In order to protect the cell against stress stimuli, heat-shock proteins function as chaperone proteins, stabilize the cytoskeleton [2], and prevent apoptosis [3–5]. Functions of HSP27 (HSPB1) are modulated by phosphorylation, although the precise mechanisms and functions of HSP27 regulation are not fully understood. Previous reports imply that phosphorylation of HSP27 appears to reflect cell protective functions [2,3,6] and seems to be the initiating step prior to intracellular translocation enabling HSP27 to interact with the cytoskeleton [7]. Increased phosphorylation of HSP27 has been observed during different modes of platelet activation [8–10] or cardiovascular risk conditions such as diabetes mellitus [11]. HSP27 directly interacts with the actin cytoskeleton and thus downregulates inflammatory cell migration [12]. Other heat-shock proteins such as HSP70 and HSP90 are also actively involved in integrin-mediated platelet adhesion [13], and HSP90 associates with the platelet glycoprotein Ib-IX-V complex [14]. Inactivation of heat-shock proteins has been associated with progression of atherosclerosis and acute vascular events. Increased expression or release of heat-shock proteins further seem to reflect systemic stress responses or ischemia [15]. Oxidized low density lipoprotein (LDL) molecules, for example, induce expression of heat-shock proteins [16], whereas decreased expression of protective HSPs was found in atherosclerotic vessel areas [17,18]. In patients with acute coronary syndrome, increased plasma levels of HSP27 and HSP70 were detected, although the source of HSP27 remained unknown [19–21]. Phosphorylated HSP27 is further released from platelets in patients with diabetes, which is a major risk factor for vascular disease [11].

In this work, we demonstrate for the first time that HSP27 (HSPB1) is upregulated and phosphorylated in platelets of patients with ST-elevation myocardial infarction compared to control patients with nonischemic chest pain. HSP27 in platelets showed a distinct intracellular translocation pattern that was not observed in the control group.

2. Results

2.1. HSP27 Is Upregulated in Human Platelets during Myocardial Infarction

Protein spots that showed differential regulation in platelets during myocardial infarction in two-dimensional gel electrophoresis were sampled and analyzed by mass spectrometry. Figure 1 shows a protein that was consistently upregulated in platelets during myocardial infarction. We identified this protein as heat-shock protein 27 (HSP27) by amino acid sequencing in mass spectrometry. Increased HSP27 levels were observed in twelve patients with myocardial infarction compared to control samples.

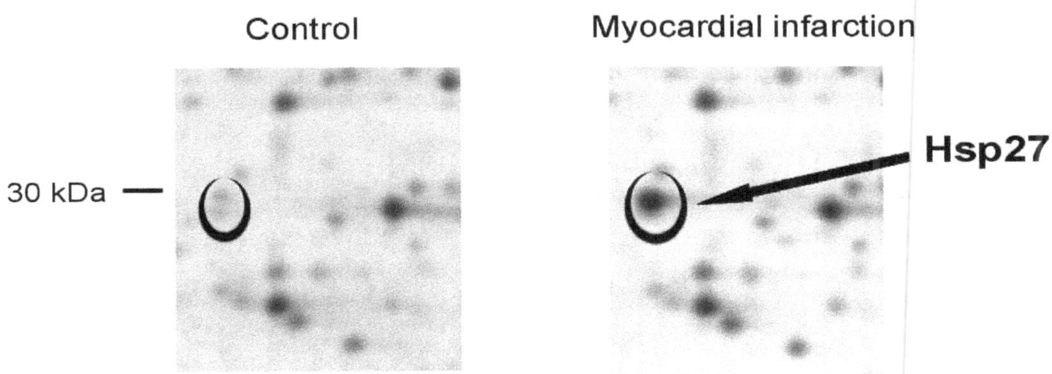

Figure 1. Heat-shock protein 27 (HSP27) is upregulated in platelets during myocardial infarction. Two-dimensional electrophoresis shows a protein spot that was upregulated in patients with ST-elevation myocardial infarction (black circles), which was identified as heat-shock protein 27 (HSP27) by mass spectrometry. Images show a representative sample of a patient with myocardial infarction and a matched control patient, which are representative of twelve independent patient pairs. Protein spots are visualized in a 2-fold magnification.

2.2. HSP27 Is Upregulated and Phosphorylated in Human Platelets during Myocardial Infarction

Increased HSP27 levels in platelets of patients with myocardial infarction compared to controls were confirmed by immunoblotting (Figure 2A). Detection of phosphorylated HSP27 (pHSP27) with a specific antibody also showed increased phosphorylation of HSP27 in myocardial infarction (Figure 2A). Four representative, independent patient pairs are shown that illustrate the increased HSP27 levels and phosphorylation in patients with myocardial infarction. Band densitometry of the Western blot results ($n = 12$ in each group) showed a significant increase of HSP27 protein and HSP27 phosphorylation in platelets from patients with myocardial infarction (MI) compared to controls (6.4 ± 1.0-fold versus 1.0 ± 0.9 and 5.9 ± 1.8-fold versus 1.0 ± 0.8; $p < 0.05$) (Figure 2B,C).

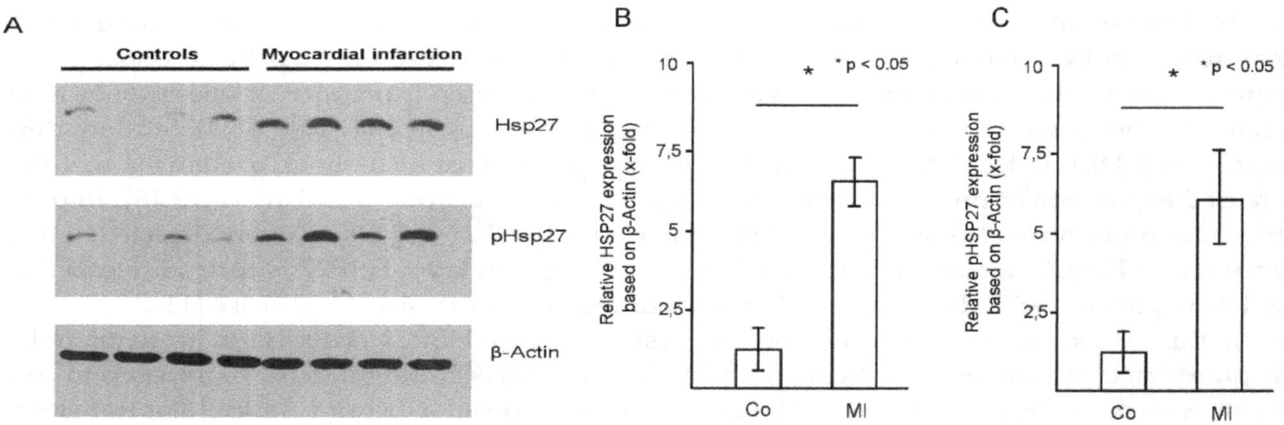

Figure 2. HSP27 is upregulated and phosphorylated in platelets during myocardial infarction. (**A**) HSP27 levels and phosphorylation (pHSP27) in four representative patients with nonischemic chest pain (controls, lanes 1–4) and four patients with myocardial infarction (MI) (lanes 5–8) are shown. β-actin served as loading control. (**B**) Quantitative analysis of HSP27 levels and phosphorylation (**C**) in the group of patients with myocardial infarction (MI) ($n = 12$) compared to controls ($n = 12$); * $p < 0.05$. Mean HSP27 and phospho-HSP27 to actin ratio for the control group was set as 1. The blot is representative of 12 independent patient pairs.

2.3. HSP27 Levels Are Increased by Thrombin Stimulation

Platelets were stimulated with thrombin (0.5 U/L) for up to two hours, and HSP27 levels were quantified by immunoblotting. After thrombin stimulation, we observed a gradual increase of HSP27 with a peak value at 30 min, which declined with longer stimulation (Figure 3). Heat-shock treatment of platelets (HS, 42 °C, 10 min) induced a robust upregulation of HSP27 and served as a positive control.

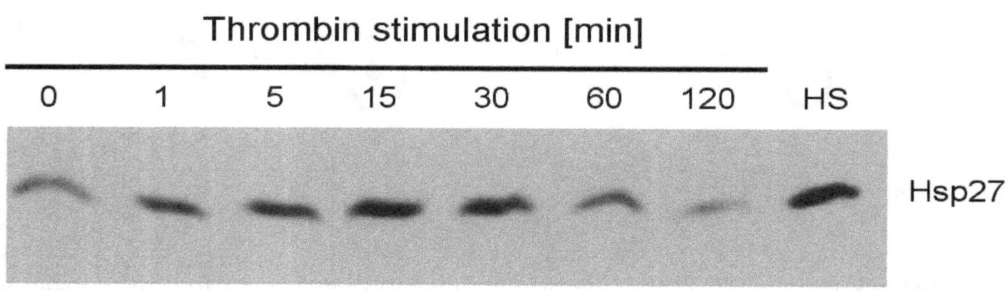

Figure 3. Thrombin activation induces upregulation of HSP27 in platelets. Platelets were stimulated with thrombin (0.5 U/mL) for 1, 5, 15, 30, 60, and 120 min, and HSP27 levels were quantified by immunoblotting. Heat activation at 42 °C for 10 min (HS) served as positive control. The immunoblot time course for HSP27 levels is representative of 3 independent experiments.

2.4. HSP27 Translocates from the Cytoskeletal into the Membrane Fraction of Platelets during Myocardial Infarction

Thrombin stimulation of platelets induced an intracellular translocation of HSP27 from the cytoskeletal into the membrane-associated fraction of platelets as illustrated by confocal microscopy

(Figure 4A). To quantify intracellular HSP27 distribution, platelet lysates were separated into the cytoskeletal and plasma membrane fraction by stepwise ultracentrifugation. Figure 4B shows a representative Western blot of a patient pair from the study population. Resting control platelets only showed small HSP27 levels in both the membrane and cytoskeletal fraction. During myocardial infarction, HSP27 was found at increased levels in the membrane fraction, which was not observed in control patients.

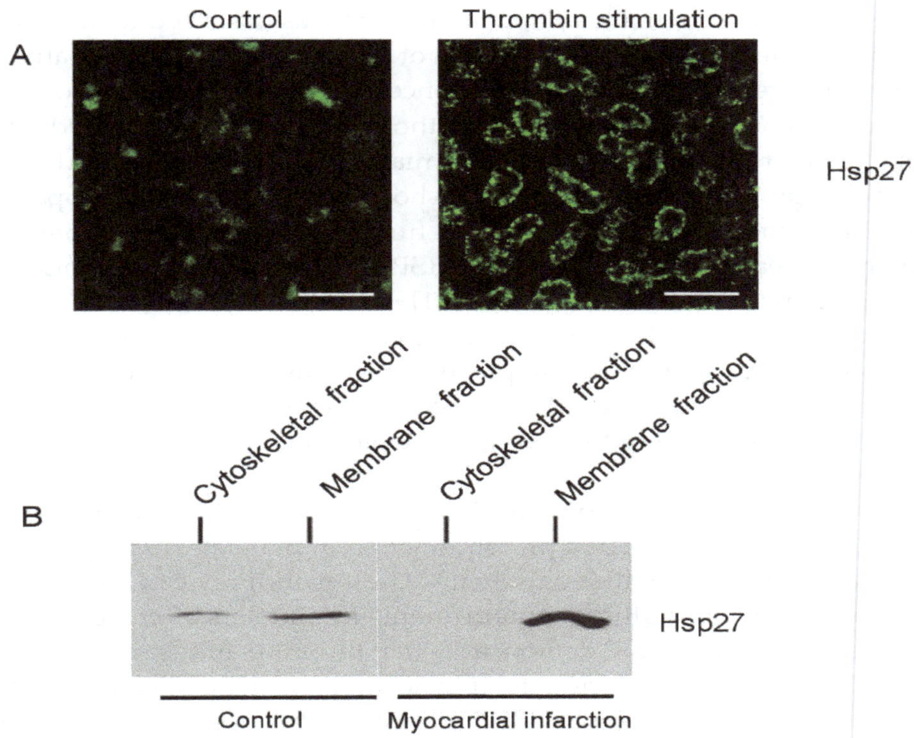

Figure 4. HSP27 translocates into the membrane fraction of platelets during myocardial infarction. (**A**) Confocal microscopy of resting (left) and thrombin activated platelets (0.5 U/L) (right) illustrates the characteristic translocation of HSP27 into the cell membrane of platelets with activation; scale bars represent 5 μm. (**B**) HSP27 distribution in platelets was further quantified in the membrane and cytoskeletal fraction of platelets from patients with nonischemic chest pain (controls, lanes 1–2) and patients with myocardial infarction (lanes 3–4). The immunoblot is representative of twelve independent patient pairs.

3. Discussion

In the present study, we used two-dimensional electrophoresis to analyze proteomic changes of platelets from patients with acute myocardial infarction and nonischemic chest pain and observed increased phosphorylation (pHSP27) and upregulation of HSP27 in platelets during myocardial infarction. So far, intracellular modulation of HSP27 has not been studied in human platelets during myocardial infarction ex vivo.

Patients with ST-elevation myocardial infarction were treated with aspirin immediately after diagnosis, and additional loading with ADP receptor antagonists such as clopidogrel was initiated immediately after blood was drawn. All control patients were also treated with aspirin by the time of blood sampling due to acute onset chest pain and suspected coronary artery disease. Therefore, there was no difference in antiplatelet therapy between groups at the time of blood sampling to make sure that the observed effects can be attributed to the acute vascular event and not to drug-related effects.

Phosphorylation of HSP27 seems to indicate that the protein is in a cell protective, anti-oxidative, and antiapoptotic functional state [2,3,22]. The role of heat-shock proteins in inflammation and its anti-oxidative capacity is of paramount interest, and animal studies have been conducted to investigate the use of heat-shock proteins in anti-atherosclerotic therapy [8,18,23]. Heat-shock proteins are activated in response to oxidized LDL and mediate an anti-inflammatory response through release of Interleukin 10 (IL-10) [24], through increased levels of gluthathione [25], and through inactivation of Nuclear Factor kappa B (NFκB) [26]. Low-density lipoproteins in return induce dephosphorylation of HSP27 [27], which alters anti-oxidative functions of HSP27. Malfunction or downregulation of heat-shock proteins has been associated with vascular disease progression and atherosclerosis [17,18,21]. Furthermore, decreased expression of heat-shock proteins was observed in atherosclerotic vascular areas and vulnerable plaques, which could be a sign of defective local anti-oxidative protection [1,2].

Heat-shock proteins were found in increased concentrations in the plasma of patients with acute myocardial infarction or coronary artery disease although the source remained unknown [8,28,29], and HSP27 levels were increased after global ischemia in coronary sinus blood samples of patients after surgical aortic clamping [15]. Altogether, heat-shock protein expression appears to be increased in response to vascular ischemia and compromised function of heat-shock proteins favors vascular disease development. Release of phosphorylated HSP27 from platelets was found in patients with diabetes who are at great risk for vascular disease [11].

Although we found a robust HSP27 phenotype in platelets during myocardial infarction, it remains to be determined if this is a result of the ischemic process or whether these changes occur prior to the acute event. Further studies suggest that HSP27 also regulates the cytoskeleton and that phosphorylation of HSP27 is a step that immediately precedes the association of HSP27 with the activated cytoskeleton of platelets [7] to stabilize actin fibers.

In this study, we consistently observed a distinct intracellular redistribution pattern from the cytoskeletal into the membrane-associated protein fraction in platelets during myocardial infarction, which was not observed in patients with nonischemic chest pain. Previous observations demonstrate that HSP27 translocates between cellular compartments in platelets after thrombin stimulation [7]. Supporting these observations, we could demonstrate that thrombin induces an upregulation of HSP27 with a peak value at 30 min and a characteristic translocation into the platelet membrane in confocal microscopy. In addition, phosphorylation of HSP27 in platelets has been closely correlated with platelet dense granule secretion [8–10], which is another characteristic of platelet activation. Plasma levels of phosphorylated HSP27 correlated closely with platelet aggregation in patients with diabetes; however, it remains unclear if this finding is functionally connected [11].

Since HSP27 directly associates with the cytoskeleton to execute signaling functions, intracellular translocation of HSP27 during cellular activation or in response to stress signals appears to make sense physiologically.

Others have reported that HSP27 interacts with the cytoskeleton through actin fiber stabilization, which has also been associated with decreased migratory capacity of inflammatory cells and anti-inflammatory functions [12]. We have previously shown that platelets also gain migratory capacity in response to inflammatory signals and that platelets can transmigrate into the vessel wall [30]. The role of heat-shock proteins in this context however is still unknown.

Besides direct regulatory effects on a cytoskeletal level, HSP27 controls translation factors such as eIF4F through binding of eIF4G, which then results in inhibition of translation [31]. Likely HSP27 prevents production of inflammatory proteins as part of an anti-inflammatory response that way. The above findings support the hypothesis and give reason to believe that phosphorylation of HSP27 may be a step that enables HSP27 to execute its cell protective effects.

An elegant study by Liu and colleagues demonstrates increased levels of CCL2 (Chemokine CC-motif ligand 2) in plasma, platelets, and thrombus material of patients with myocardial infarction. Activation of platelets with CCL2 in return resulted in increased phosphorylation of HSP27 and other signaling proteins [32].

It is tempting to speculate that increased phosphorylation of HSP27 in platelets from patients with myocardial infarction could be a sign of an ischemic stress response. We do not know however if this phenotype is a direct result of the acute vascular events or a consequence of platelet activation. The question whether heat-shock protein phosphorylation is an anti-inflammatory reaction in platelets while unphosphorylated HSP27 may contribute to accelerated aggregation or if it is a passive result of the stress stimulus remains to be determined in future studies. Further research will additionally be necessary to investigate the regulatory mechanisms of HSP27 protein and its phosphorylation during myocardial infarction.

In summary, we observed a significant increase of HSP27 (HSPB1) protein levels and phosphorylation of HSP27 in human platelets during myocardial infarction compared to matched controls with nonischemic chest pain in this study. Supporting previous observations in platelets, a characteristic intracellular translocation of HSP27 from the cytoskeletal into the membrane associated protein fraction was also observed during myocardial infarction. This platelet phenotype was distinctly different in platelets from patients with myocardial infarction compared to controls. It will be interesting to extend the observed characteristics of the HSP27 phenotype in platelets during myocardial infarction to other acute vascular events such as ischemic stroke or peripheral artery disease.

4. Material and Methods

4.1. Study Design

Twelve patients with acute ST-elevation myocardial infarction (mean age 62 years) and obstruction of a proximal left dominant or right coronary artery as well as twelve age-matched control patients with acute onset typical chest pain but rule-out for acute myocardial infarction by electrocardiogram (ECG) and cardiac enzymes were enrolled in the study. Due to suspected coronary artery disease, control patients underwent coronary angiography in which coronary artery disease was ruled out. According to guideline recommendations, patients with ST-elevation myocardial infarction were treated with aspirin immediately after diagnosis. Blood for this study was drawn in the cath lab before other antiplatelet agents such as clopidogrel or tirofiban were administered for acute myocardial infarction. All control patients were also on aspirin treatment by the time of blood sampling. Written informed consent was obtained from all patients. The study was approved by the local ethics committee of the University of Tübingen (No. 264/2007BO2, approved 25 October 2007) in accordance with the Declaration of Helsinki. Blood was drawn on admission after indication for coronary angiography had been established and blood samples were processed immediately.

4.2. Platelet Isolation

Platelets were isolated according to standard protocols in our lab as previously described [30,33]. For some experiments, platelets were stimulated with thrombin (Sigma-Aldrich, Taufkirchen, Germany) 0.1 U/mL for different time points or were heat-activated at 42 °C for 10 min prior to lysis. A total of 1×10^9 platelets was lysed in 100 μL of CHAPS lysis buffer (8 M Urea, 4% CHAPS, and 2% DTT) and stored for two-dimensional gel analysis and immunoblotting at −20 °C.

4.3. Fractionation of Cytoskeleletal and Membrane Proteins of Platelets

Freshly isolated platelets (1×10^9) were lysed in 1 mL Triton X-lysis buffer (containing complete protease inhibitor, Roche, Penzberg, Germany) for 10 min on ice. The pellet containing the cytoskeletal cell fraction was collected after centrifugation at 15,600× *g* for 4 min after supernatants had been carefully removed. Supernatants were centrifugated again at 100,000× *g*, and pellets containing the membrane fraction were collected. Both fractions were resuspended in SDS loading buffer before electrophoresis.

4.4. Two-Dimensional Electrophoresis

Platelets from patients with myocardial infarction and nonischemic chest pain were lysed in 100 μL of CHAPS lysis buffer (8 M Urea, 4% CHAPS, and 2% DTT) and purified using a 2D clean-up kit (GE Healthcare, Freiburg, Germany) according to the manufacturer's instructions. Comparative 2D gel analysis of the proteomes was performed as described previously with slight modifications [34,35]. First dimension isoelectric focusing was performed using a Protean IEF Cell focusing unit (BioRad, Hercules, CA, USA) with pH 3–10 NL gel strips (11 cm, GE Healthcare, for some subsequent gels pH 4–7). For the second dimension equilibrated (5.7 M Dithiotreitol and 1.5 M Iodoacetamide) gel strips were applied to 12% polyacrylamide gels. Proteins were stained with Lava purple fluorescent staining (Fluorotechnics, Sydney, Australia). Protein spots of interest were excised manually from the gels, digested with trypsin. and analyzed by LC ESI-MS/MS (Applied Biosystems/MDS Sciex, Darmstadt, Germany) as described previously [35].

4.5. Immunoblotting

Platelets lysed in Laemmli buffer were separated by SDS PAGE gel electrophoresis and probed for HSP27, phospho-HSP27, and β-actin. After protein transfer to a nitrocellulose membrane, membranes were blocked in 5% nonfat milk in Tris-buffered saline with Tween-20 (TBS-T, Sigma-Aldrich, Taufkirchen, Germany) for 1 h. Afterwards, primary monoclonal antibody to HSP27 or phospho-HSP27 (1:1000, Santa Cruz Biotechnologies, Dallas, TX, USA) were added and incubated overnight at 4 °C with constant agitation. Membranes were washed in TBS-T repeatedly, and HRP-conjugated secondary antibody (1:10000) (GE Healthcare, Freiburg, Germany) was added for 1 h at room temperature. After washing, chemiluminescent substrate (ECL reagent, Amersham Biosciences, Munich, Germany) was added for 1–5 min and bands were visualized on plain film. β-Actin served as loading control (1:000, rabbit anti-human β-Actin, Cell Signaling Technology, Dallas, TX, USA). HSP27 and phospho-HSP27 levels were quantified by band densitometry analysis with the help of HSP27- and pHSP27-to-actin ratios. Mean HSP27- and pHSP27-to-actin ratios for the myocardial infarction group were set as 1 and relative increases in HSP27 and pHSP27 levels in the control group were compared accordingly. Relative HSP27 and pHSP27 levels between groups based on actin loading are shown.

4.6. Confocal Microscopy

Immunofluorescence staining was performed as previously described [30]. In brief, washed platelets were allowed to adhere to a fibrinogen surface (50 μg/mL) on a chamber slide for 20 min and were then fixed with paraformaldehyde (2%) and permeabilized with Triton-X-100 (TX-100, 0.025%). The adherent platelets were washed and blocked with 2% bovine serum albumin for 30 min followed by incubation with the primary antibody for 2 h at room temperature. Primary antibody against HSP27 (Santa Cruz Biotechnologies, Dallas, TX, USA) was used in a 1:100 dilution in TBS with 1% BSA (Bovine Serum Albumine). Slides were then washed and incubated with an Alexa488-linked secondary antibody (Dianova, Hamburg, Germany) for 1 h. Confocal microscopy was performed using a Zeiss LSM 5 EXCITER confocal laser scanning microscope (Carl Zeiss Micro Imaging, Jena, Germany).

4.7. Statistical Analysis

All data are presented as means ± SEM. Statistical analyses were performed with Sigma Plot 10.0. For comparisons between two groups of normal distributed data, the student´s t-test was used. For the comparison of two groups without normal distributed data, a rank-sum test was performed. For multiple comparisons between groups of normal distributed data, the one-way analysis of variance (1-way ANOVA) was used. Differences were considered significant at an error probability level of $p < 0.05$.

5. Conclusions

We observed a significant increase of HSP27 (HSPB1) levels and phosphorylation of HSP27 in platelets during myocardial infarction compared to matched controls with nonischemic chest pain in this study. Supporting previous observations in platelets, a characteristic intracellular translocation of

HSP27 from the cytoskeletal into the membrane-associated protein fraction was also observed during myocardial infarction. This platelet phenotype was distinctly different in platelets from patients with myocardial infarction compared to controls. It will be interesting to extend the observed characteristics of the HSP27 phenotype in platelets during myocardial infarction to other acute vascular events such as ischemic stroke or peripheral artery disease.

Author Contributions: Conceptualization and project administration: S.L.; supervision: S.L.; investigation and data analysis: S.L., T.L., and M.F.-W.; methodology: S.L., T.L., and M.F.-W.; visualization: S.L. and B.F.K.; writing—original draft preparation: S.L., H.M., and B.F.K.

Acknowledgments: The authors wish to thank the Proteome Center Tübingen for supporting the 2D gel analysis. The Proteome Center Tübingen was supported by the Ministerium für Wissenschaft und Kunst, Landesregierung Baden-Württemberg.

References

1. Madrigal-Matute, J.; Martin-Ventura, J.L.; Blanco-Colio, L.M.; Egido, J.; Michel, J.B.; Meilhac, O. Heat-shock proteins in cardiovascular disease. *Adv. Clin. Chem.* **2011**, *54*, 1–43. [PubMed]
2. Wettstein, G.; Bellaye, P.S.; Micheau, O.; Bonniaud, P. Small heat shock proteins and the cytoskeleton: An essential interplay for cell integrity? *Int. J. Biochem. Cell Biol.* **2012**, *44*, 1680–1686. [CrossRef] [PubMed]
3. Acunzo, J.; Katsogiannou, M.; Rocchi, P. Small heat shock proteins HSP27 (HspB1), alphaB-crystallin (HspB5) and HSP22 (HspB8) as regulators of cell death. *Int. J. Biochem. Cell Biol.* **2012**, *44*, 1622–1631. [CrossRef] [PubMed]
4. Gorman, A.M.; Szegezdi, E.; Quigney, D.J.; Samali, A. Hsp27 inhibits 6-hydroxydopamine-induced cytochrome c release and apoptosis in PC12 cells. *Biochem. Biophys. Res. Commun.* **2005**, *327*, 801–810. [CrossRef]
5. Paul, C.; Simon, S.; Gibert, B.; Virot, S.; Manero, F.; Arrigo, A.P. Dynamic processes that reflect anti-apoptotic strategies set up by HspB1 (Hsp27). *Exp. Cell Res.* **2010**, *316*, 1535–1552. [CrossRef]
6. Stetler, R.A.; Gao, Y.; Zhang, L.; Weng, Z.; Zhang, F.; Hu, X.; Wang, S.; Vosler, P.; Cao, G.; Sun, D.; et al. Phosphorylation of HSP27 by protein kinase D is essential for mediating neuroprotection against ischemic neuronal injury. *J. Neurosci.* **2012**, *32*, 2667–2682. [CrossRef]
7. Zhu, Y.; O'Neill, S.; Saklatvala, J.; Tassi, L.; Mendelsohn, M.E. Phosphorylated HSP27 associates with the activation-dependent cytoskeleton in human platelets. *Blood* **1994**, *84*, 3715–3723. [CrossRef]
8. Harats, D.; Yacov, N.; Gilburd, B.; Shoenfeld, Y.; George, J. Oral tolerance with heat shock protein 65 attenuates Mycobacterium tuberculosis-induced and high-fat-diet-driven atherosclerotic lesions. *J. Am. Coll. Cardiol.* **2002**, *40*, 1333–1338. [CrossRef]
9. Bennett, W.F.; Belville, J.S.; Lynch, G. A study of protein phosphorylation in shape change and Ca++-dependent serotonin release by blood platelets. *Cell* **1979**, *18*, 1015–1023. [CrossRef]
10. Lyons, R.M.; Stanford, N.; Majerus, P.W. Thrombin-induced protein phosphorylation in human platelets. *J. Clin. Invest.* **1975**, *56*, 924–936. [CrossRef]
11. Tokuda, H.; Kuroyanagi, G.; Tsujimoto, M.; Enomoto, Y.; Matsushima-Nishiwaki, R.; Onuma, T.; Kojima, A.; Doi, T.; Tanabe, K.; Akamatsu, S.; et al. Release of Phosphorylated HSP27 (HSPB1) from Platelets Is Accompanied with the Acceleration of Aggregation in Diabetic Patients. *PLoS ONE* **2015**, *10*, e0128977. [CrossRef] [PubMed]
12. Rayner, K.; Chen, Y.X.; McNulty, M.; Simard, T.; Zhao, X.; Wells, D.J.; de Belleroche, J.; O'Brien, E.R. Extracellular release of the atheroprotective heat shock protein 27 is mediated by estrogen and competitively inhibits acLDL binding to scavenger receptor-A. *Circ. Res.* **2008**, *103*, 133–141. [CrossRef] [PubMed]
13. Gear, A.R.; Simon, C.G.; Polanowska-Grabowska, R. Platelet adhesion to collagen activates a phosphoprotein complex of heat-shock proteins and protein phosphatase 1. *J. Neural Transm.* **1997**, *104*, 1037–1047. [CrossRef] [PubMed]
14. Staron, M.; Wu, S.; Hong, F.; Stojanovic, A.; Du, X.; Bona, R.; Liu, B.; Li, Z. Heat-shock protein gp96/grp94 is an essential chaperone for the platelet glycoprotein Ib-IX-V complex. *Blood* **2011**, *117*, 7136–7144. [CrossRef]

15. Jin, C.; Cleveland, J.C.; Ao, L.; Li, J.; Zeng, Q.; Fullerton, D.A.; Meng, X. Human myocardium releases heat shock protein 27 (HSP27) after global ischemia: The proinflammatory effect of extracellular HSP27 through toll-like receptor (TLR)-2 and TLR4. *Mol. Med.* **2014**, *20*, 280–289. [CrossRef]
16. Dupont, A.; Chwastyniak, M.; Beseme, O.; Guihot, A.L.; Drobecq, H.; Amouyel, P.; Pinet, F. Application of saturation dye 2D-DIGE proteomics to characterize proteins modulated by oxidized low density lipoprotein treatment of human macrophages. *J. Proteome Res.* **2008**, *7*, 3572–3582. [CrossRef]
17. Lepedda, A.J.; Cigliano, A.; Cherchi, G.M.; Spirito, R.; Maggioni, M.; Carta, F.; Turrini, F.; Edelstein, C.; Scanu, A.M.; Formato, M. A proteomic approach to differentiate histologically classified stable and unstable plaques from human carotid arteries. *Atherosclerosis* **2009**, *203*, 112–118. [CrossRef]
18. Martin-Ventura, J.L.; Duran, M.C.; Blanco-Colio, L.M.; Meilhac, O.; Leclercq, A.; Michel, J.B.; Jensen, O.N.; Hernandez-Merida, S.; Tunon, J.; Vivanco, F.; et al. Identification by a differential proteomic approach of heat shock protein 27 as a potential marker of atherosclerosis. *Circulation* **2004**, *110*, 2216–2219. [CrossRef]
19. Dybdahl, B.; Slordahl, S.A.; Waage, A.; Kierulf, P.; Espevik, T.; Sundan, A. Myocardial ischaemia and the inflammatory response: Release of heat shock protein 70 after myocardial infarction. *Heart* **2005**, *91*, 299–304. [CrossRef]
20. Ghayour-Mobarhan, M.; Rahsepar, A.A.; Tavallaie, S.; Rahsepar, S.; Ferns, G.A. The potential role of heat shock proteins in cardiovascular disease: Evidence from In Vitro and in vivo studies. *Adv. Clin. Chem.* **2009**, *48*, 27–72.
21. Park, H.K.; Park, E.C.; Bae, S.W.; Park, M.Y.; Kim, S.W.; Yoo, H.S.; Tudev, M.; Ko, Y.H.; Choi, Y.H.; Kim, S.; et al. Expression of heat shock protein 27 in human atherosclerotic plaques and increased plasma level of heat shock protein 27 in patients with acute coronary syndrome. *Circulation* **2006**, *114*, 886–893. [CrossRef] [PubMed]
22. Gawad, A.; Ptak-Belowska, A.; Brzozowski, T.; Pawlik, W.W. Monocytes and vascular endothelial cells apoptosis. Role of p-HSP27. *J. Physiol. Pharmacol.* **2009**, *60*, 55–61. [PubMed]
23. Jun, L.; Jie, L.; Dongping, Y.; Xin, Y.; Taiming, L.; Rongyue, C.; Jie, W.; Jingjing, L. Effects of nasal immunization of multi-target preventive vaccines on atherosclerosis. *Vaccine* **2012**, *30*, 1029–1037. [CrossRef] [PubMed]
24. De, A.K.; Kodys, K.M.; Yeh, B.S.; Miller-Graziano, C. Exaggerated human monocyte IL-10 concomitant to minimal TNF-alpha induction by heat-shock protein 27 (Hsp27) suggests Hsp27 is primarily an antiinflammatory stimulus. *J. Immunol.* **2000**, *165*, 3951–3958. [CrossRef] [PubMed]
25. Preville, X.; Salvemini, F.; Giraud, S.; Chaufour, S.; Paul, C.; Stepien, G.; Ursini, M.V.; Arrigo, A.P. Mammalian small stress proteins protect against oxidative stress through their ability to increase glucose-6-phosphate dehydrogenase activity and by maintaining optimal cellular detoxifying machinery. *Exp. Cell Res.* **1999**, *247*, 61–78. [CrossRef] [PubMed]
26. Rinaldi, B.; Romagnoli, P.; Bacci, S.; Carnuccio, R.; Maiuri, M.C.; Donniacuo, M.; Capuano, A.; Rossi, F.; Filippelli, A. Inflammatory events in a vascular remodeling model induced by surgical injury to the rat carotid artery. *Br. J. Pharmacol.* **2006**, *147*, 175–182. [CrossRef]
27. Garcia-Arguinzonis, M.; Padro, T.; Lugano, R.; Llorente-Cortes, V.; Badimon, L. Low-density lipoproteins induce heat shock protein 27 dephosphorylation, oligomerization, and subcellular relocalization in human vascular smooth muscle cells. *Arterioscler. Thromb. Vasc. Biol.* **2010**, *30*, 1212–1219. [CrossRef]
28. Satoh, M.; Shimoda, Y.; Akatsu, T.; Ishikawa, Y.; Minami, Y.; Nakamura, M. Elevated circulating levels of heat shock protein 70 are related to systemic inflammatory reaction through monocyte Toll signal in patients with heart failure after acute myocardial infarction. *Eur. J. Heart Fail.* **2006**, *8*, 810–815. [CrossRef]
29. Zhang, X.; He, M.; Cheng, L.; Chen, Y.; Zhou, L.; Zeng, H.; Pockley, A.G.; Hu, F.B.; Wu, T. Elevated heat shock protein 60 levels are associated with higher risk of coronary heart disease in Chinese. *Circulation* **2008**, *118*, 2687–2693. [CrossRef]
30. Kraemer, B.F.; Borst, O.; Gehring, E.M.; Schoenberger, T.; Urban, B.; Ninci, E.; Seizer, P.; Schmidt, C.; Bigalke, B.; Koch, M.; et al. PI3 kinase-dependent stimulation of platelet migration by stromal cell-derived factor 1 (SDF-1). *J. Mol. Med.* **2010**, *88*, 1277–1288. [CrossRef]
31. Cuesta, R.; Laroia, G.; Schneider, R.J. Chaperone hsp27 inhibits translation during heat shock by binding eIF4G and facilitating dissociation of cap-initiation complexes. *Genes Dev.* **2000**, *14*, 1460–1470. [PubMed]
32. Liu, D.; Cao, Y.; Zhang, X.; Peng, C.; Tian, X.; Yan, C.; Liu, Y.; Liu, M.; Han, Y. Chemokine CC-motif ligand 2 participates in platelet function and arterial thrombosis by regulating PKCalpha-P38MAPK-HSP27 pathway. *Biochim. Biophys. Acta Mol. Basis Dis.* **2018**, *1864*, 2901–2912. [CrossRef] [PubMed]

33. Weyrich, A.S.; Elstad, M.R.; McEver, R.P.; McIntyre, T.M.; Moore, K.L.; Morrissey, J.H.; Prescott, S.M.; Zimmerman, G.A. Activated platelets signal chemokine synthesis by human monocytes. *J. Clin. Invest.* **1996**, *97*, 1525–1534. [CrossRef] [PubMed]
34. Tiffert, Y.; Franz-Wachtel, M.; Fladerer, C.; Nordheim, A.; Reuther, J.; Wohlleben, W.; Mast, Y. Proteomic analysis of the GlnR-mediated response to nitrogen limitation in Streptomyces coelicolor M145. *Appl. Microbiol. Biotechnol.* **2011**, *89*, 1149–1159. [CrossRef]
35. Hala, M.; Cole, R.; Synek, L.; Drdova, E.; Pecenkova, T.; Nordheim, A.; Lamkemeyer, T.; Madlung, J.; Hochholdinger, F.; Fowler, J.E.; et al. An exocyst complex functions in plant cell growth in Arabidopsis and tobacco. *Plant Cell* **2008**, *20*, 1330–1345. [CrossRef]

10

Platelet Features and Derivatives in Osteoporosis: A Rational and Systematic Review on the Best Evidence

Francesca Salamanna, Melania Maglio *, Maria Sartori, Matilde Tschon and Milena Fini

Laboratory of Preclinical and Surgical Studies, IRCCS Istituto Ortopedico Rizzoli, Via di Barbiano 1/10, 40136 Bologna, Italy; francesca.salamanna@ior.it (F.S.); maria.sartori@ior.it (M.S.); matilde.tschon@ior.it (M.T.); milena.fini@ior.it (M.F.)

* Correspondence: melania.maglio@ior.it

Abstract: *Background*: With the increase in aging population, the rising prevalence of osteoporosis (OP) has become an important medical issue. Accumulating evidence showed a close relationship between OP and hematopoiesis and emerging proofs revealed that platelets (PLTs), unique blood elements, rich in growth factors (GFs), play a critical role in bone remodeling. The aim of this review was to evaluate how PLT features, size, volume, bioactive GFs released, existing GFs in PLTs and PLT derivatives change and behave during OP. *Methods*: A systematic search was carried out in PubMed, Scopus, Web of Science Core Collection and Cochrane Central Register of Controlled Trials databases to identify preclinical and clinical studies in the last 10 years on PLT function/features and growth factor in PLTs and on PLT derivatives during OP. The methodological quality of included studies was assessed by QUIPS tool for assessing risk of bias in the clinical studies and by the SYRCLE tool for assessing risk of bias in animal studies. *Results*: In the initial search, 2761 studies were obtained, only 47 articles were submitted to complete reading, and 23 articles were selected for the analysis, 13 on PLT function/features and growth factor in PLTs and 10 on PLT derivatives. Risk of bias of almost all animal studies was high, while the in the clinical studies risk of bias was prevalently moderate/low for the most of the studies. The majority of the evaluated studies highlighted a positive correlation between PLT size/volume and bone mineralization and an improvement in bone regeneration ability by using PLTs bioactive GFs and PLT derivatives. *Conclusions*: The application of PLT features as OP markers and of PLT-derived compounds as therapeutic approach to promote bone healing during OP need to be further confirmed to provide clear evidence for the real efficacy of these interventions and to contribute to the clinical translation.

Keywords: platelet function; platelet derivatives; osteoporosis; bone

---

1. Introduction

Although the bone tissue has the exclusive ability to self-repair and regenerate, in several situations this capability results inadequate or linked to complications. Osteoporosis (OP) is defined by the World Health Organization (WHO) as a "*progressive systemic skeletal disease characterized by low bone mass and microarchitectural deterioration of bone tissue, with a consequent increase in bone fragility and susceptibility to fracture*" [1] (Figure 1).

Affecting about 200 million people in the world, with considerable morbidity and mortality, OP is one of the main epidemics of the 21st century (International Osteoporosis Foundation [2]). Fractures resulting from OP are the main cause of morbidity and mortality, bringing elevated social-economic burden on both families and health care system [3].

Several risk factors, i.e., clinical, medical, behavioral, nutritional and genetic, are related to OP [4]. One of the main causes of OP is the postmenopausal state which involves increased degree of imbalance of bone resorption and formation in favor of bone resorption [5]. Osteoclasts, originating from hematopoietic cells, are mainly responsible for bone resorption. Despite the hematopoietic origin of osteoclasts, the hematological changes occurring during OP are not yet well elucidated. In the last decade, some studies found that platelet (PLT), fragments of cytoplasm derived from the megakaryocytes of the bone marrow, have a critical role in skeletal homeostasis, modulating bone formation and resorption [6–8].

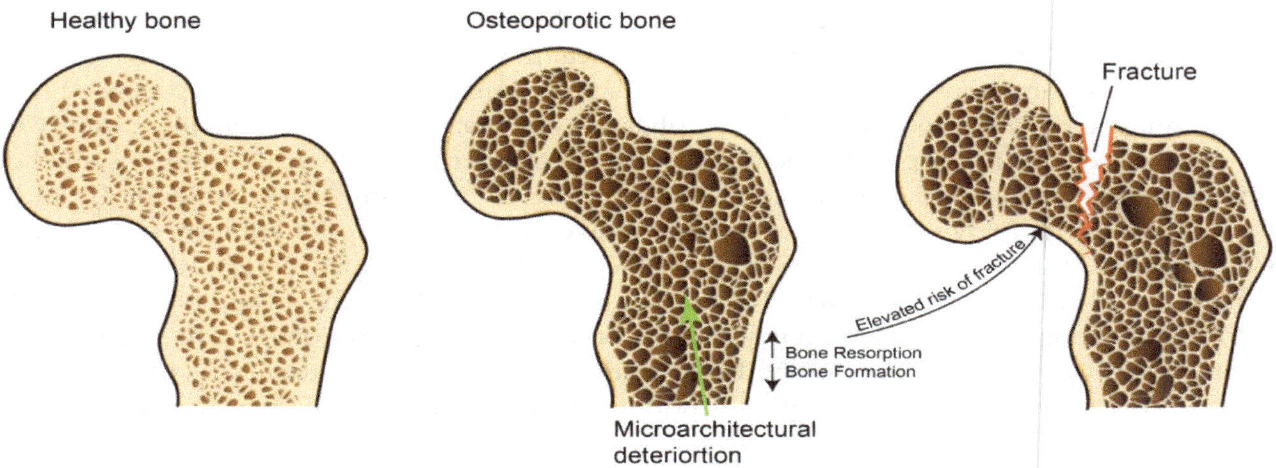

Figure 1. Schematic representation of osteoporosis disease.

PLTs are 2–3 μm in diameter and around their periphery a contractile microtubules ring containing actin and myosin is present. PLTs have several intracellular structures, i.e., lysosomes and two types of granules, dense granule organelles, containing adenosine triphosphate (ATP), adenosine diphosphate (ADP), serotonin, and calcium, and the alpha (α) granules, containing growth factors (GFs), clotting factors, and other proteins (Figure 2) [9]. PLTs also act as a reserve for glycogen [9].

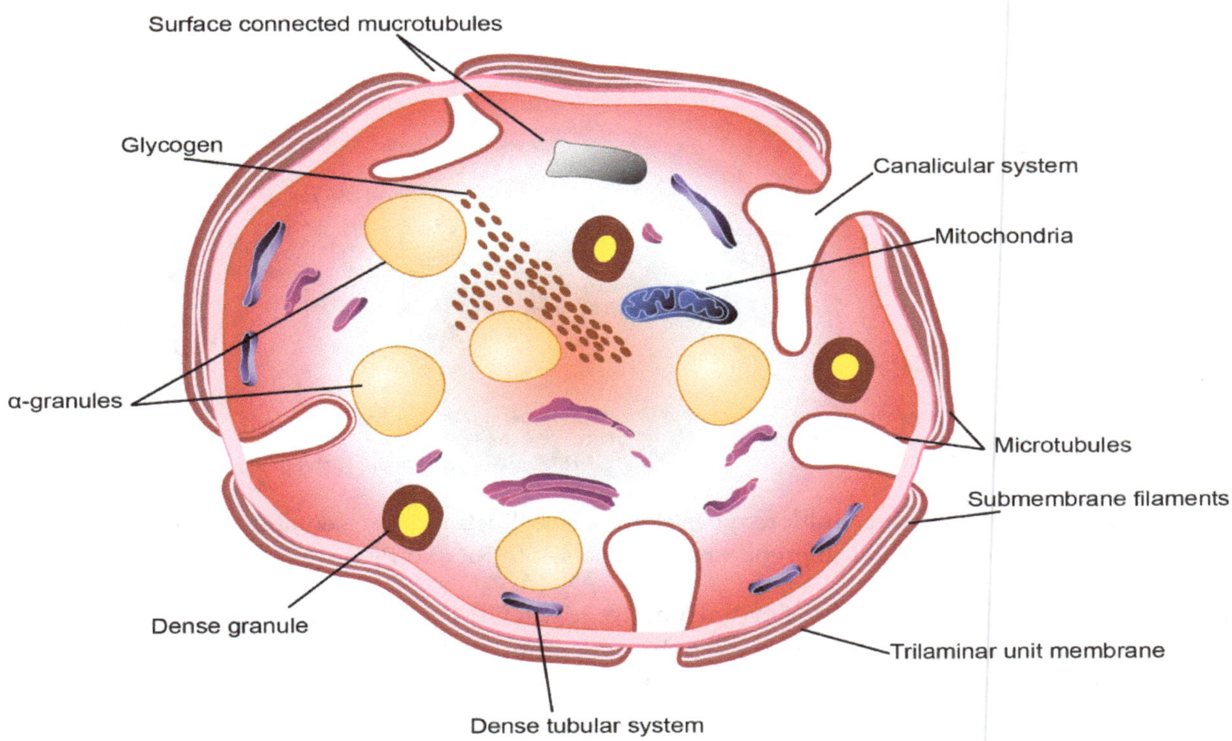

Figure 2. Schematic overview of PLT structure (diagrammatic representation) in the equatorial plane.

These GFs play a central role in the healing process and tissue regeneration, being used as messengers to regulate various processes [10]. Tissue repair begins with PLT clot formation, activation of the coagulation cascade and PLT degranulation, and release of platelets growth factors (PGFs). These PGFs join to specific target tyrosine growth factor receptors, which then activate intracellular signal transduction pathways [11,12]. Several preclinical and clinical studies highlighted the supportive effect of PLTs on bone formation showing that platelet-derived growth factors (PDGFs) favor bone formation by affecting cell proliferation, chemotaxis differentiation, and extracellular matrix synthesis [13,14]. On the other hand, preclinical in vitro studies showed the role of PLTs in osteoclastogenesis and bone resorption, but the exact mechanism has not been yet proposed [13,14]. These unique biological properties of PLTs emphasize why their derivatives were increasingly used in the clinical scenario to support the healing process in different pathological conditions, including musculoskeletal diseases [15–17]. In comparison to the use of a single recombinant GF in high concentrations, the employment of PLT derivatives have the advantage of offering several synergistic GFs able to cooperate in a specific site and for a specific goal. For this reason PLT derivatives (i.e., platelet poor plasma, PPP; platelet-rich plasma, PRP; platelet-rich fibrin, PRF; leucocyte and platelet-rich fibrin, L-PRF) [18] are considered an attractive option for bone tissue regeneration (Figure 3), containing a high concentration of local GFs including PDGF, transforming growth factor (TGF), platelet-derived angiogenesis factor (PDAF), platelet-derived endothelial growth factor (PDEGF), vascular endothelial growth factor (VEGF) and many others able to modulate the regenerative process [19].

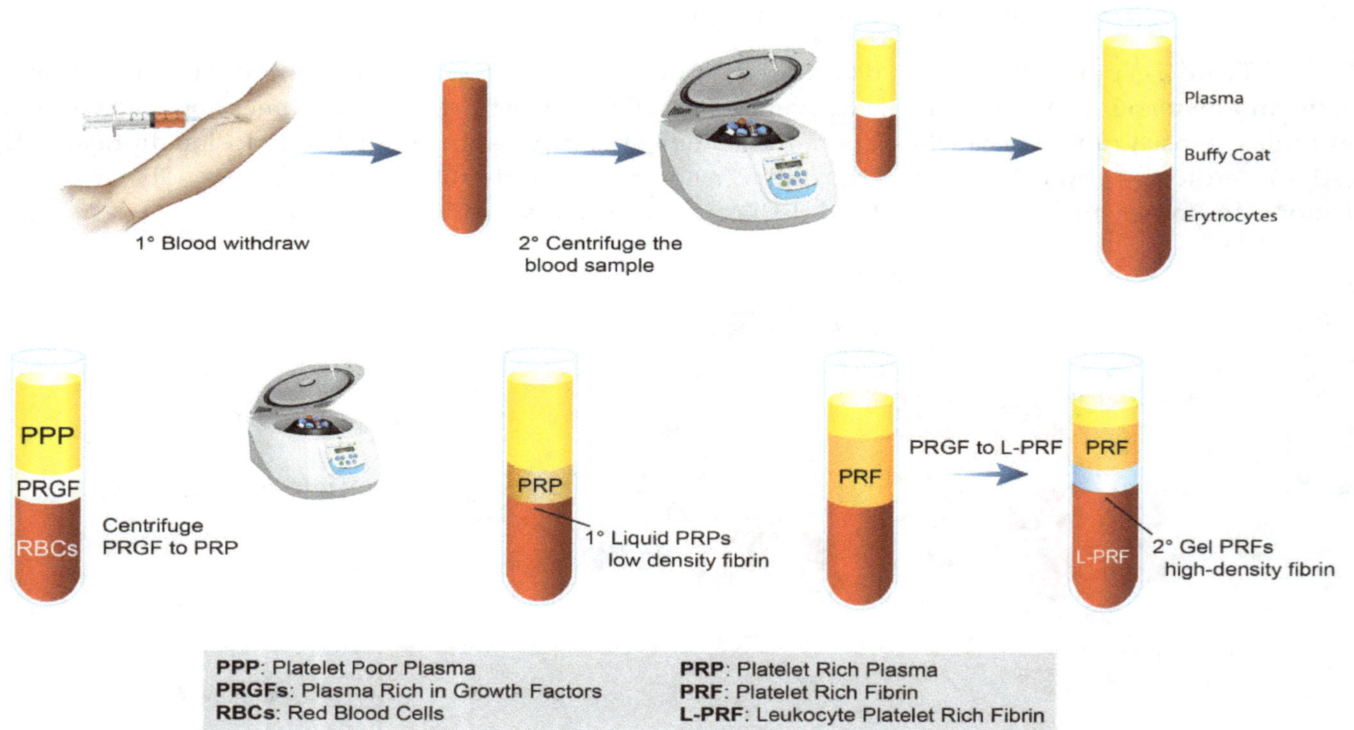

Figure 3. PLT concentrates preparation, types/classes, and illustration/presentation of PLT derivatives. Schematic drawing of the classical preparation protocols of PRP and PRF.

It has been observed that PLT derivatives (e.g., PRF, PPP, PRF) improved proliferation and osteogenic activity of bone marrow mesenchymal stem cells (BMSCs) and osteoblasts [20–22]. Additionally, in vivo studies have revealed that clots of PLT derivatives, also in combination with different materials/scaffolds, improved bone regeneration by promoting the expression of TGF-β and bone morphogenic protein-2 (BMP-2) [23–27]. Several clinical studies have also applied PLT derivatives, alone and in association with natural and synthetic biomaterials, in patients with different grades of bone defects, reporting improved bone regeneration, early bone formation, bone-depth reduction and more mature bone [28–31]. However, the exact function of PLTs and its derivatives on bone resorption and bone formation is still complex to understand because of the multifaceted interactions between GFs, inflammatory mediators, and cytokines. Consequently PLTs role and skills during OP and their relationship with bone loss are even more complex to understand and conflicting results have been obtained [32–34].

To date, many key questions remain unanswered and controversial, in particular concerning PLT function, size, volume, role of bioactive GFs released and use of PLT derivatives during OP pathogenesis. Thus, we carried out a systematic review in which we wondered: How do PLTs work and what changes occur in their function, features and/or structure (volume, size, number) during OP? How do GFs released by PLTs or GFs existing in PLTs and PLT derivatives "work" during OP? Which are the main derivatives used in OP? and How are they used? In the present systematic review we tried to highlight and answer to these points, attempting to give an up-to-date tool for researchers and clinicians involved in PLT-mediated bone tissue regenerative applications in OP condition.

2. Methods

2.1. Eligibility Criteria

The PICOS model was used to formulate the questions for this study: (1) studies that considered cells, animals and patients with OP (Population), (2) studies where one of the primary aims were to evaluate PLTs and PLTs derivatives during OP (Interventions), (3) studies that presented a control interventions (Comparisons), (3) studies that reported the effects/functions/roles of PLTs and PLTs derivatives during OP (Outcomes) and (4) preclinical (in vitro and in vivo) and clinical studies (Study design). Studies from 27 July 2009 to 27 July 2019 were included in this review if they met the PICOS criteria.

We excluded studies investigating (1) PLTs functions and/or PLTs derivatives in pathological conditions different from OP, (2) pathological conditions where OP is a bone manifestation of another disease (i.e., diabetes, Gaucher disease, cancer, rheumatic diseases), (3) osteonecrosis of the jaw due to OP therapy, (4) PLT functions and/or PLT derivatives during the administration of drugs active on bone metabolism (e.g., alendronate, zolendronate, denosumab, raloxifene), (5) drug (e.g., glucocorticoid)-induced osteoporosis. Additionally, we excluded case reports, abstracts, editorials, letters, comment to Editor, reviews, meta-analysis, book chapters and articles not written in English.

2.2. Information Source and Search Strategies

Our literature review involved a systematic search conducted on 27 July 2019. We performed our review according to the Preferred Reporting Items for Systematic Reviews and Meta-Analyses (PRISMA) statement [35]. The search was carried out on PubMed. MEDLINE, ProQuest, Scopus, Web of Science Core Collection and Cochrane Central Register of Controlled Trials databases to identify preclinical and clinical studies on PLT functions/features and use of PLT derivatives in OP condition. Search was conducted combining the terms "Osteoporosis" AND "Platelets"; for each of these terms, free words and controlled vocabulary specific to each bibliographic database were combined using the operator "OR". The combination of free-vocabulary and/or MeSH terms for the identification of studies in PubMed/MEDLINE, ProQuest, Scopus, Web of Science Core Collection and Cochrane Central Register of Controlled Trials were reported in Table 1. In addition, reference lists of relevant studies were searched for other potentially appropriate publications.

Table 1. Search terms used in the PubMed, ProQuest, Scopus, Web of Science Core Collection and Cochrane Central Register of Controlled Trials.

Database	Search Items
PubMed	((((((“blood platelets”[MeSH Terms] OR (“blood”[All Fields] AND “platelets”[All Fields]) OR “blood platelets”[All Fields] OR “platelet”[All Fields]) OR (“blood platelets”[MeSH Terms] OR (“blood”[All Fields] AND “platelets”[All Fields]) OR “blood platelets”[All Fields] OR “platelets”[All Fields])) OR ((“blood platelets”[MeSH Terms] OR (“blood”[All Fields] AND “platelets”[All Fields]) OR “blood platelets”[All Fields] OR “platelet”[All Fields]) AND functions[All Fields])) OR ((“blood platelets”[MeSH Terms] OR (“blood”[All Fields] AND “platelets”[All Fields]) OR “blood platelets”[All Fields] OR “platelet”[All Fields]) AND (“Changes”[Journal] OR “changes”[All Fields]))) OR ((“blood platelets”[MeSH Terms] OR (“blood”[All Fields] AND “platelets”[All Fields]) OR “blood platelets”[All Fields] OR “platelet”[All Fields]) AND (“Structure”[Journal] OR “structure”[All Fields]))) OR ((“blood platelets”[MeSH Terms] OR (“blood”[All Fields] AND “platelets”[All Fields]) OR “blood platelets”[All Fields] OR “platelet”[All Fields]) AND size[All Fields])) AND ((((((((“osteoporosis, postmenopausal”[MeSH Terms] OR (“osteoporosis”[All Fields] AND “postmenopausal”[All Fields]) OR “postmenopausal osteoporosis”[All Fields] OR “osteoporosis”[All Fields] OR “osteoporosis”[MeSH Terms]) OR (“bone demineralization, pathologic”[MeSH Terms] OR (“bone”[All Fields] AND “demineralization”[All Fields] AND “pathologic”[All Fields]) OR “pathologic bone demineralization”[All Fields] OR (“bone”[All Fields] AND “demineralization”[All Fields]) OR “bone demineralization”[All Fields])) OR (“bone density”[MeSH Terms] OR (“bone”[All Fields] AND “density”[All Fields]) OR “bone density”[All Fields])) OR (“osteoporotic fractures”[MeSH Terms] OR (“osteoporotic”[All Fields] AND “fractures”[All Fields]) OR “osteoporotic fractures”[All Fields])) OR (“bone diseases, metabolic”[MeSH Terms] OR (“bone”[All Fields] AND “diseases”[All Fields] AND “metabolic”[All Fields]) OR “metabolic bone diseases”[All Fields] OR “osteopenia”[All Fields])) OR (“bone diseases, metabolic”[MeSH Terms] OR (“bone”[All Fields] AND “diseases”[All Fields] AND “metabolic”[All Fields]) OR “metabolic bone diseases”[All Fields] OR (“bone”[All Fields] AND “loss”[All Fields]) OR “bone loss”[All Fields])) OR (“bone density”[MeSH Terms] OR (“bone”[All Fields] AND “density”[All Fields]) OR “bone density”[All Fields])) OR bmd[All Fields]) AND (“2009/07/27”[PDAT]: “2019/07/27”[PDAT])
ProQuest	(platelet AND (bdl(1007527) AND pd(20090727-20190727))) AND (osteoporosis AND (bdl(1007527) AND pd(20090727-20190727))) Applied limits: Database: Biological Science Collection British Nursing Database Health Research Premium Collection *Part of the search defined by the query is performed in these databases.* Restricted based on: Database: Biological Science Collection; Health Research Premium Collection; Biological Science Index; MEDLINE®; TOXLINE
Web of Science Core Collection	(TS = platelet OR TS = platelets OR TS = platelet functions OR TS = platelet changes OR TS = platelet structure OR TS = platelet size) AND (TS = osteoporosis OR TS = bone demineralization OR TS = bone density OR TS = Osteoporotic Fractures OR TS = osteopenia OR TS = bone loss OR TS = bone density OR TS = bmd)—with Publication Year from 2009 to 2019
Scopus	(TITLE-ABS-KEY (platelet) OR TITLE-ABS-KEY (platelets) OR TITLE-ABS-KEY (platelet AND functions) OR TITLE-ABS-KEY (platelet AND changes) OR TITLE-ABS-KEY (platelet AND structure) OR TITLE-ABS-KEY (platelet AND size) AND TITLE-ABS-KEY (osteoporosis) OR TITLE-ABS-KEY (bone AND demineralization) AND TITLE-ABS-KEY (bone AND density) OR TITLE-ABS-KEY (osteoporotic AND fractures) OR TITLE-ABS-KEY (osteopenia) OR TITLE-ABS-KEY (bone AND loss) OR TITLE-ABS-KEY (bone AND density) OR TITLE-ABS-KEY (bmd) OR TITLE-ABS-KEY (bone AND mass)) AND DOCTYPE (ar) AND PUBYEAR > 2008
Cochrane Central Register of Controlled Trials	((((((platelet) OR platelets) OR platelet functions) OR platelet changes) OR platelet structure) OR platelet size)) AND ((((((((osteoporosis) OR bone demineralization) OR bone density) OR Osteoporotic Fractures) OR osteopenia) OR bone loss) OR bone density) OR bmd) in All Text—with Publication Year from 2009 to 2019

2.3. Study Selection and Data Extraction

Possible relevant articles were screened using title and abstract by one reviewer (FS) and articles that did not meet the inclusion criteria were excluded. After screening the title and abstract, articles were submitted to a public reference manager (Mendeley; "www.mendeley.com") to eliminate duplicates. Subsequently, the remaining full text articles were retrieved and examined by two reviewers (FS, MM). Any disagreement was resolved through discussion until a consensus was reached, or with the involvement of a third reviewer (MF).

Data from the retrieved studies were tabulated taking into consideration studies that evaluated PLT functions/features and growth factor in PLTs (Table 2) during OP and studies that evaluated PLT derivatives in OP (Table 3). Each table was split-up based on preclinical and clinical studies. We extracted the following data from the articles on PLT functions/features and growth factor in PLTs during OP: Reference, Aim, Study design, Methodology, Platelet function, Link between platelet and OP, Main results (Table 2). The extracted data for the studies on PLT derivatives in OP were: Reference, Aim, Study type, Platelet formulation, Platelets concentration, White blood cells content, Activation method, Platelet application, Experimental design, Main results.

2.4. Assessment of Methodological Quality

Two reviewers (FS and MM) independently assessed the methodological quality of selected studies (Tables 4 and 5). In case of disagreement, they attempted to reach consensus; if this failed, a third reviewer (MF) made the final decision. The methodological quality of the clinical studies was assed using the Quality in Prognosis Studies (QUIPS) tool [36,37]. Studies were assessed on six domains: study participation, study attrition, prognostic factor measurement, outcome measurement, study confounding, and statistical analysis and reporting. Methodological quality appraisal of included in vivo studies was performed according to the Systematic Review Centre for Laboratory Animal Experimentation (SYRCLE) tool [38], which has been specifically designed to assess the risk of bias of animal studies. We have not assessed risk of bias for in vitro studies because, to our knowledge, no standard quality assessment tool exists for the type of in vitro studies included in this review.

3. Results

3.1. Study Selection and Characteristics

The initial literature search retrieved 2928 studies. Of those, 1169 studies were identified using PubMed/MEDLINE, 167 using ProQuest, 134 using Scopus, 1141 were found in Web of Science Core Collection and 317 using Cochrane Central Register of Controlled Clinical Trials. After screening the title and abstract107 articles were run through Mendeley to eliminate duplicate articles. The resulting 49 complete articles were then reviewed to establish whether the publication met the inclusion criteria and 23 were considered eligible for this review. From the reference lists of the selected articles no additional publications were found. Search strategy and study inclusion and exclusion criteria are detailed in Figure 4.

We divided the extracted data in two tables, taking into consideration studies that evaluated PLT functions/features and growth factor in PLTs during OP (n = 13) (Table 2) and studies that used PLT derivatives in OP (n = 10) (Table 3). Each table was split-up based on preclinical and clinical studies.

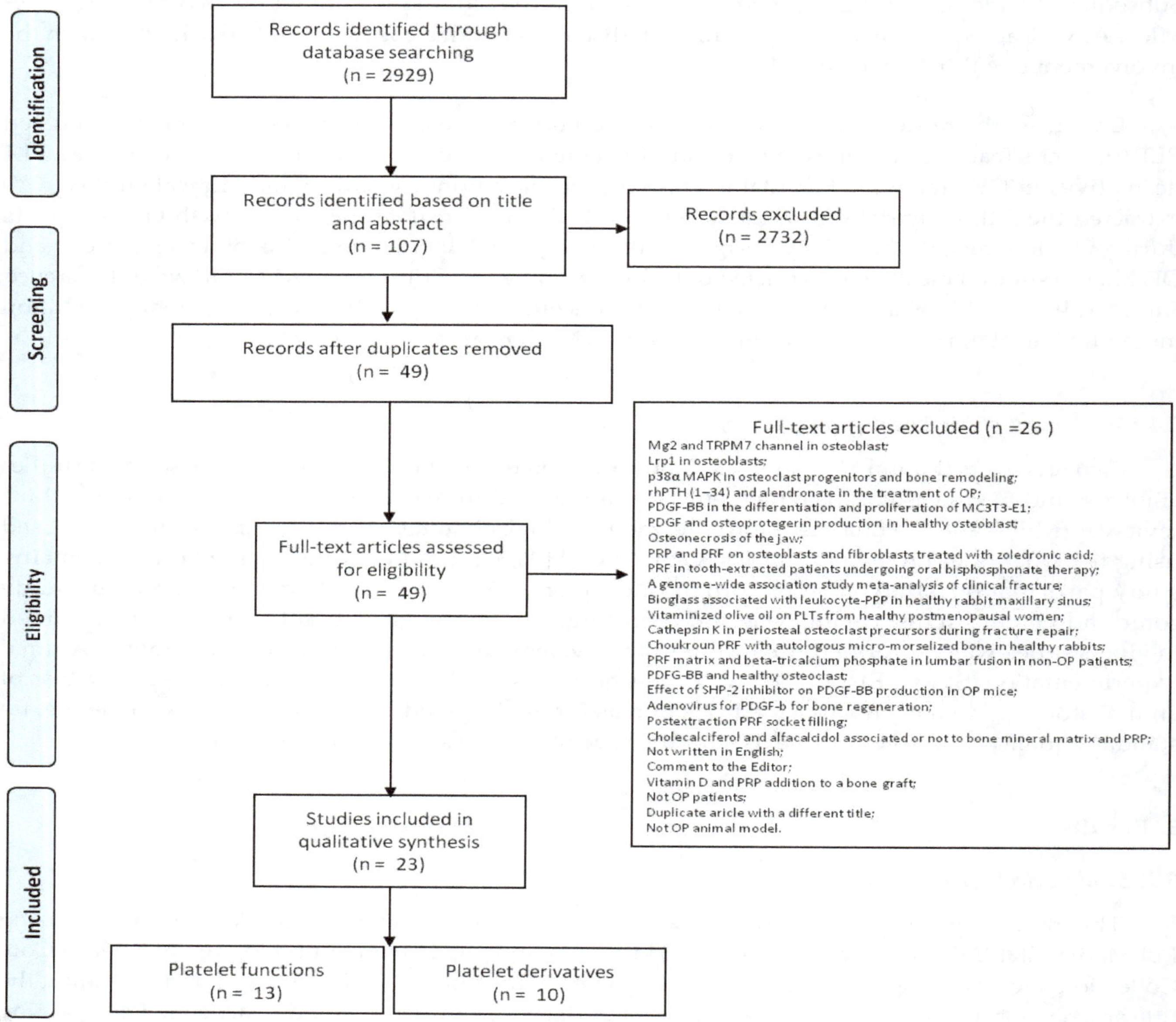

Figure 4. PRISMA flowchart for the selection of studies.

Table 2. Preclinical (in vitro and in vivo) and clinical studies on PLT functions/features and growth factor in PLTs during osteoporosis.

Reference	Aim	Study Design	Methodology	Platelet Function	Link Between Platelet and Osteoporosis	Main Results
In vitro studies						
Pountos et al. 2010 [39]	Effect of BMP-2, BMP-7, PTH, and PDGF on proliferation and osteogenic differentiation of OP MSCs	MSCs isolated from trabecular bone of 10 OP patients (4 male and 6 female) treated with a 10^6 range of concentrations (0.001 to100 ng/mL) of PDGF-BB	Functional assays of proliferation and osteogenic differentiation	PDGF-BB	PDGF-BB have a positive effect on osteogenic differentiation of OP MSCs	MSC proliferation stimulated by BMP-7 and PDGF-BB
In vivo studies						
Xie et al. 2014 [40]	Role of PDGF-BB in OVX mice	OVX C57BL/6 female mice injected with 1 µg PDGF-BB into the bone marrow cavity	Micro-CT, immunocytochemistry, immunofluorescence and histomorphometry.	PDGF-BB	Local PDGF-BB administration can temporally increase angiogenesis and spatially promote bone formation to couple angiogenesis with osteogenesis in bone modeling and remodeling	↑PDGF-BB concentrations, VEGF concentrations, vessel volume, CD31hi Emcnhi cells, proliferation of endothelial cells in metaphysis, trabecular bone volume, thickness and number, cortical bone thickness, serum osteocalcin concentration in OVX mice treated with PDGF-BB
Zhang et al. 2014 [41]	Effect on osteointegration of nanotube arrays loaded with rhPDGF-BB	OVX rat femur implantation: - oxalic acid-etched titanium rods - titanium rods modified with TiO_2 nanotube arrays - PDGF group (titanium rods immersed in 100 µg/mL rhPDGF-BB) - PDGF + Vacuum extraction (vacuum pump -PDGF + Vacuum group-for 10 min)	Static and dynamic histomorphometry and biomechanical test	PDGF-BB	Immobilization of rhPDGF-BB on nanotube arrays as implant surface modification strategy in orthopedic applications in osteoporotic patients	rhPDGF-BB immobilized on the nanotube surface ↑ new bone formation and osseointegration
Tang et al. 2017 [42]	Association between low plasma PDGF-BB levels and oestradiol	Sprague–Dawley rat: -Sham -OVX -OVX+oestradiol (100 mg/kg/d) -OVX+PDGF-BB (1mg/3 d/wk)	Plasma oestradiol and PDGF-BB levels measured using ELISA kits	PDGF-BB	Plasma PDGF-BB levels play a major role in OVX rats	↓PDGF-BB levels in OVX rats than SHAM group. Oestradiol replacement ↑plasma PDGF-BB levels, while PDGF-BB systematic treatment not affect plasma estradiol levels

Table 2. *Cont.*

Reference	Aim	Study Design	Methodology	Platelet Function	Link Between Platelet and Osteoporosis	Main Results
			Clinical studies			
Kim et al. 2011 [43]	Association between peripheral blood cell (PLT, WBC, RBC) counts BMD	Case-control study 17 OP patients 167 osteopenic patients 154 control subjects	DXA, biochemical parameters	PLTs count	Positive relationship between blood cell counts and BMD	WBC, RBC and PLT counts significantly associated with BMD
Li et al. 2012 [44]	Relationship between PLT count, MPV, and BMD	Case-control study 111 OP patients 171 osteopenic patients 128 control subjects	DXA, biochemical parameters	PLTs count and MPV	MPV negatively correlated with BMD	Negative correlation between MPV and the lumbar and femoral neck BMD. Univariate and multivariate analysis: MPV significantly associated with lumbar spine L2–L4 BMD and femoral neck BMD
D'Amelio et al. 2012 [45]	Correlation between PLTs vitamin D receptor expression and OP	Case-control study 77 postmenopausal OP patients 33 healthy control of childbearing age 49 healthy control men 11 healthy women matched with patients for age and postmenopausal period	DXA, markers of bone metabolism and vitamin D receptor levels	PLTs vitamin D receptor expression	Reduced level of PLT vitamin D receptor is correlate to OP	↓PLTs vitamin D receptor expression in OP patients respect to healthy postmenopausal controls. PLTs vitamin D receptor not influenced by gender. PLTs vitamin D receptor predict 65% of the BMD variation.
Akbal et al. 2014 [32]	Correlation between BMD and MPV and PDW	Case-control study 30 OP patients 30 osteopenic patients 20 control subjects	DXA, full laboratory test	MPV PDW	Significant role of PDW and MPV in the postmenopausal OP development	↓MPV and PDW in OP than the normal BMD patients. PDW positively correlated with FTT and L1–4T scores. Age and PDW independently related to FTT and LTT scores.
Kim et al. 2015 [34]	Association between plasma PAF, OP vertebral fracture and BMD	Case-control study 73 OP patients with vertebral fracture 73 OP patients without vertebral fracture	Radiography, DXA, biochemical parameters, plasma PAF concentration	PAF	Plasma PAF levels inversely correlated with BMD	34.6% ↑ plasma PAF levels in postmenopausal women with vertebral fracture than subjects without vertebral fracture

Table 2. *Cont.*

Reference	Aim	Study Design	Methodology	Platelet Function	Link Between Platelet and Osteoporosis	Main Results
Aypak et al. 2016 [46]	Correlation between BMD and MPV	Case-control study 126 OP patients 37 osteopenic patients 12 control subjects	DXA, laboratory tests including complete blood count (CBC), calcium, phosphorus, serum 25 hydroxyvitamin D (25OHD), and intact parathormone (iPTH)	MPV	MPV correlated with BMD in postmenopausal OP women.	MPV significantly associated with BMD in normal weight and overweight-obese OP patients.
Tang et al. 2017 [42]	Association between low plasma PDGF-BB levels and oestradiol in postmenopausal OP	Case-control study 28 postmenopausal OP patients 69 control young woman 24 age-matched women	DXA, plasma oestradiol and PDGF-BB levels	PDGF-BB	Plasma PDGF-BB levels maintained by oestrogen in normal young women and play a major role in postmenopausal OP	↓plasma oestradiol and PDGF-BB levels in postmenopausal women, especially in OP patients. PDGF-BB levels were positively correlated with oestradiol levels and inversely correlated with age
Vural et al. 2017 [47]	Correlation between PLT functions, vitamin D and BMD	Case-control study 124 OP patients 151 osteopenic patients 87 control subjects	DXA, biochemical parameters	MPV	No correlation between MPV and OP. MPV considered a less important indicator in serum 25-hydroxyvitamin D levels and OP	No difference in MPV and PLT counts between groups. No correlation between MPV and serum 25-hydroxyvitamin D levels. Correlation between PLT count and lumbar spine (L1–4) T score
Koseoglu et al. 2017 [48]	Correlation between PLT/lymphocyte *ratio* and low BMD in postmenopausal woman	Case-control study 179 OP and osteopenic patients 32 control subjects	DXA, biochemical parameters	PLT/lymphocyte *ratio*	PLT/lymphocyte *ratio* as new inflammatory marker for bone loss and low BMD	↑PLT/lymphocyte *ratio* in OP and osteopenic patients than in the control subjects. Negative correlation between lumbar and femur neck BMD and PLT/lymphocyte *ratio*
Eroglu et al. 2019 [33]	Correlation between PLT/lymphocyte and BMD	Case-control study 48 OP patients 112 osteopenic patients 92 control subjects	DXA, biochemical parameters	PLT/lymphocyte *ratio*	Negative correlation Between PLT/lymphocyte and BMD	↑PLT/lymphocyte *ratio* in OP and osteopenic patients.

MPV: mean platelet volume; PDW platelet distribution width; FTT: femur total T; L1–4T: lumbar 1–4T; PAF: platelet-activating factor; WBC: peripheral blood white blood cell RBC: red blood cell.

Table 3. Preclinical studies on PLT derivatives.

Reference	Aim	Study Type	Platelet Formulation	Platelets Concentration	White Blood Cells Content	Activation Method	Platelets Application	Experimental Design	Main Results
Lo et al. 2009 [49]	Transplantation of PRP/NIH3T3-G cells to induced bone regeneration in OP	in vitro and in vivo	Human PRP	NS	NS	Exogenous-bovine thrombin	NIH3T3-G alone, BMCs alone, and NIH3T3 G/BMC co-culture Bone marrow cavity of the tibia	OVX-SAMP8 mice treated with PRP/NIH3T3-G	PRP/NIH3T3-G treatment prevent OP development
Liu et al. 2011 [50]	Balance between adipogenesis and osteogenesis in bone regeneration by PRP for age-related OP	in vitro and in vivo	Human PRP	NS	NS	Exogenous-bovine thrombin	Mouse pre-adipocytes (3T3-L1) and osteoblast cell line (7F2) co-culture Bone marrow cavity of the hind femur	OVX-SAMP8 mice treated with PRP	PRP treatment exert its action promoting bone regeneration and suppressing adipogenesis within the marrow
Clafshenkel et al. 2012 [51]	Incorporation of melatonin and/or PRP into CA scaffolds to enhance bone regeneration in OP	in vivo	OVX rat PRP	NS	NS	NS	Calvaria critical-sized defect	OVX rats treated with CA scaffold with PRP and melatonin, associated or not	PRP not improves bone formation
Chen et al. 2013 [52]	PRP to promote healing of OP fractures	in vitro and in vivo	OVX rat PRP	High: $8.21 \pm 0.4 \times 10^9$ Medium:$2.65 \pm 0.2 \times 10^9$ Low: $0.85 \pm 0.16 \times 10^9$ PPP:$8 \pm 0.5 \times 10^6$ (PLTs/mL)	NS	Exogenous-thrombin/ $CaCl_2$	BMSCs culture Femoral fracture	OVX rats treated with high-, medium- and low-concentration PRP and with PPP	Medium-concentration of PRP is the more suitable in promoting fracture healing
Cho et al. 2014[53]	Incorporation of PRP into CPC to enhance bone regeneration in OP	in vivo	OVX rat PRP	4.12×10^9 (PLTs/mL)	NS	NS	Vertebral body critical-size defects	OVX rats treated with CPC associated to PRP	PRP accelerates osteoconduction and improves trabecular bone microarchitecture and BMD
Jiang et al. 2016 [54]	PRP treatment and TiO_2 nanoporous modification on the stability of titanium implants in OP	in vivo	Human PRP	2×10^9 (PLTs/mL)	NS	Exogenous-calcium enriched batroxobin	Bone marrow cavity of the hind tibia	OVX rats treated with TiO_2 associated to PRP	PRP treatment improves implant biomechanical stability

Table 3. *Cont.*

Reference	Aim	Study Type	Platelet Formulation	Platelets Concentration	White Blood Cells Content	Activation Method	Platelets Application	Experimental Design	Main Results
Wei et al. 2016 [55]	PRP in combination with BMSCs for the treatment of OP defect	in vivo	Rats PRP	NS	NS	Exogenous-thrombin	Tibia critical size defects defect	OVX rats treated with allogenic BMSC associated to PRP	PRP combined with BMSCs promotes bone defects healing
Rocha et al. 2017 [56]	PRP and MSCs, associated or not, in the repair of bone failures in secondary OP	in vivo	Equine PRP	200×10^3 (PLTs/μL)	NS	NS	Tibia failures	Rabbits submitted to ovariosalpingohysterectomy and hypercortisolism treated with allogeneic BMSCs and PRP, associated or not	PRP contributes positively to the repair of bone failure, but less than the treatment with MSCs and similarly to the association of both
Sakata et al. 2018 [57]	Bone regeneration of OP defects by PRP and β-TCP	in vivo	Rats PRP	NS	NS	NS	Vertebral body critical-size defects	OVX rats treated with β-TCP associated to PRP	PRP associated to β-TCP sponge facilitates bone regeneration in OVX lumbar vertebral bone defect
Engler-Pinto et al. 2019 [58]	L-PRF associated or not with bovine bone graft on the healing of OP bone defects	in vivo	Rats L-PRF	NS	NS	NS	Calvaria critical size defects	OVX rats treated with bovine bone graft associated to PRP	L-PRF clot improves bone formation but less than the use of L-PRF associated to bovine bone graft

NS: not specified; PRP/NIH3T3-G: NIH3T3-G pre-differentiated into osteoblast-like cells using PRP; OVX-SAMP8: ovariectomized senescence-accelerated mice; CA: calcium aluminate; CPC: calcium phosphate cement; β-TCP: β-tricalciumphosphate.

3.2. Assessment of Methodological Quality

Risks of bias assessments for each clinical study were summarized in Table 4. The overall risk of bias across studies was low to moderate for the majority of the studies (n = 8; 75%), with the exception of two studies that have high risk due the lack of information [33,45], for at least one aspect of study attrition [45], prognostic factor measurement [33] and outcome measurement [45].

Risks of bias assessments for each in vivo study were summarized in Table 5. Risk of bias of animal studies was high for almost all the examined studies. Among the 13 included in vivo studies, three for PLT functions during OP and 10 for PLT derivatives in OP, 10 of them have not declared the method of sequence generation [41,42,49–53,56,57], in one study the method was unclear [55] and in the remaining two studies the method of sequence generation was clearly declared [40,58]. The majority (n = 7) of the studies showed that groups were similar concerning baseline characteristics (i.e., age, weight, sex) and two studies showed that allocation was adequately concealed [40,58]. One study reported that animals were housed randomly during the experiment [40] and another reported the blinding of investigators [52]. Only one study reported that the animals were selected at random for outcome assessment [41] and another one reported the blinding of outcome assessors [52]. Almost all the studies included all the animals in the analyses (n = 10), reported and detailed the primary outcome (n = 12) and were apparently free of other problems that could result in high risk of bias (n = 8) (Table 5).

Table 4. QUIPS tool for assessing risk of bias in the clinical studies.

Study	QUIPS					
	Study Participation	Study Attrition	Prognostic Factor Measurement	Outcome Measurement	Confounding Measurement and Account	Analysis
PLT functions/features and growth factor in PLTs during osteoporosis						
Kim et al.2011 [43]	Low	Low	Moderate	Moderate	Moderate	Low
Li et al. 2012 [44]	Low	Low	Low	Moderate	Low	Low
D'Amelio et al. 2012 [45]	Moderate	High	Moderate	High	Moderate	Moderate
Akbal et al. 2014 [32]	Low	Low	Low	Moderate	Low	Low
Kim et al. 2015 [34]	Low	Low	Low	Low	Low	Low
Aypak et al. 2016 [46]	Low	Low	Low	Moderate	Moderate	Moderate
Tang et al. 2017 [42]	Low	Moderate	Moderate	Moderate	Moderate	Moderate
Vural et al. 2017 [47]	Low	Low	Low	Moderate	Low	Moderate
Koseoglu et al. 2017 [48]	Low	Low	Low	Moderate	Moderate	Low
Eroglu et al. 2019 [33]	Low	Moderate	High	Moderate	Moderate	Low

low (good) indicator, moderate indicator, high (bad) indicator.

Table 5. SYRCLE's tool for assessing risk of bias in the in vivo studies.

Study	SYRCLE									
	Selection Bias			Performance Bias		Detection Bias		Attrition Bias	Reporting Bias	Other
	Sequence Generation	Baseline Characteristics	Allocation Concealment	Random Housing	Blinding	Random Outcome Assessment	Blinding	Incomplete Outcome Data	Selective Outcome Reporting	Other Sources of Bias
PLT functions/features and growth factor in PLTs during osteoporosis										
Xie et al. 2014 [40]	Yes	Yes	Yes	Yes	No	Unclear	No	No	Yes	No
Zhang et al. 2014 [41]	No	Yes	Unclear	Unclear	Unclear	Yes	No	No	Yes	No
Tang et al. 2017 [42]	No	Unclear	Unclear	Unclear	Unclear	Unclear	Unclear	No	Yes	Yes
PLT derivatives in osteoporosis										
Lo et al. 2009 [49]	No	Unclear	No	Unclear	Unclear	No	No	No	Yes	No
Liu et al. 2011 [50]	No	No	No	No	No	No	No	No	Unclear	No
Clafshenkel et al. [51]	No	Yes	No	No	No	No	No	Unclear	Yes	Unclear
Chen et al. 2013 [52]	No	Unclear	No	No	Yes	No	Yes	No	Yes	No
Cho et al. 2014 [53]	No	No	No	No	No	No	Unclear	Yes	Yes	Yes
Jiang et al. 2016 [54]	No	Yes	No	No	No	No	No	No	Yes	Unclear
Wei et al. 2016 [55]	Unclear	Yes	Unclear	No	No	No	No	No	Yes	No
Rocha et al. 2017 [56]	No	No	No	No	No	No	No	Unclear	Yes	Yes
Sakata et al. 2018 [57]	No	Yes	Unclear	No	No	No	No	No	Yes	No
Engler-Pinto et al. 2019 [58]	Yes	Yes	Yes	No	No	No	Unclear	No	Yes	No

positive (good) indicator, unclear, negative (bad) indicator.

4. Study Results

4.1. Platelet Functions/Features and Growth Factor in PLTs in Osteoporosis

Articles that evaluated PLT functions/features and growth factor in PLTs during OP were prevalently clinical ($n = 9$), two were in vivo studies, one was both in vivo and clinical and another was exclusively in vitro (Table 2). In this last [39], starting from the evidence of the reduced healing capability of MSCs during OP, a comparison between the effect of BMP-2, BMP-7, parathyroid hormone (PTH) and PDGF on proliferation and osteogenic differentiation of MSCs derived from OP patients was performed. MSCs isolated from trabecular bone showed to be more sensitive to high dose of PDGF-BB, as well of BMP-7 in comparison to the other stimulations, in terms of alkaline phosphatase (ALP) activity and calcium release, in a dose dependent manner. The critical role of the PDGF-BB was also investigated in two in vivo studies [40,41]. Zhang et al. showed that the immobilization of PDGF-BB on titanium nanotube arrays was effective in stimulating osteogenesis both in vitro, on BMSCs isolated from OVX rats, and in an in vivo model of osteointegration in OVX rat femurs [41]. Exploiting the use of a Ctsk$^{-/-}$ mouse model, in which higher levels of PDGF-BB have been found to be secreted by pre-osteoclasts, Xie et al. [40] investigated instead the bone remodeling rate in relationship with angiogenesis stimulation. Results showed that PDGF-BB promotes angiogenesis, recruiting MSCs and endothelial progenitor cells (EPCs) and stimulating CD31hiEmcnhi vessel and bone formation in OVX mice [40]. Additionally, in vitro tube formation assays performed with conditioned medium from pre-osteoclasts and osteoclasts isolated after bone marrow flushing confirmed the results. Always focusing on the PDGF-BB function Tang et al. [42], in an in vivo study on OVX rats and in a prospective clinical study involving young woman, postmenopausal and OP postmenopausal woman, showed that the levels of oestradiol and PDGF-BB correlate with patients age and that the lowest levels are found in the postmenopausal OP cohort.

Several clinical studies evaluated the relationship between PLT and OP status based on Bone Mineral Density (BMD) value considering different cohorts of woman, i.e., healthy, osteopenic and OP [44,46,47]. The investigation on mean PLT volume (MPV) [46,47] and PLT distribution width (PDW) [46] showed that the levels of these markers were reduced in OP patients and that they correlated with BMD T-score. In addition, in a bigger clinical study (175 patients, 72% osteoporotic) it was also seen that MPV inversely correlate with body mass index [46]. On the contrary, performing the same evaluation on the same cohorts of patients, Vural et al. found no differences in MPV and PLT, neither founded any possible relationship between vitamin D levels and MPV [47]. However, D'Amelio et al. [45] evaluating woman with postmenopausal OP and using as control healthy man and woman matched for age and postmenopausal period, showed that PLT vitamin D receptor was less expressed in OP patients. Additionally PLT vitamin D receptor level can be related to the variation of BMD independently form the health status of patients. The lower expression of the receptor also induced a worst response to vitamin D and a consequent increase in PTH levels [45]. In addition, evaluating postmenopausal OP patients, the PLT/lymphocyte ratio was found to correlate with low BMD [33,48], in particular in reference to the femoral and lumbar district [48], which is also related to low vitamin D levels, supporting the hypothesis that inflammation correlates with vitamin D levels [48]. The relationship between peripheral blood cell count and BMD in OP was investigated also by Kim et al. which observed that PLTs count, as well as white and red blood cells counts, correlate with BMD in OP patients [43]. Finally, Kim et al. evaluated the level of plasma PLT activating factor (PAF) in OP woman with radiological evident vertebral fracture. After assessing BMD and serum calcium levels, results showed that PAF levels correlated with the presence of vertebral fracture, as well as with BMD in all sites except for femoral neck, and that PAF levels increase in parallel with ALP levels [34].

4.2. Platelet Derivatives in Osteoporosis

All 10 articles on PLT derivatives in OP were about vivo or both in vivo and in vitro studies. With the exception of the study by Rocha et al. [56] that used rabbits submitted to elective ovariosalpingohysterectomy as OP animal model all the other studies employed ovariectomized (OVX) mice and rats to induce OP (Table 3). All studies used PRP as PLT products with the exception of a study which used L-PRF [58]. Six of these studies (60%) also employed a scaffold/biomaterial (calcium phosphate, calcium aluminate/calcium aluminate-melatonin, β-tricalcium phosphate, bovine bone graft, nanoporous TiO_2, hemostatic sponge) to incorporate the PLTs derivatives and subsequently evaluated bone regeneration in calvaria [51,58], tibia [54,56] and in caudal and lumbar vertebrae [53,57]. Clafshenkel et al. [51] evaluating a calcium aluminate and calcium aluminate-melatonin scaffolds implanted in a critical size calvaria defect of OVX rats, showed that the addition of PRP did not significantly improve degree, intensity and abundance of osteoid tissue mineralization and bone formation in either of the two scaffolds [51]. Differently, Engler-Pinto et al. [58] using the same animal model to evaluate L-PRF alone, blood clot alone, bovine bone graft alone, or a combination of L-PRF with bovine bone graft, showed that the association of L-PRF to bovine bone graft potentiate the bone healing and the production of VEGF, osteocalcin (OCN) and BMP-2/4 [58]. Increased osteogenetic efficiency were also observed evaluating the healing of a long bone defect (tibia) in OVX animals treated with TiO_2 nanoporous implant associated with PRP [54]. Additionally, it was seen that PRP in association with TiO_2 not only promoted the osteogenesis but also increased the expression of RUNX2 and COL1 genes and suppressed osteoclastogenesis with increased expression of OPG and decreased levels of RANKL [54]. Despite calvaria defects and long bones defects are the most frequently models used to evaluate bone regeneration and healing during OP, also vertebrae, despite the higher cancellous bone content and the different anatomical and biomechanical properties, were used to evaluate the role of scaffold/biomaterial in association to PRP [53,57]. Two in vivo studies in OVX animals showed that incorporating PRP into calcium phosphate cement [53] and β-TCP sponge [57] accelerated osteoconduction in the caudal [53] and lumbar [57] site, also demonstrating an improvement of the trabecular bone microarchitecture. Additionally, calcium phosphate cements associated to PRP improve the bone mineral density [53] and increase the stiffness of the affected vertebral bodies [57]. Finally, Rocha et al. [56] used a hydrolyzed collagen sponge made from freeze-dried sterile porcine gelatine as carrier for allogenic BMSCs and PRP, alone or in combination, to evaluate the repair of bone failure in tibiae of osteoporotic rabbits secondary to estrogenic deprivation and iatrogenic hypercortisolism [56]. Results suggested that PRP contributed positively to repair of bone failure, but less than the group treated with BMSC and similarly to the association of both [56].

Differently from the above mentioned studies, four studies (40%) used PRP without scaffold/biomaterial to analyze the balance between adipogenesis and osteogenesis in bone regeneration [50], to induce bone regeneration from embryonic fibroblasts [49], to evaluate PRP association to BMSCs [55], to analyze PRP effect in the treatment of OP fractures and to clarify PRP best concentration of use [52]. By using an OVX senescence-accelerated mice (SAMP8) model in which genetically modified NIH3T3 embryonic fibroblasts (pre-differentiated into osteoblast-like cells using PRP) were injected into the bone marrow cavity, an improvement in BMD scores and in the skeletal bone architecture were detected [49]. Using the same animal model it was also seen that PRP alone exerted its action by promoting bone regeneration and suppressing adipogenesis within the marrow [50]. PRP-induced osteogenesis was confirmed by simultaneously up-regulating osteogenesis-promoting genes RUNX2, OPN and OCN and down-regulating adipogenesis regulators such as PPAR-γ2 and leptin in bone marrow cells of PRP treated animals [50]. Allogenic BMSCs in combination with PRP were also used for the treatment of OP bone defects in an OVX rat model showing that bone defects of OVX rats treated with PRP and BMSCs were completely repaired, whereas those treated with PRP or BMSCs alone exhibited slower healing [55]. In addition, higher expression levels of RUNX2, OSX, and OPN were found in rats treated with PRP and BMSCs [55]. Autologous BMSCs cultured with high-, medium-, low-concentration PRP and with PPP from OVX rats were also used to treat fracture healing in an OVX animal model [52]. Results highlighted that the medium-concentration of PRP showed faster healing than the other groups, with a faster bridging of the fracture gaps and higher bridging rate [52].

5. Discussion

To date OP management still remains a difficult task for clinicians and based on the progressively increase of aging population the global implications of OP and impaired bone healing are considerable. In the past decade, an increasing number of studies explored the use of new and advanced markers as well as of bioactive factors able to promote bone formation/regeneration during OP [59,60]. Although promising results have been documented both for OP markers and bioactive factors, the available evidence does not yet support their use and further investigation for their clinical use, in particular for PLTs and their derivatives, are mandatory [61]. Thus, the aim of the present systematic review was to evaluate the PLT function, i.e., size, volume, bioactive GFs released, and the usage of their derivatives during OP in order to understand the potential of PLT function as OP markers and the physiopathological mechanisms that underlie the regenerative effects of their derivatives.

In this review, preclinical studies on PLTs function/features and growth factor in PLTs during OP mainly deal with PDGF, a naturally molecule released from the α-granules of PLTs, as part of the clotting process that occurs in response to injury [62]. Homodimer BB constitutes a dimeric glycoprotein of PDGF and is considered the universal PDGF isoform, rendering it the most logical form of the protein to develop as a therapeutic [62]. The ability to simultaneously influence cellular chemotaxis, mitogenesis and angiogenesis gives to PDGF a fundamental role in musculoskeletal repair and regeneration also in OP condition [63]. In this review, PDGF-BB administered in vitro in OP MSCs and in vivo in OVX small animal models respectively stimulated osteogenesis, proliferation and improve angiogenesis and implant osteointegration. In addition, it was found that in OVX animal model, bone marrow levels of PDGF-BB, which was partially produced by preosteoclasts, were drastically decreased. This probably occurred because an increase in mature bone resorption by osteoclasts reduced preosteoclasts and consequently PDGF-BB secretion in OVX animals. Therefore, PDGF-BB is likely mediated by oestrogen in bone metabolism. In fact, it was demonstrated that plasma PDGF-BB levels are maintained by oestrogen in normal young women and play a major role in postmenopausal OP [42]. However, despite numerous studies suggested PDGF-BB as potential therapeutic target during OP, before moving toward the next step, further studies will ascertain the exact mechanisms of PDGF-BB on increasing new bone formation and improving angiogenesis in OP conditions. In addition, clinical studies for dose, delivery site and mode optimization will be mandatory in order to examine the side effects, overall safety and effectiveness of PDGF-BB. In this review the examined clinical studies also found a positive correlation between PLT size, distribution width, volume changes and low BMD due to OP with also a correlation with low levels of PLT vitamin D receptor that underlined a lower ability to respond to vitamin D in OP condition. On the other hand, some other studies demonstrated that PLT size cannot be used as a predictive marker of vitamin D status and BMD during OP. However, as reported by Varol et al. [64] accurate measurements of PLT count and volume are fundamental factors for diagnostic, therapeutic, and research purposes, thus to avoid artefactual results. Unfortunately, not all the studies analyzed detail the procedures used to obtain PLT count and volume and not all used a standardized procedure. An additional mechanism always available with routine blood counts, which explains the relation between PLTs and OP, was found in the correlation between PLT/lymphocyte ratio and OP, since PLT/lymphocyte ratio seems to be a discriminative factor for low BMD. The possibility to exploit data related to PLT size, distribution width, volume changes and PLT/lymphocyte ratio, obtained from a simple and routine investigation, to diagnose and correlate a specific pathological condition is undoubtedly fascinating. However, despite in this review the clinical studies suggested that these parameters may be used as potential OP predictors, a consensus has not been reached and there are still limited results. To date, these aspects strongly restrict the clinical translation and further studies, including larger patient groups, are mandatory and could allow identifying a subset of patients who are at greater risk for developing OP and who may benefit from early screening, intervention, and additional research.

Another key question concerning PLTs regards the role and use of their derivatives during OP. The rationale for PLT derivatives use in bone healing process is due to the abundance and accessibility of key GFs and other signaling molecules in PLTs [65–67]. To date, PLT derivatives have been used for the improvement of bone fracture healing, such as common fracture healing, diabetic fracture healing, and nonunion [16,68]. Although not yet definite, it appears that most research supports a positive role for PLT derivatives in bone regeneration. However, the majority of these studies were done in non-OP condition and consequently it is not clear whether the impact of PLT derivatives would be compromised by OP. Thus, in the present review we searched preclinical and clinical studies on PLT derivatives employed in OP condition. However, our search strategy provided only preclinical studies and this is probably due to the fact that the use PLTs derivatives in OP still requires a better understanding of the physiopathological mechanisms that underlie their real regenerative effects. All preclinical studies examined in this review used PRP as PLT products with the exception of one study which used L-PRF and most of them used a scaffold/biomaterial to incorporate the PLTs derivatives. Most of the examined studies demonstrated that PRP improve overall bone quality in OP animal models by promoting osteogenesis while suppressing adipogenesis in bone marrow. Moreover, PRP seems to stimulate the differentiation of embryonic fibroblasts into osteoblast-like cells; the transplantation of these PRP-treated cells also significantly improved bone architecture in OP animal models. It has also been demonstrated that PRP treatment combined with BMSCs may enhance the formation of new bone. However, a minority of studies (2/10) reported that the use of PRP associated to a scaffold and/or to BMSCs did not improve degree, intensity, mineralization and bone formation. Thus, in spite of numerous experimental evidences showed in this review, the use of PLT derivatives during OP is still subject of controversy also considering the high risk of bias of most animal studies. Several explanations for this dispute could be first of all due to the different interval between implantation and investigation as well as to the volume of whole blood and final volume of PLT derivative, final PLT and GFs concentration, methods that PLT derivatives is produced, activator agents, presence or absence of leukocytes and red blood cells, the origin of platetelet derivative used (autologous, allogeneic or xenogeneic). Additionally, specific factors associated to the surgical approach, i.e., size of the bone defect, type and nature of the bone implant, bone graft substitute and bone fixation device could also affect the efficacy. Thus, despite the use of PLT derivatives increased in the past years mainly due to the easy use and biosafety that facilitates the translation in humans, to date further research should be performed to fully reveal the characteristics of the relationship between PLT derivatives and OP. These researches would be of fundamental importance as they would allow a rapid clinical translation of the PLT derivatives in the clinical theatre, leading to an improvement in the patient's quality of life and a reduction in the ever-increasing financial burden for governments and society due to OP.

6. Conclusions

Given the fundamental role of PLT features (size, volume, width distribution, GFs released and growth factor that exists in the PLTs) and PLT derivatives in musculoskeletal repair and regeneration, their future role in OP is expected to expand. Additional researches are under way to further improve our understanding on PLT as markers for OP and on PLT derivatives as therapeutic treatment to enhance bone healing and control inflammation during OP. These future investigations will hopefully continue to shed more light on how PLTs could best used to further improve the outcomes of OP patients in the clinical *scenario*.

Author Contributions: F.S. and M.F. designed the review. F.S. and M.M. performed the literature search. F.S. and M.M. analyzed the obtained articles. F.S., M.M. and M.F. wrote the paper. F.S., M.M. and M.S. collected and assembled the data. M.T. revised the manuscript critically. F.S., M.F., M.M. and M.S. finally approved the article. All authors have read and agreed to the published version of the manuscript.

Acknowledgments: This work was supported by grants from IRCCS Istituto Ortopedico Rizzoli (Ricerca Corrente) and by the 5 × 1000 Project, year 2017. The authors gratefully acknowledge Silvia Bassini for the support in the realization of the figures of this review.

References

1. Curtis, E.M.; Harvey, N.C.; Cooper, C. The burden of osteoporosis. In *Osteoporosis: A Life Course Epidemiology Approach to Skeletal Health*; Harvey, N.C., Cooper, C., Eds.; CRC Press: Boca Raton, FL, USA, 2018; pp. 1–20.
2. Johnell, O.; Kanis, J.A. An estimate of the worldwide prevalence and disability associated with osteoporotic fractures. *Osteoporos. Int.* **2006**, *17*, 1726–1733. [CrossRef] [PubMed]
3. Kanis, J.A.; Cooper, C.; Rizzoli, R.; Reginster, J.Y.; Scientific Advisory Board of the European Society for Clinical and Economic Aspects of Osteoporosis and Osteoarthritis (ESCEO) and the Committees of Scientific Advisors and National Societies of the International Osteoporosis Foundation (IOF). Executive summary of European guidance for the diagnosis and management of osteoporosis in postmenopausal women. *Aging Clin. Exp. Res.* **2019**, *31*, 15–17. [CrossRef] [PubMed]
4. Al Anouti, F.; Taha, Z.; Shamim, S.; Khalaf, K.; Al Kaabi, L.; Alsafar, H. An insight into the paradigms of osteoporosis: From genetics to biomechanics. *Bone Rep.* **2019**, *11*, 100216. [CrossRef] [PubMed]
5. Black, D.M.; Rosen, C.J. Postmenopausal Osteoporosis. *N. Engl. J. Med.* **2016**, *374*, 2096–2097. [CrossRef] [PubMed]
6. Bord, S.; Frith, E.; Ireland, D.C.; Scott, M.A.; Craig, J.I.; Compston, J.E. Megakaryocytes modulate osteoblast synthesis of type-l collagen, osteoprotegerin, and RANKL. *Bone* **2005**, *36*, 812–819. [CrossRef]
7. Ciovacco, W.A.; Goldberg, C.G.; Taylor, A.F.; Lemieux, J.M.; Horowitz, M.C.; Donahue, H.J.; Kacena, M.A. The role of gap junctions in megakaryocyte-mediated osteoblast proliferation and differentiation. *Bone* **2009**, *44*, 80–86. [CrossRef]
8. Kacena, M.A.; Nelson, T.; Clough, M.E.; Lee, S.K.; Lorenzo, J.A.; Gundberg, C.M.; Horowitz, M.C. Megakaryocyte-mediated inhibition of osteoclast development. *Bone* **2006**, *39*, 991–999. [CrossRef]
9. Maynard, D.M.; Heijnen, H.F.; Horne, M.K.; White, J.G.; Gahl, W.A. Proteomic analysis of platelet alpha-granules using mass spectrometry. *J. Thromb. Haemost.* **2007**, *5*, 1945–1955. [CrossRef]
10. Singh, A.; Ali, S.; Srivastava, R.N. Platelet-rich plasma in osteoporotic fractures: A review of literature. *J. Orthop. Traumatol. Rehabil.* **2014**, *7*, 123–138. [CrossRef]
11. Lieberman, J.R.; Daluiski, A.; Einhorn, T.A. The role of growth factors in the repair of bone. Biology and clinical applications. *J. Bone Jt. Surg. Am.* **2002**, *84*, 1032–1044. [CrossRef]
12. Tabata, Y. Tissue regeneration based on growth factor release. *Tissue Eng.* **2003**, *9* (Suppl. 1), 5–15. [CrossRef]
13. Sharif, P.S.; Abdollahi, M. The role of platelets in bone remodeling. *Inflamm. Allergy Drug Targets* **2010**, *9*, 393–399. [CrossRef] [PubMed]
14. Khan, F.A.; Parayaruthottam, P.; Roshan, G.; Menon, V.; Fidha, M.; Fernandes, A.K. Platelets and Their Pathways in Dentistry: Systematic Review. *J. Int. Soc. Prev. Community Dent.* **2017**, *7* (Suppl. 2), 55–60. [CrossRef] [PubMed]
15. Lang, S.; Loibl, M.; Herrmann, M. Platelet-Rich Plasma in Tissue Engineering: Hype and Hope. *Eur. Surg. Res.* **2018**, *59*, 265–275. [CrossRef]
16. Oryan, A.; Alidadi, S.; Moshiri, A. Platelet-rich plasma for bone healing and regeneration. *Expert Opin. Biol. Ther.* **2016**, *16*, 213–232. [CrossRef] [PubMed]
17. Salamanna, F.; Veronesi, F.; Maglio, M.; Della Bella, E.; Sartori, M.; Fini, M. New and emerging strategies in platelet-rich plasma application in musculoskeletal regenerative procedures: General overview on still open questions and outlook. *Biomed Res. Int.* **2015**, *2015*, 846045. [CrossRef]
18. Dohan Ehrenfest, D.M.; Rasmusson, L.; Albrektsson, T. Classification of platelet concentrates: From pure platelet-rich plasma (P-PRP) to leucocyte- and platelet-rich fibrin (L-PRF). *Trends Biotechnol.* **2009**, *27*, 158–167. [CrossRef]
19. Camargo, P.M.; Lekovic, V.; Weinlaender, M.; Vasilic, N.; Madzarevic, M.; Kenney, E.B. Platelet-rich plasma and bovine porous bone mineral combined with guided tissue regeneration in the treatment of intrabony defects in humans. *J. Periodontal. Res.* **2002**, *37*, 300–306. [CrossRef]
20. Uggeri, J.; Belletti, S.; Guizzardi, S.; Poli, T.; Cantarelli, S.; Scandroglio, R.; Gatti, R. Dose-dependent effects of platelet gel releasate on activities of human osteoblasts. *J. Periodontol.* **2007**, *78*, 1985–1991. [CrossRef]

21. Man, Y.; Wang, P.; Guo, Y.; Xiang, L.; Yang, Y.; Qu, Y.; Gong, P.; Deng, L. Angiogenic and osteogenic potential of platelet-rich plasma and adipose-derived stem cell laden alginate microspheres. *Biomaterials* **2012**, *33*, 8802–8811. [CrossRef]
22. Parsons, P.; Butcher, A.; Hesselden, K.; Ellis, K.; Maughan, J.; Milner, R.; Scott, M.; Alley, C.; Watson, J.T.; Horner, A. Platelet-rich concentrate supports human mesenchymal stem cell proliferation, bone morphogenetic protein-2 messenger RNA expression, alkaline phosphatase activity, and bone formation in vitro: A mode of action to enhance bone repair. *J. Orthop. Trauma* **2008**, *22*, 595–604. [CrossRef] [PubMed]
23. Mariano, R.; Messora, M.; de Morais, A.; Nagata, M.; Furlaneto, F.; Avelino, C.; Paula, F.; Ferreira, S.; Pinheiro, M.; de Sene, J.P. Bone healing in critical-size defects treated with platelet-rich plasma: A histologic and histometric study in the calvaria of diabetic rat. *Oral. Surg. Oral. Med. Oral. Pathol. Oral. Radiol. Endod.* **2010**, *109*, 72–78. [CrossRef] [PubMed]
24. Simman, R.; Hoffmann, A.; Bohinc, R.J.; Peterson, W.C.; Russ, A.J. Role of platelet-rich plasma in acceleration of bone fracture healing. *Ann. Plast. Surg.* **2008**, *61*, 337–344. [CrossRef] [PubMed]
25. Torres, J.; Tamimi, F.M.; Tresguerres, I.F.; Alkhraisat, M.H.; Khraisat, A.; Lopez-Cabarco, E.; Blanco, L. Effect of solely applied platelet-rich plasma on osseous regeneration compared to Bio-Oss: A morphometric and densitometric study on rabbit calvaria. *Clin. Implant. Dent. Relat. Res.* **2008**, *10*, 106–112. [CrossRef] [PubMed]
26. Giovanini, A.F.; Gonzaga, C.C.; Zielak, J.C.; Deliberador, T.M.; Kuczera, J.; Göringher, I.; de Oliveira Filho, M.A.; Baratto-Filho, F.; Urban, C.A. Platelet-rich plasma (PRP) impairs the craniofacial bone repair associated with its elevated TGF-β levels and modulates the co-expression between collagen III and α-smooth muscle actin. *J. Orthop. Res.* **2011**, *29*, 457–463. [CrossRef] [PubMed]
27. Cheng, X.; Lei, D.; Mao, T.; Yang, S.; Chen, F.; Wu, W. Repair of critical bone defects with injectable platelet rich plasma/bone marrow-derived stromal cells composite: Experimental study in rabbits. *Ulus. Travma. Acil. Cerrahi. Derg.* **2008**, *14*, 87–95.
28. Célio-Mariano, R.; de Melo, W.M.; Carneiro-Avelino, C. Comparative radiographic evaluation of alveolar bone healing associated with autologous platelet-rich plasma after impacted mandibular third molar surgery. *J. Oral. Maxillofac. Surg.* **2012**, *70*, 19–24. [CrossRef]
29. Poeschl, P.W.; Ziya-Ghazvini, F.; Schicho, K.; Buchta, C.; Moser, D.; Seemann, R.; Ewers, R.; Schopper, C. Application of platelet-rich plasma for enhanced bone regeneration in grafted sinus. *J. Oral. Maxillofac. Surg.* **2012**, *70*, 657–664. [CrossRef]
30. Pappalardo, S.; Guarnieri, R. Efficacy of Platelet-Rich-Plasma (PRP) and Highly Purified Bovine Xenograft (Laddec®) Combination in Bone Regeneration after Cyst Enucleation: Radiological and Histological Evaluation. *J. Oral. Maxillofac. Res.* **2013**, *4*, 3. [CrossRef]
31. Gupta, G. Clinical and radiographic evaluation of intra-bony defects in localized aggressive periodontitis patients with platelet rich plasma/hydroxyapatite graft: A comparative controlled clinical trial. *Contemp. Clin. Dent.* **2014**, *5*, 445–451. [CrossRef]
32. Akbal, A.; Gökmen, F.; Gencer, M.; Inceer, B.S.; Kömürcü, E. Mean platelet volume and platelet distribution width can be related to bone mineralization. *Osteoporos. Int.* **2014**, *25*, 2291–2295. [CrossRef]
33. Eroglu, S.; Karatas, G. Platelet/lymphocyte ratio is an independent predictor for osteoporosis. *Saudi Med. J.* **2019**, *40*, 360–366. [CrossRef] [PubMed]
34. Kim, H.; Kim, B.J.; Ahn, S.H.; Lee, S.H.; Koh, J.M. Higher plasma platelet-activating factor levels are associated with increased risk of vertebral fracture and lower bone mineral density in postmenopausal women. *J. Bone Miner. Metab.* **2015**, *33*, 701–707. [CrossRef] [PubMed]
35. Moher, D.; Liberati, A.; Tetzlaff, J.; Altman, D.G.; PRISMA Group. Preferred reporting items for systematic reviews and meta-analyses: The PRISMA statement. *PLoS Med.* **2009**, *6*, e1000097. [CrossRef] [PubMed]
36. Hayden, J.A.; Côté, P.; Bombardier, C. Evaluation of the quality of prognosis studies in systematic reviews. *Ann. Intern. Med.* **2006**, *144*, 427–437. [CrossRef]
37. Hayden, J.A.; van der Windt, D.A.; Cartwright, J.L.; Côté, P.; Bombardier, C. Assessing bias in studies of prognostic factors. *Ann. Intern. Med.* **2013**, *158*, 280–286. [CrossRef]
38. Hooijmans, C.R.; Rovers, M.M.; de Vries, R.B.; Leenaars, M.; Ritskes-Hoitinga, M.; Langendam, M.W. SYRCLE's risk of bias tool for animal studies. *BMC Med. Res. Methodol.* **2014**, *14*, 43. [CrossRef]

39. Pountos, I.; Georgouli, T.; Henshaw, K.; Bird, H.; Jones, E.; Giannoudis, P.V. The effect of bone morphogenetic protein-2, bone morphogenetic protein-7, parathyroid hormone, and platelet-derived growth factor on the proliferation and osteogenic differentiation of mesenchymal stem cells derived from osteoporotic bone. *J. Orthop. Trauma* **2010**, *24*, 552–556. [CrossRef]
40. Xie, H.; Cui, Z.; Wang, L.; Xia, Z.; Hu, Y.; Xian, L.; Li, C.; Xie, L.; Crane, J.; Wan, M.; et al. PDGF-BB secreted by preosteoclasts induces angiogenesis during coupling with osteogenesis. *Nat. Med.* **2014**, *20*, 1270–1278. [CrossRef]
41. Zhang, W.; Jin, Y.; Qian, S.; Li, J.; Chang, Q.; Ye, D.; Pan, H.; Zhang, M.; Cao, H.; Liu, X.; et al. Vacuum extraction enhances rhPDGF-BB immobilization on nanotubes to improve implant osseointegration in ovariectomized rats. *Nanomedicine* **2014**, *10*, 1809–1818. [CrossRef]
42. Tang, L.; Xia, Z.; Luo, Z.; Long, H.; Zhu, Y.; Zhao, S. Low plasma PDGF-BB levels are associated with estradiol in postmenopausal osteoporosis: PDGF-BB mediated by estradiol in women. *J. Int. Med. Res.* **2017**, *45*, 1332–1339. [CrossRef] [PubMed]
43. Kim, H.L.; Cho, H.Y.; Park, I.Y.; Choi, J.M.; Kim, M.; Jang, H.J.; Hwang, S.M. The positive association between peripheral blood cell counts and bone mineral density in postmenopausal women. *Yonsei Med. J.* **2011**, *52*, 739–745. [CrossRef] [PubMed]
44. Li, X.S.; Zhang, J.R.; Meng, S.Y.; Li, Y.; Wang, R.T. Mean platelet volume is negatively associated with bone mineral density in postmenopausal women. *J. Bone Miner. Metab.* **2012**, *30*, 660–665. [CrossRef] [PubMed]
45. D'Amelio, P.; Cristofaro, M.A.; De Vivo, E.; Ravazzoli, M.; Grosso, E.; Di Bella, S.; Aime, M.; Cotto, N.; Silvagno, F.; Isaia, G.C.; et al. Platelet vitamin D receptor is reduced in osteoporotic patients. *Panminerva Med.* **2012**, *54*, 225–231. [PubMed]
46. Aypak, C.; Türedi, Ö.; Bircan, M.A.; Civelek, G.M.; Araz, M. Association between mean platelet volume and bone mineral density in postmenopausal women. *J. Phys. Ther. Sci.* **2016**, *28*, 1753–1758. [CrossRef] [PubMed]
47. Vural, M.; Mert, M.; Erhan, B.; Gunduz, B.; Keles, B.Y.; Erdem, A.E.; Bozan, A.; Arslan, H. Is there any relationship between mean platelet volume, bone mineral density and vitamin d in postmenopausal women? *Acta Med. Mediterr.* **2017**, *33*, 443.
48. Koseoglu, S.B. Bone loss & platelet-to-lymphocyte ratio. *Biomark. Med.* **2017**, *11*, 5–10.
49. Lo, W.C.; Chiou, J.F.; Gelovani, J.G.; Cheong, M.L.; Lee, C.M.; Liu, H.Y.; Wu, C.H.; Wang, M.F.; Lin, C.T.; Deng, W.P. Transplantation of embryonic fibroblasts treated with platelet-rich plasma induces osteogenesis in SAMP8 mice monitored by molecular imaging. *J. Nucl. Med.* **2009**, *50*, 765–773. [CrossRef]
50. Liu, H.Y.; Wu, A.T.; Tsai, C.Y.; Chou, K.R.; Zeng, R.; Wang, M.F.; Chang, W.C.; Hwang, S.M.; Su, C.H.; Deng, W.P. The balance between adipogenesis and osteogenesis in bone regeneration by platelet-rich plasma for age-related osteoporosis. *Biomaterials* **2011**, *32*, 6773–6780. [CrossRef]
51. Clafshenkel, W.P.; Rutkowski, J.L.; Palchesko, R.N.; Romeo, J.D.; McGowan, K.A.; Gawalt, E.S.; Witt-Enderby, P.A. A novel calcium aluminate-melatonin scaffold enhances bone regeneration within a calvarial defect. *J. Pineal Res.* **2012**, *53*, 206–218. [CrossRef]
52. Chen, L.; Yang, X.; Huang, G.; Song, D.; Ye, X.S.; Xu, H.; Li, W. Platelet-rich plasma promotes healing of osteoporotic fractures. *Orthopedics* **2013**, *36*, 687–694. [CrossRef] [PubMed]
53. Cho, A.R.; Kim, H.K.; Kwon, J.Y.; Kim, T.K.; Choi, Y.M.; Kim, K.H. The incorporation of platelet-rich plasma into calcium phosphate cement enhances bone regeneration in osteoporosis. *Pain Physician* **2014**, *17*, 737–745.
54. Jiang, N.; Du, P.; Qu, W.; Li, L.; Liu, Z.; Zhu, S. The synergistic effect of TiO_2 nanoporous modification and platelet-rich plasma treatment on titanium-implant stability in ovariectomized rats. *Int. J. Nanomed.* **2016**, *11*, 4719–4733.
55. Wei, B.; Huang, C.; Zhao, M.; Li, P.; Gao, X.; Kong, J.; Niu, Y.; Huang, R.; Quan, J.; Wei, J.; et al. Effect of Mesenchymal Stem Cells and Platelet-Rich Plasma on the Bone Healing of Ovariectomized Rats. *Stem Cells Int.* **2016**, *2016*, 9458396. [CrossRef] [PubMed]
56. Rocha, M.A.C.; Silva, L.M.C.; Oliveira, W.A.; Bezerra, D.O.; Silva, G.C.D.; Silva, L.D.S.; Medeiros, B.L.D.N.; Baêta, S.A.F.; Carvalho, M.A.M.; Argôlo, N.M. Allogeneic mesenchymal stem cells and xenogenic platelet rich plasma, associated or not, in the repair of bone failures in rabbits with secondary osteoporosis. *Acta Cir. Bras.* **2017**, *32*, 767–780. [CrossRef] [PubMed]

57. Sakata, M.; Tonomura, H.; Itsuji, T.; Ishibashi, H.; Takatori, R.; Mikami, Y.; Nagae, M.; Matsuda, K.I.; Tabata, Y.; Tanaka, M.; et al. Bone Regeneration of Osteoporotic Vertebral Body Defects Using Platelet-Rich Plasma and Gelatin β-Tricalcium Phosphate Sponges. *Tissue Eng. Part A* **2018**, *24*, 1001–1010. [CrossRef]
58. Engler-Pinto, A.; Siéssere, S.; Calefi, A.; Oliveira, L.; Ervolino, E.; de Souza, S.; Furlaneto, F.; Messora, M.R. Effects of leukocyte- and platelet-rich fibrin associated or not with bovine bone graft on the healing of bone defects in rats with osteoporosis induced by ovariectomy. *Clin. Oral. Implant. Res.* **2019**, *30*, 962–976. [CrossRef]
59. Goltzman, D. The Aging Skeleton. *Adv. Exp. Med. Biol.* **2019**, *1164*, 153–160.
60. Compston, J.E.; McClung, M.R.; Leslie, W.D. Osteoporosis. *Lancet* **2019**, *393*, 364–376. [CrossRef]
61. Dede, A.D.; Callan, M. Treatment of osteoporosis: Whom, how and for how long? *Br. J. Hosp Med. (Lond.)* **2018**, *79*, 259–264. [CrossRef]
62. Caplan, A.I.; Correa, D. PDGF in bone formation and regeneration: New insights into a novel mechanism involving MSCs. *J. Orthop. Res.* **2011**, *29*, 1795–1803. [CrossRef] [PubMed]
63. Graham, S.; Leonidou, A.; Lester, M.; Heliotis, M.; Mantalaris, A.; Tsiridis, E. Investigating the role of PDGF as a potential drug therapy in bone formation and fracture healing. *Expert Opin. Investig. Drugs* **2009**, *18*, 1633–1654. [CrossRef] [PubMed]
64. Varol, E. The relationship between platelet indices and postmenopausal osteoporosis. *Osteoporos. Int.* **2015**, *26*, 1871–1872. [CrossRef] [PubMed]
65. Barrientos, S.; Stojadinovic, O.; Golinko, M.S.; Brem, H.; Tomic-Canic, M. Growth factors and cytokines in wound healing. *Wound Repair Regen.* **2008**, *16*, 585–601. [CrossRef]
66. Suzuki, S.; Morimoto, N.; Ikada, Y. Gelatin gel as a carrier of platelet-derived growth factors. *J. Biomater. Appl.* **2013**, *28*, 595–606. [CrossRef]
67. Veronesi, F.; Pagani, S.; Torricelli, P.; Filardo, G.; Cavallo, C.; Grigolo, B.; Fini, M. PRP and MSCs on tenocytes artificial wound healing: An in vitro study comparing fresh and frozen PRP. *Histol. Histopathol.* **2018**, *33*, 1323–1334.
68. Scully, D.; Naseem, K.M.; Matsakas, A. Platelet biology in regenerative medicine of skeletal muscle. *Acta Physiol. (Oxf.)* **2018**, *223*, 13071. [CrossRef] [PubMed]

11

Feedback Regulation of Syk by Protein Kinase C in Human Platelets

Stephanie Makhoul [1], Stephanie Dorschel [1], Stepan Gambaryan [1,2], Ulrich Walter [1] and Kerstin Jurk [1,*]

1 Center for Thrombosis and Hemostasis (CTH), University Medical Center Mainz of the Johannes Gutenberg University Mainz, 55131 Mainz, Germany; stephanie.makhoul@uni-mainz.de (S.M.); stephanie.dorschel@gmail.com (S.D.); s.gambaryan@klin-biochem.uni-wuerzburg.de (S.G.); ulrich.walter@uni-mainz.de (U.W.)

2 Sechenov Institute of Evolutionary Physiology and Biochemistry, Russian Academy of Sciences, 194223 St. Petersburg, Russia

* Correspondence: kerstin.jurk@unimedizin-mainz.de

Abstract: The spleen tyrosine kinase (Syk) is essential for immunoreceptor tyrosine-based activation motif (ITAM)-dependent platelet activation, and it is stimulated by Src-family kinase (SFK)-/Syk-mediated phosphorylation of Y352 (interdomain-B) and Y525/526 (kinase domain). Additional sites for Syk phosphorylation and protein interactions are known but remain elusive. Since Syk S297 phosphorylation (interdomain-B) was detected in platelets, we hypothesized that this phosphorylation site regulates Syk activity via protein kinase C (PKC)-and cyclic adenosine monophosphate (cAMP)-dependent pathways. ADP, the GPVI-agonist convulxin, and the GPIbα-agonist echicetin beads (EB) were used to stimulate human platelets with/without effectors. Platelet aggregation and intracellular messengers were analyzed, along with phosphoproteins, by immunoblotting using phosphosite-specific antibodies or phos-tags. ADP, convulxin, and EB upregulated Syk S297 phosphorylation, which was inhibited by iloprost (cAMP pathway). Convulxin-stimulated Syk S297 phosphorylation was stoichiometric, transient, abolished by the PKC inhibitor GF109203X, and mimicked by the PKC activator PDBu. Convulxin/EB stimulated Syk S297, Y352, and Y525/526 phosphorylation, which was inhibited by SFK and Syk inhibitors. GFX and iloprost inhibited convulxin/EB-induced Syk S297 phosphorylation but enhanced Syk tyrosine (Y352/Y525/526) and substrate (linker adaptor for T cells (LAT), phospholipase γ2 (PLC γ2)) phosphorylation. GFX enhanced convulxin/EB-increases of inositol monophosphate/Ca^{2+}. ITAM-activated Syk stimulates PKC-dependent Syk S297 phosphorylation, which is reduced by SFK/Syk/PKC inhibition and cAMP. Inhibition of Syk S297 phosphorylation coincides with enhanced Syk activation, suggesting that S297 phosphorylation represents a mechanism for feedback inhibition in human platelets.

Keywords: spleen tyrosine kinase (Syk); protein kinase C; cyclic adenosine monophosphate (cAMP); platelets; glycoprotein VI; glycoprotein Ibα

---

1. Introduction

Human platelets are small circulating blood cells, which control, monitor, and preserve the integrity of the vessel wall. These anucleate cells prevent blood loss during vascular injury and, on the other hand, have prominent roles in thrombotic, inflammatory, and immune pathologies [1,2]. After vascular injury, distinct platelet receptors sense and bind newly exposed proteins immobilized within the extracellular matrix or on activated endothelial cells. This is followed by receptor-mediated platelet activation resulting in a plethora of cellular responses, including cytoskeletal remodeling, integrin activation (e.g., integrin $\alpha_{IIb}\beta_3$), degranulation, synthesis/release of thromboxane A2 (TxA2), and surface exposure of anionic phospholipids, leading to platelet shape change, adhesiveness, aggregation, and coagulant activity to form a protective thrombus [2,3]. These essential physiological processes are often impaired in diseases and are tightly controlled by numerous hormones, vasoactive factors, and adhesive proteins, which activate, modulate, or inhibit these platelet functions. Two major classes of platelet activators include soluble agonists and adhesion molecules, e.g., von Willebrand factor (vWF), collagen, fibrin, and podoplanin, which stimulate specific G-protein-coupled receptors (GPCRs) [4,5] and cell membrane-spanning adhesion receptors, respectively [6–8]. GPCR-coupled agonists activate phospholipase (PLC) β, elevate cytosolic Ca^{2+} concentration, and activate Ca^{2+}-dependent protein kinases, such as protein kinase C (PKC), thereby inducing platelet activation [4,5].

The adhesion molecules collagen, vWF, podoplanin, fibrinogen, and fibrin bind to and activate platelet membrane receptors such as glycoprotein (GP) VI (GPVI), GPIb-V-IX, CLEC-2, integrin $\alpha_{IIb}\beta_3$ [6,7], and subsequently Src-family tyrosine protein kinases (SFKs) causing platelet activation [8]. The SFKs tyrosine-phosphorylate other proteins/protein kinases including membrane proteins with the immunoreceptor tyrosine-based activation motif (ITAM) [6,8,9], which then recruits src homology 2 (SH2) domain-containing proteins, in particular the spleen tyrosine kinase (Syk). In human platelets, ITAM-mediated Syk activation is mediated by the Fc receptor γ-chain and the low-affinity IgG receptor FcγRIIa [10,11], while it is only mediated by the Fc receptor γ-chain in murine platelets [12–15]. Mice embryos presenting with a homozygous targeted mutation in the *Syk* gene (by deletion of one exon on *Syk* gene encoding for 41 residues in the Syk kinase domain in embryonic stem cells) die from severe hemorrhages before birth [16], and mice lacking platelet Syk were protected from arterial thrombosis and ischemic stroke [17], highlighting the important role of Syk in platelets. Tyrosine-phosphorylated ITAM proteins recruit Syk from the cytosol to the cell membrane and activate Syk via two distinct overlapping mechanisms, the described ITAM-dependent process and a tyrosine phosphorylation-dependent process [15,18–20]. The Syk Y-phospho-sites closely associated with activation, Y348/Y352 and Y525/Y526, are two pairs within the interdomain-B and kinase domains, respectively. Syk activation is initiated when these Y-sites are phosphorylated by SFKs or when dually Y-phosphorylated ITAM-containing membrane proteins recruit the two Syk-SH2 domains followed by Syk autophosphorylation, leading to the activation of the LAT-signalosome [18,19]. However, in addition to these Syk tyrosine phosphorylation sites involved in kinase activation, it was demonstrated, primarily with murine and human B-cells, that Syk contains multiple tyrosine, serine, and threonine phosphorylation sites, and that some of them are important for recruiting additional regulatory binding proteins [21–23]. Syk serine phosphorylation at S297 (S291 in murine cells) is observed in B-cells [23,24]. While Syk S291 phosphorylation in murine B-cell lines was reported to enhance Syk coupling to the B-cell antigen receptor (BCR) [24], Syk S297 phosphorylation diminished antigen–receptor signaling in human B-cell lines [23]. However, the role of Syk S297 phosphorylation in human platelets remains unknown. In our recent phosphoproteomic studies with human platelets, the cyclic adenosine monophosphate (cAMP)-elevating platelet inhibitor and stable prostacyclin analog iloprost (cAMP/protein kinase A (PKA) pathway), as well as adenosine diphosphate (ADP), affected the phosphorylation of many protein kinases including several tyrosine protein kinases such as Janus kinase (JAK) 3, activated CDC42 kinase 1(ACK1), Bruton-tyrosine kinase (BTK), and Syk [25,26]. Interestingly, ADP, which activates platelet Ca^{2+}/calmodulin-dependent protein kinases such as PKC, but not iloprost, stimulated Syk S297 phosphorylation. Very recently, we established methods for the selective quantitative assessment of GPVI-and GPIbα-mediated activation and function of human platelet Syk [27,28]. We observed that cAMP-and cyclic guanosine monophosphate (cGMP)-elevating platelet inhibitors strongly inhibited GPIbα-/GPVI-mediated platelet activation but enhanced the initial Syk activation [28]. These phosphoproteomic and functional approaches suggest that there is a network of interacting protein kinases at the level of Syk in platelets [29,30].

Based on previously published data and our own findings on Syk S297 phosphorylation in human platelets, and considering the crucial Syk interdomain location of S297 [20], we hypothesized that this serine site is phosphorylated in response to the activation of several signaling pathways. In particular, we hypothesized that PKC-and cAMP-dependent pathways, via their respective protein kinases, regulate the phosphorylation of Syk S297, thereby affecting activation and/or activity of Syk in human platelets. With this approach, we aimed to show that phosphorylation of Syk S297 in platelets modulates Syk activity and, subsequently, further Syk substrates important for platelet function.

2. Results

2.1. ADP, Convulxin, and Echicetin Beads Upregulate Syk S297 Phosphorylation, Which Is Inhibited by Iloprost

Our previous phosphoproteomic studies with human platelets showed that ADP induced Syk serine phosphorylation at S297, which is located in the interdomain-B of Syk [26]. Using a phosphospecific antibody against this site, we investigated the regulation of this phosphorylation site by ADP, by its functional inhibitor (iloprost) and by agonists, which activate platelets via ITAM-/Syk-dependent mechanisms. As expected, ADP-induced platelet aggregation was completely inhibited by iloprost (Supplementary Materials Figure S1a). ADP increased Syk S297 phosphorylation 3–4-fold within 4 min of stimulation, which was strongly inhibited by iloprost (Figure 1a). Then, a rapid (within 1 min) but transient phosphorylation of S297 was observed upon platelet stimulation with the selective GPVI agonist convulxin (cvx), as well as with the specific GPIbα agonist echicetin beads (EB) (Figure 1b,c), which was also strongly inhibited by iloprost. Furthermore, cvx/EB-induced S297 phosphorylation was significantly downregulated by the blockage of the TxA2 receptor and the ADP receptor $P2Y_{12}$ (Supplementary Materials Figure S2a,b). These data demonstrate that Syk S297 is upregulated by distinct signaling pathways in human platelets and, with both GPIbα and GPVI, is significantly dependent on the secondary mediators ADP and TxA2.

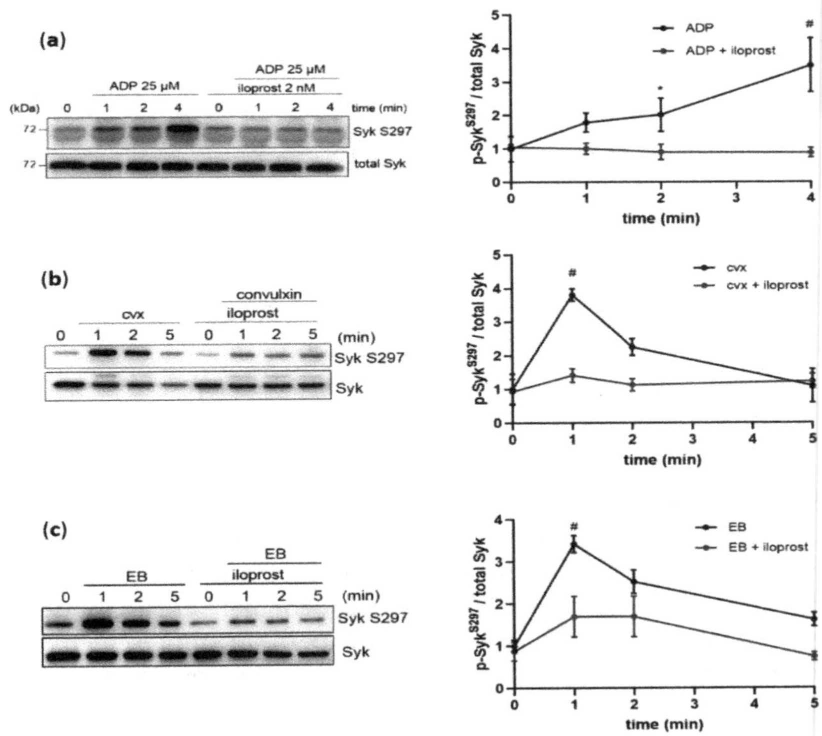

Figure 1. Platelet Syk S297 is upregulated by ADP, convulxin, and echicetin beads. Human washed platelets were pre-incubated with iloprost (2 nM; 3 min) at 37 °C prior to stimulation with (**a**) 25 µM ADP, (**b**) 50 ng/mL convulxin (cvx), or (**c**) 0.5% (*v*/*v*) echicetin beads (EB). Platelet aggregation was stopped after 1, 2, or 5 min of stimulation under stirring conditions by adding Laemmli buffer. Syk S297

phosphorylation was analyzed in the presence or absence of iloprost by immunoblotting compared to total protein. Quantitative data are represented as means ± SD from three independent experiments with platelets from three healthy donors. * $p < 0.05$, untreated versus iloprost-treated platelets in response to ADP at 2 min; # $p < 0.0001$, untreated versus iloprost-treated platelets at 4 min in response to ADP and at 1 min in response to cvx or EB.

2.2. *Transient GPVI/GPIbα-Stimulated Syk S297 Phosphorylation Parallels Syk Tyrosine Phosphorylation*

To better understand a possible link between the phosphorylation of S297 and Syk activation, we studied simultaneously the phosphorylation of Syk on S297, on Y525/526 (a Syk activation marker), and on Syk Y352 (important for Syk activity enhancement and for binding other proteins). Syk S297 showed a transient phosphorylation, which increased within 1 min of stimulation and then decreased both with convulxin (Figure 2a,ai) and EB (Figure 2b,bi) stimulation. A very similar transient time course was observed with Syk Y525/526 and Y352 phosphorylation in response to cvx (Figure 2a,aii) and EB (Figure 2b,bii). In contrast, ADP did not induce Syk Y525/526 or Y352 phosphorylation [26]. The observed Syk phosphorylation declined after initial stimulation, suggesting that a protein phosphatase is active.

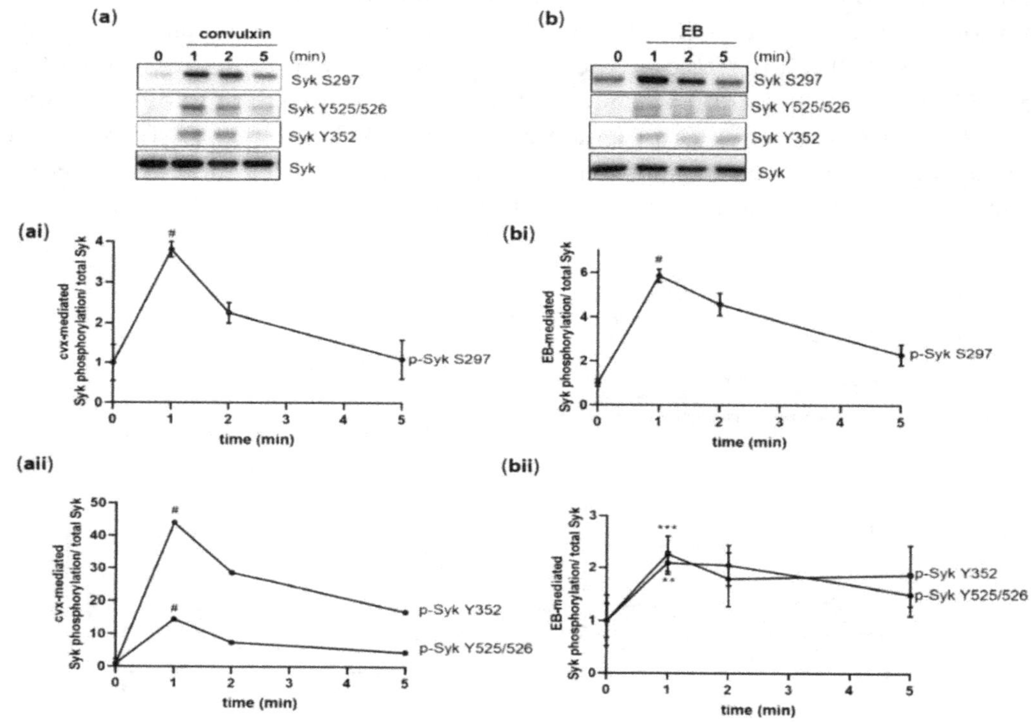

Figure 2. Cvx and EB induce a transient Syk S297 phosphorylation in parallel with Syk tyrosine phosphorylation. Washed human platelets were stimulated under stirring conditions with (**a**) 50 ng/mL cvx or (**b**) 0.5% (*v*/*v*) EB. (**ai**,**bi**) Syk phosphorylation on S297 and tyrosine sites (525/526 and 352) mediated by cvx or by EB were analyzed in a time-dependent manner (1, 2, and 5 min) compared to total Syk. (**aii**,**bii**) The kinetics of the phosphorylation patterns are represented as means ± SD from three independent experiments with platelets from three healthy donors. # $p < 0.0001$, *** $p < 0.001$, ** $p < 0.01$, untreated versus cvx-or EB-stimulated platelets at 1 min.

2.3. *Syk S297 Phosphorylation Is Dependent on Src Family Kinases (SFKs) and Syk Kinase Activity*

EB-/cvx-mediated Syk tyrosine phosphorylation is mediated by SFKs and Syk autophosphorylation [18,19,28]. Therefore, we evaluated the role of SFKs and Syk for Syk S297 phosphorylation. The SFK inhibitor PP2 abolished Syk S297 phosphorylation induced by cvx and EB (Figure 3a,b), which was also observed with the two different Syk inhibitors OXSI-2 and PRT-060318

(Figure 3c,d). These data indicate that GPVI-/GPIbα–increased Syk S297 phosphorylation is highly dependent on SFK-stimulated Syk kinase activity.

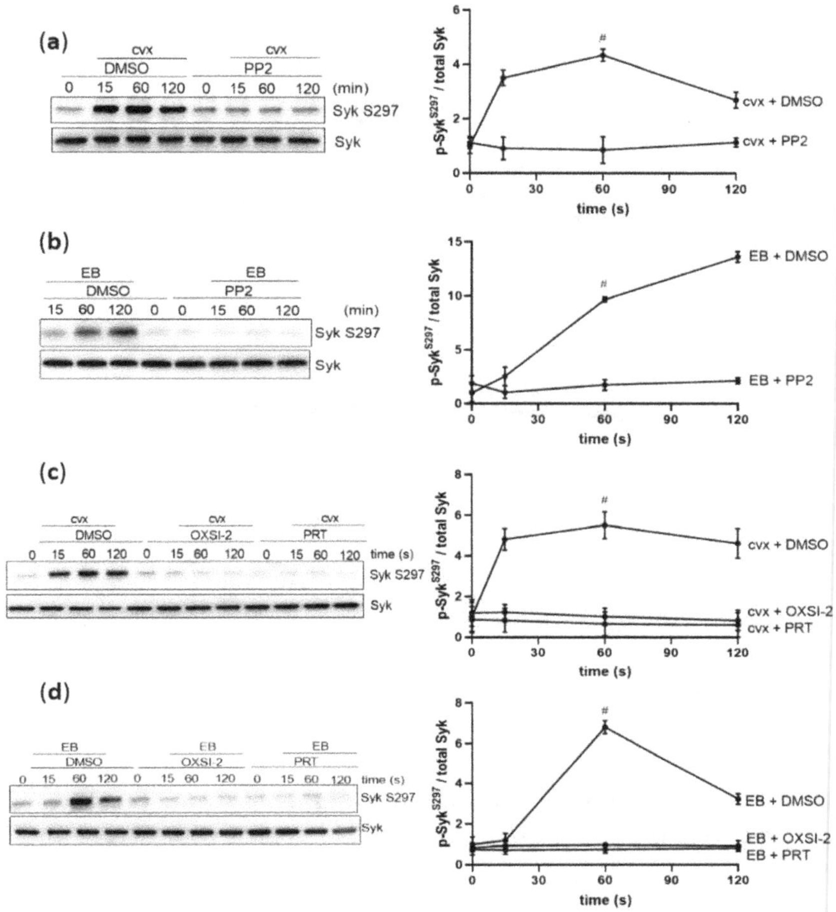

Figure 3. Cvx-and EB-mediated Syk S297 upregulation is dependent on SFKs and Syk activation. Washed human platelets were pre-incubated with vehicle control (DMSO) or with the SFK inhibitor PP2 (10 μM) for 5 min at 37 °C prior to stimulation, under stirring conditions with (**a**) 50 ng/mL cvx or (**b**) 0.5% (*v/v*) EB or with two different Syk inhibitors, OXSI-2 (2 μM) and PRT-060318 (1 μM) for 5 min at 37 °C prior to stimulation with (**c**) cvx or (**d**) EB. Phosphorylation of Syk S297 was analyzed by immunoblotting compared to total Syk. Data are represented as means ± SD from three independent experiments with platelets from three healthy donors. # $p < 0.0001$, DMSO versus inhibitor-treated (PP2, OXSI-2, or PRT-060318) platelets at 1 min in response to cvx or EB.

2.4. Convulxin-Induced Syk S297 Phosphorylation Is Stoichiometric and Abolished by Inhibition of PKC

Upon BCR stimulation in murine B-cells, Syk S291, in addition to several tyrosine sites, was a major PKC phosphorylation site in the interdomain-B of Syk which corresponds to S297 in human Syk [24]. Therefore, we tested the role of PKC in Syk S297 regulation in human platelets using a global and potent PKC inhibitor, GF109203X (GFX). Pre-incubation of platelets with GFX completely inhibited cvx-stimulated Syk S297 phosphorylation (Figure 4a).

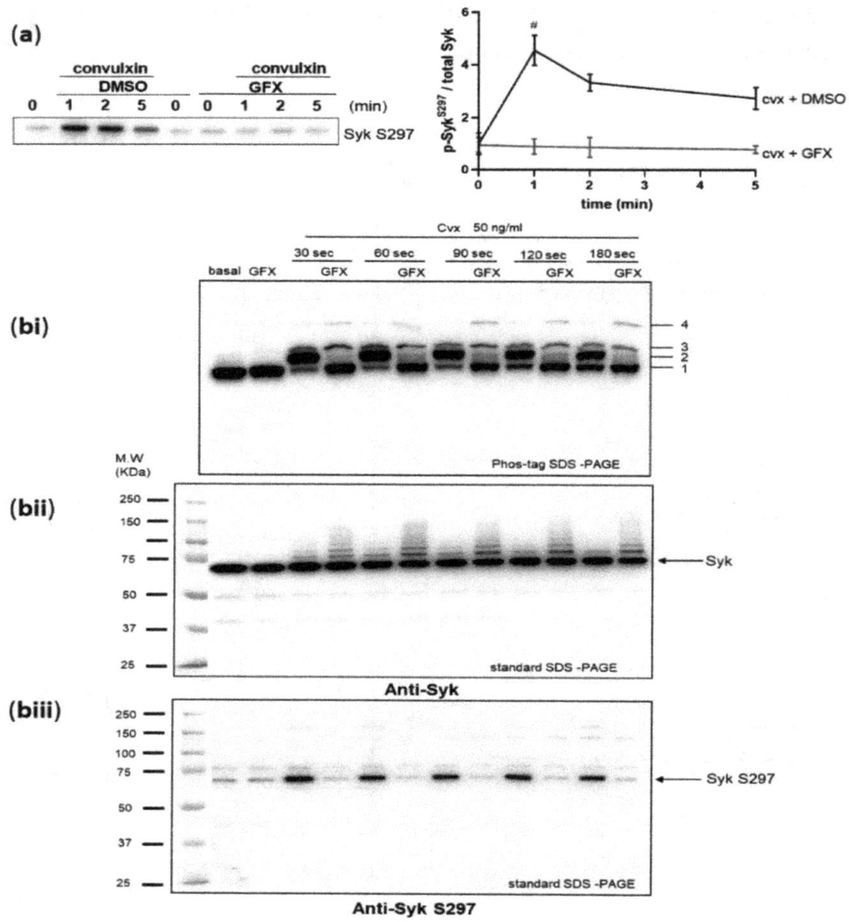

Figure 4. Syk S297 phosphorylation is stoichiometric and completely blocked by PKC inhibition. Washed human platelets were pre-incubated with DMSO as vehicle control or with pan-PKC inhibitor GF109203X (GFX) (5 μM) at 37 °C for 5 min prior to stimulation with 50 ng/mL cvx. (**a**) Platelet aggregation was stopped by adding Laemmli buffer directly in the cuvettes after 1, 2, and 5 min of stimulation under stirring conditions, and Syk S297 phosphorylation was analyzed by standard SDS-PAGE/Western blot analysis compared to total Syk. Quantitative analyses are represented as means ± SD from three independent experiments with platelets from three healthy donors. # $p < 0.0001$, DMSO versus GFX-treated platelets at 1 min. (**bi–biii**) Washed human platelets were stimulated under non-stirring conditions at 37 °C, and platelets were lysed after the indicated time points by using Laemmli buffer for phos-tag SDS-PAGE. (**bi**) Samples were analyzed by phos-tag SDS-PAGE followed by immunoblotting using total Syk antibody. Same samples were analyzed by standard SDS-PAGE followed by immunoblotting using (**bii**) total Syk antibody or (**biii**) anti-Syk S297. Blots are representative of two independent experiments from two healthy donors.

To analyze quantitative aspects of Syk phosphorylation, we used the phos-tag SDS- polyacrylamide gel-electrophoresis (PAGE) method developed by the group of Koike [31,32]. This method separates phosphorylated and non-phosphorylated forms of proteins including tyrosine kinases depending on their degree of phosphorylation. Immunoblotting using anti-Syk antibody revealed the presence of four migration bands (Figure 4bi). At basal conditions, one major Syk band was detected (band 1), which was not affected by the PKC inhibitor GFX. After 30 sec of activation with cvx, a complete shift of Syk was detected with several bands (2–4), while the major band 1 disappeared and a major band 2 appeared. With GFX pre-treatment, the major band 2 completely disappeared and the basal band 1 reappeared, while the other minor bands (3, 4) were not downregulated. These data indicate that the cvx-induced appearance of band 2 is highly dependent on PKC activity.

In parallel, we analyzed total Syk with the standard SDS-PAGE method and detected only one constant major Syk band in all samples but a row of very minor bands (higher than 72 kDa), especially GFX-treated samples (Figure 4bii), which may be due to Syk ubiquitination [33].

These samples from cvx-treated platelets were also analyzed by standard SDS-PAGE blots using phosphoantibodies. The phosphorylation of Syk S297 was strongly stimulated by cvx and completely inhibited by the PKC inhibitor GFX (Figure 4biii), which closely resembled the appearance/disappearance of band 2 in the phos-tag gel analysis. Overall, the major cvx-induced shift of Syk in phos-tag gels is completely prevented by PKC inhibition and due to Syk S297 phosphorylation.

2.5. PKC Activator, PDBu, Stimulates a Stoichiometric Syk S297 Phosphorylation, Which Is Prevented by PKC Inhibition

To obtain further evidence for PKC-mediated phosphorylation of Syk S297 in human platelets, we tested the effects of a global PKC activator, the phorbol ester phorbol 12, 13-dibutyrate (PDBu). PDBu induced strong platelet aggregation (Supplementary Materials Figure S3) and a very fast and strong phosphorylation of Syk on S297, which appeared within 15 s of stimulation and was stable for the first two minutes (Figure 5a). In addition, we incubated washed human platelets under non-stirring conditions with PDBu and analyzed the Syk phosphorylation profile by phos-tag SDS-PAGE/immunoblotting. At basal conditions, unstimulated Syk showed only one band (band 1), which was completely shifted (band 2) upon PDBu stimulation in a time-dependent manner (Figure 5bi). In parallel, the standard SDS-PAGE/Western blot analysis showed one Syk band for all samples (Figure 5bii) and a time-dependent PDBu stimulated Syk S297 phosphorylation (Figure 5bii,biii). These data support the role of PKC in regulating Syk through S297 phosphorylation.

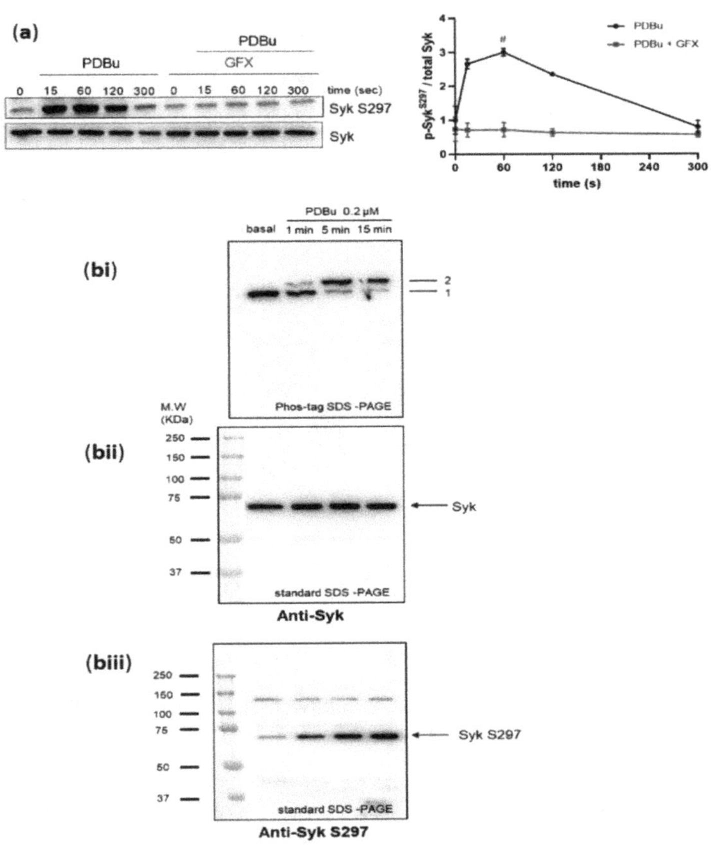

Figure 5. Syk S297 phosphorylation is upregulated by PKC activation in a stoichiometric manner. Washed human platelets were pre-incubated with GF109203X (5 µM) for 5 min at 37 °C prior to stimulation with

the phorbol ester PDBu, a global PKC activator (0.2 μM). (**a**) Platelet aggregation was stopped by adding Laemmli buffer directly in the cuvettes after 15, 60, 120, and 300 s of stimulation under stirring conditions, and Syk S297 phosphorylation was analyzed by standard SDS-PAGE/Western blot analysis compared to total Syk. Quantitative analyses are represented as means ± SD from three independent experiments with platelets from three healthy donors. # $p < 0.0001$, untreated versus GFX treated platelets at 1 min. (**bi–biii**) Washed human platelets were stimulated under non-stirring conditions at 37 °C, and platelets were lysed after the indicated time points by using Laemmli buffer for phos-tag SDS-PAGE/immunoblotting analysis. (**bi**) Samples were analyzed by phos-tag SDS-PAGE followed by immunoblotting using total Syk antibody. Same samples were analyzed by standard SDS-PAGE followed by immunoblotting using (**bii**) total Syk antibody or (**biii**) anti-pSyk S297. Blots are representative of two independent experiments from two healthy donors.

2.6. *PDBu-Mediated S297 Phosphorylation Is Independent of Syk and Only Partially Inhibited by PKA*

We then investigated the role of Syk kinase activity on the phosphorylation of S297 mediated by PKC activation. Firstly, we validated the effects of PDBu and GFX by analyzing myristoylated alanine-rich C-*kinase* substrate (MARCKS) phosphorylation, a well-established PKC substrate, also in platelets [34,35]. MARCKS was significantly phosphorylated on S159/163 in response to PDBu and convulxin, which was completely abolished by GFX (Figure 6a). Secondly, Syk inhibition by PRT-060318 did not affect PDBu-induced Syk S297 or MARCKS phosphorylation (Figure 6b), showing that PDBu-caused PKC activation does not require Syk, in contrast to S297 phosphorylation caused by cvx and EB (Figure 3c,d). Thirdly, PDBu/PKC-induced Syk S297 phosphorylation was only partially inhibited by iloprost, and PDBu/PKC-induced MARCKS phosphorylation was even increased by iloprost (Figure 6c). These data suggest that PKC per se was not inhibited by iloprost in intact platelets, but iloprost (PKA) inhibited signaling pathways upstream of PKC activation.

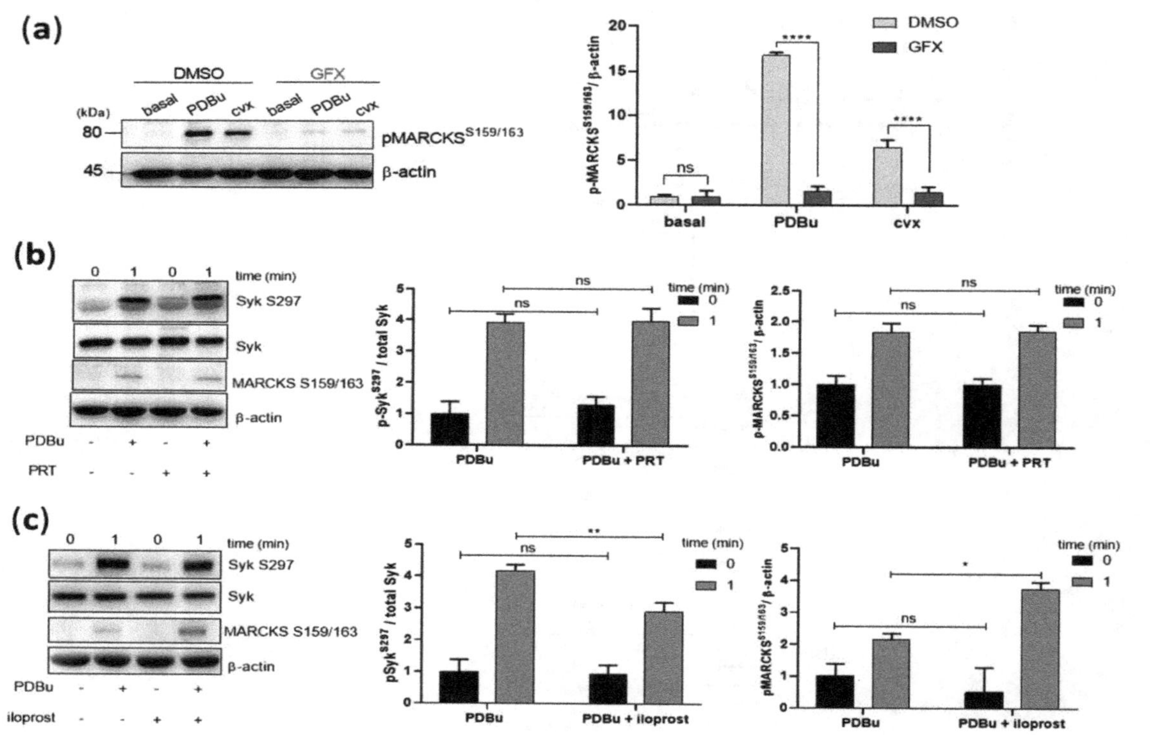

Figure 6. PKC activation induces Syk S297 phosphorylation, which is independent from Syk and only partially inhibited by PKA. Washed human platelets were pre-incubated with DMSO (vehicle control) or with pan-PKC inhibitor, GF109203X (5 μM) at 37 °C for 5 min prior to stimulation with 0.2 μM PDBu or 50 ng/mL cvx. (**a**) Platelet aggregation was stopped by adding Laemmli buffer after 1 min of stimulation under stirring conditions. MARCKS S159/163 phosphorylation was analyzed by standard

SDS-PAGE analysis compared to the loading control β-actin. Washed human platelets were pre-incubated with (**b**) Syk inhibitor, PRT-060318 (1 μM) for 5 min or (**c**) iloprost (2 nM) for 3 min prior to stimulation with PDBu. Syk S297 and MARCKS S159/163 phosphorylation was analyzed compared to total Syk or β-actin, respectively. The corresponding quantifications are represented as means ± S.D from three independent experiments with platelets from three healthy donors. **** $p < 0.0001$, ** $p < 0.01$, * $p < 0.05$, ns: not significant.

2.7. PKC Inhibition Abolished cvx/EB-Induced Syk S297 Phosphorylation But Increased Syk Tyrosine 352 and 525/526 Phosphorylation (Hyperphosphorylation)

Next, we investigated the role of PKC on Syk activation in GPVI-and GPIbα-mediated platelet activation. PKC inhibition by GFX completely prevented the S297 phosphorylation induced by cvx (Figure 7a,ai) and EB (Figure 7b,bi). In contrast, Syk tyrosine phosphorylation on 525/526 was significantly increased upon stimulation for 1 min compared to the control and subsequently prolonged in both GPVI-(Figure 7a,aii) and GPIbα-mediated signaling (Figure 7b,bii). Syk Y352 phosphorylation showed a similar hyperphosphorylation profile in GPVI-(Figure 7a,aiii), and EB/GPIbα-mediated signaling (Figure 7b,biii). These data demonstrate a striking differential regulation of Syk tyrosine phosphorylation and serine phosphorylation by PKC.

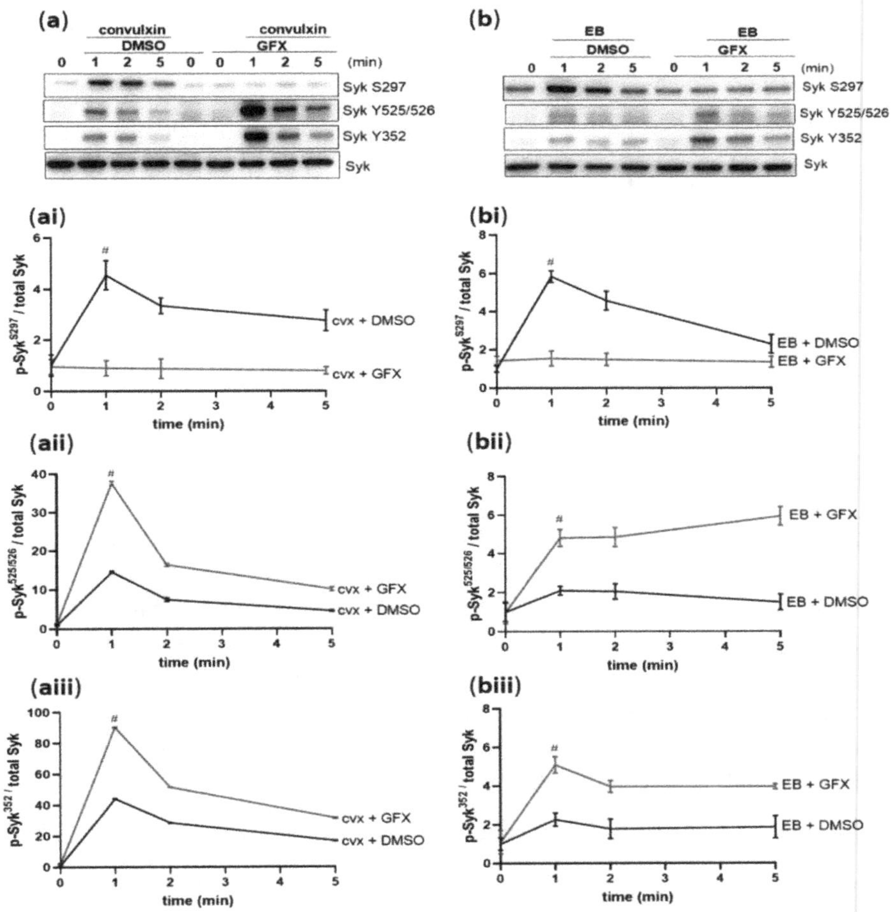

Figure 7. Syk phosphorylation is differentially regulated by PKC. Washed human platelets were pre-incubated with DMSO (vehicle control) or with pan-PKC inhibitor GF109203X (5 μM) at 37 °C for 5 min prior to stimulation with (**a**) 50 ng/mL cvx or (**b**) 0.5% (*v*/*v*) EB. Platelet aggregation was stopped by adding Laemmli buffer after 1, 2, or 5 min of stimulation under stirring conditions. The kinetics of the phosphorylation of Syk on (**ai**,**bi**) S297, (**aii**,**bii**) Y525/526, and (**aiii**,**biii**) Y352 compared to total Syk are represented as means ± SD from three independent experiments with platelets from three healthy donors. # $p < 0.0001$, DMSO-versus GFX-treated platelets at 1 min in response to cvx or EB.

2.8. PKC Inhibition Caused Hyperphosphorylation of Y-Sites of Syk Substrates

To validate the possible functional consequence of Syk tyrosine hyperphosphorylation, we investigated the phosphorylation of Syk downstream substrates linker adaptor for T cells (LAT) and PLCγ2 on Y191 and Y759, respectively. Cvx-and EB-induced LAT phosphorylation was remarkably increased in the presence of the PKC inhibitor compared to the control (Figure 8a,b,ai,bi). In addition, PLCγ2 also showed an increase and prolongation in the Y759 phosphorylation in both signaling pathways cvx/GPVI (Figure 8a,aii) and EB/GPIbα (Figure 8b,bii). These data demonstrate that Syk hyperphosphorylation is associated with direct increased tyrosine phosphorylation of LAT and PLCγ2.

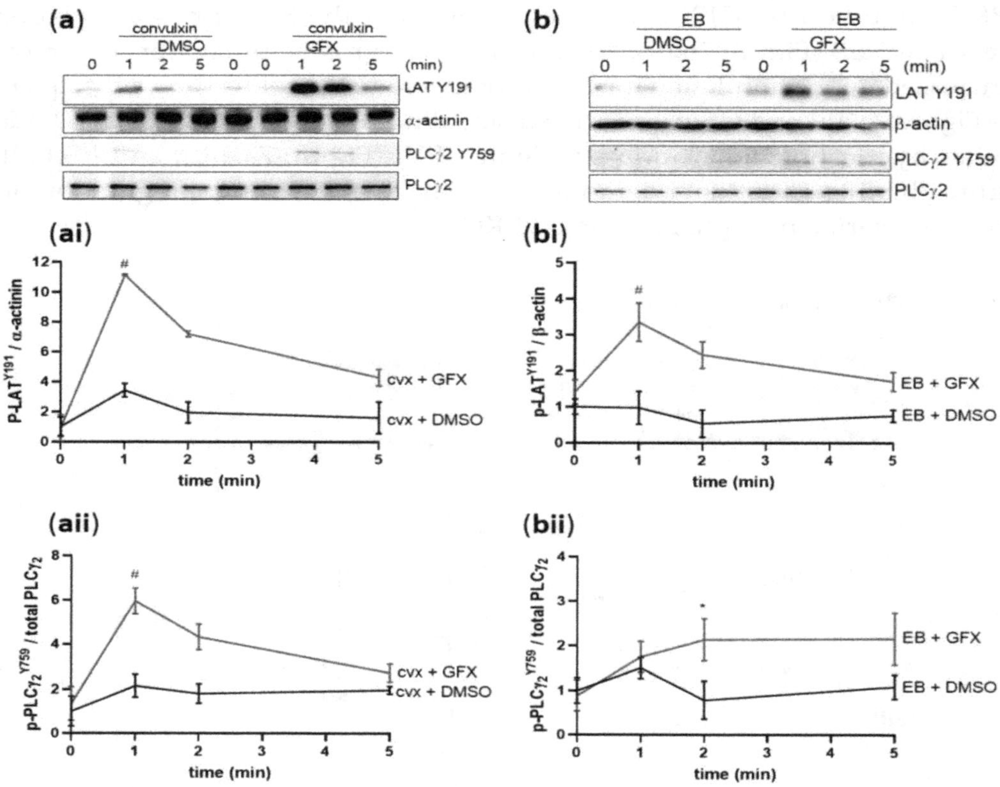

Figure 8. PKC inhibition induces hyperphosphorylation of Y-sites of Syk substrates. Washed human platelets were pre-incubated with DMSO (vehicle control) or with pan-PKC inhibitor GF109203X (5 μM) at 37 °C for 5 min prior to stimulation with (**a**) 50 ng/mL cvx or (**b**) 0.5% (*v*/*v*) EB. Platelet aggregation was stopped by adding Laemmli buffer after 1, 2, or 5 min of stimulation under stirring conditions. The kinetics of the phosphorylation of (**ai**,**bi**) LAT on Y191 and (**aii**,**bii**) PLCγ2 on Y759 compared to the adequate loading control are represented as means ± SD from three independent experiments with platelets from three healthy donors. # $p < 0.0001$, * $p < 0.05$, DMSO-versus GFX-treated platelets at 1 min or 2 min in response to cvx or EB, respectively.

2.9. PKC Inhibition Enhances cvx/EB-Mediated InsP1 Production and Intracellular Ca^{2+} Mobilization

Furthermore, we aimed to measure the activity of PLCγ2 by monitoring intracellular inositol triphosphate (InsP3) production (via inositol monophosphate (InsP1) accumulation) and intracellular Ca^{2+} mobilization. In the presence of GFX, cvx-and EB-evoked InsP3 production was significantly enhanced compared to the control in the absence of the PKC inhibitor (Figure 9a). Moreover, this InsP3 production, mediated by cvx and EB, led to a significant increase of intracellular Ca^{2+} mobilization, which was even more pronounced in the presence of GFX (Figure 9b). These data show that Syk hyperphosphorylation is followed by an increase in LAT/PLCγ2 phosphorylation leading to an increase of InsP3 production and subsequent Ca^{2+} mobilization.

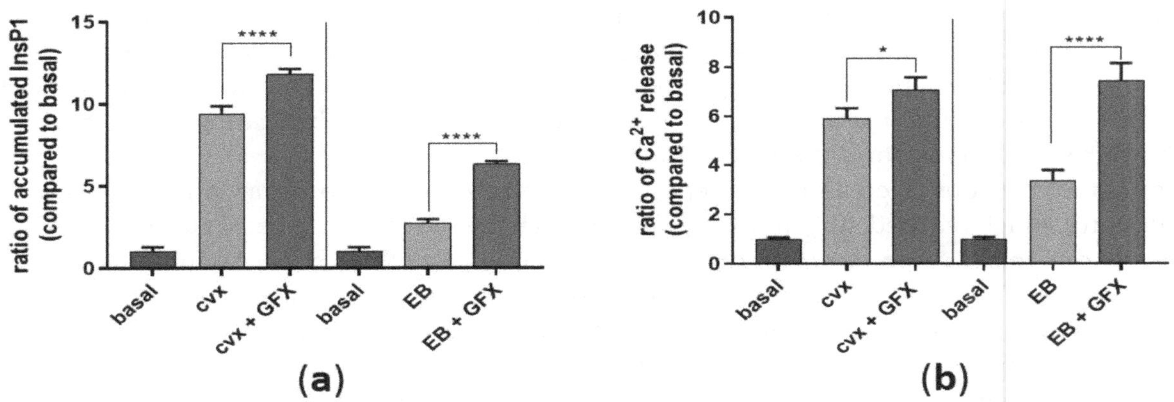

Figure 9. PKC inhibition significantly enhances InsP1 accumulation and intracellular Ca^{2+} mobilization. (**a**) Washed human platelets were pre-treated with GF109203X (5 μM) as previously described in the presence of 1 mM LiCl. Platelet aggregation was stopped after 5 min using lysis buffer provided by the manufacturer. (**b**) Washed human platelets were pre-incubated with Fluo-3 AM (5 μM) for 30 min at 37 °C, and platelet stimulation was performed directly before measurement. $Ca^{2+}{}_i$ mobilization was monitored for 120 s by flow cytometry. Quantitative data are represented as means ± SD from three independent experiments with platelets from three healthy donors. **** $p < 0.0001$, * $p < 0.05$.

3. Discussion

In this study, we show (1) that activation of human platelets via GPVI and GPIbα, as well as by pharmacological activation of PKC, caused phosphorylation of Syk S297, and (2) that pharmacological PKC inhibition or receptor-linked inhibition of PKC by cAMP/PKA abolished Syk S297 phosphorylation, but enhanced Syk tyrosine phosphorylation/activity and phosphorylation of Syk substrates. These data suggest that Syk S297 phosphorylation is mediated by PKC and represents a mechanism of feedback inhibition of Syk in human platelets.

Several lines of evidence show that a classical, diacylgycerol/phorbolester-responsive protein kinase C (PKC) phosphorylates Syk S297 in human platelets:

Increased Syk S297 phosphorylation is observed when ADP receptors, GPVI, and GPIbα are stimulated by selective agonists, which also stimulate PLC and PKC activation.

- GPVI-and GPIbα-caused Syk S297 phosphorylation is Syk-dependent and requires, at least predominantly, the release of secondary mediators ADP and TxA2.
- GPVI-/GPIbα-induced Syk S297 phosphorylation is blocked by a selective PKC inhibitor (GFX) and mimicked by an established, membrane permeable PKC activator (PDBu).
- This agonist-induced Syk S297 phosphorylation is also inhibited by the cAMP/PKA pathway, which is known to strongly inhibit receptor-stimulated $Ca2^{+}$ responses and PKC activation [30,36,37].
- The peptide sequence around Syk S297* 9ISRIKSYS*FPKPGHR) in humans (in the murine form S291, ISRIKSYS*FPKPGHK) agrees well with the PKC phosphosite consensus sequence [24] (see also other phosphosite databases).

While the evidence for a conventional PKC as a responsible Syk S297 kinase in intact human platelets is quite strong, we do not know which PKC isoform (or perhaps more than one) is involved. PKC isoforms consist of conventional forms (α, βI, βII, γ) activated by Ca^{2+}/diacylglycerol, novel forms (δ, ε, η, θ) activated by diacylglycerol alone, and atypical forms, which are diacylglycerol-independent. Human and mouse platelets highly express the conventional PKC isoforms α/β and the novel isoforms δ and θ [36,38], which all have important but diverse roles in platelets, which may be inhibitory or activating [36,39]. Therefore, we cannot rule out that more than one member of the PKC family is capable of phosphorylating platelet Syk S297. Previously, a functional interaction of PKCα with both

Src and Syk was reported for human platelets, which occurred together with a Syk-dependent tyrosine phosphorylation of PKCα, but not the other way around, i.e., PKCα-dependent serine phosphorylation of Syk [40]. An Src/SFK-dependent phosphorylation and activation of PKCδ was observed in platelets, as well as a PKCδ-dependent activation of Syk in endothelial cells [41], indicating interactions between Syk, other tyrosine kinases, and the PKC family.

In our study, we observed a rapid receptor-mediated Syk S297 phosphorylation (onset within 15 s of stimulation), which reached stoichiometry and then declined back to the basal status. This conclusion was possible due to the combination of regular Western blot/pSyk S297 phosphoantibody analysis with phos-tag analysis. The initial major Syk band in phos-tag gels was almost completely shifted to another major band, which was completely shifted back to the basal state when platelets where pre-treated with the PKC inhibitor GFX. This major band was the only Syk band with this property, a pattern also observed by regular Syk pS297 phosphoantibody analysis. The conclusion that these effects are PKC-mediated is strongly supported by observations made with the direct PKC activator PDBu. Treatment of platelets with PDBu caused, as detected in phos-tag gels, a time-dependent, complete shift of this major Syk band, which was also completely prevented by the PKC inhibitor. Compared to cvx treatment, no minor additional shifted band appeared in the PDBu treatment detected in phos-tag gels. The phos-tag data with PDBu additionally suggest that PKC phosphorylates only one major site (S297) in Syk, although Syk contains a number of serine/threonine protein kinase phosphosites [23]. The other, minor cvx-shifted Syk bands, detected in phos-tags, are not downregulated by GFX and correspond to tyrosine-phosphorylated sites of Syk. Altogether, Syk S297 phosphorylation in human platelets is mediated by PKC, in a stoichiometric and reversible manner.

While the phosphorylation of Syk S297 can reach stoichiometry in our studies with human platelets, the corresponding stoichiometry of Syk tyrosine phosphorylation appears to be much smaller. Interestingly, Syk phosphorylation data obtained with murine B-cells suggested that phosphorylation of Syk at tyrosines is restricted to a subset of the kinase physically associated with ITAM-containing membrane proteins, whereas the interdomain serine sites are also phosphorylated in Syk, when Syk is not membrane-bound [24].

Another feature of GPVI/GPIbα-mediated Syk phosphorylation and activation in platelets is the transient nature of both tyrosine (Y352, Y525/Y526) and serine (S297) phosphorylation. Onset and decline of Cvx/EB stimulated SykS297 phosphorylation resemble the time course of Syk Y352/Y525/526 phosphorylation, which indicates Syk activation [18,19]. The state of protein phosphorylation is determined, for each individual site, by both protein kinases and protein phosphatases, and it can be decreased, for example, by kinase inhibition, phosphatase activation, or both. When murine B-cell Syk activation is discontinued, both Syk and its substrates are rapidly tyrosine-dephosphorylated, and the downstream signaling effects end [22,42]. Several protein tyrosine phosphatases were identified to dephosphorylate tyrosine phosphorylated Syk including the SH2 domain-containing tyrosine protein phosphatase-1 (SH-1; PTPN6), e.g., T-cell ubiquitin ligand 2 (TULA-2; UBASH3B) [22], and most of them are present in human platelets [29,38]. TULA-2 is an unconventional protein tyrosine phosphatase with a ubiquitin-associated domain (UAB), a SH3 domain and a special histidine phosphatase domain, and TULA-2 binds and uses Syk as substrate [43,44]. This is of functional significances since the state of Syk tyrosine phosphorylation is reduced in cells overexpressing TULA-2, but enhanced when TULA-2 is impaired or deficient. In both murine and human platelets, strong evidence was obtained that TULA-2 suppresses the GPVI-induced, FcRγ-mediated Syk activation by dephosphorylating primarily Syk Y346 (Y352 in human) [45], which also belongs to the interdomain-B. Compared to the tyrosine dephosphorylation of Syk, there is no information available on serine/threonine dephosphorylation, which is most likely catalyzed by one or more of the many serine/threonine protein phosphatases present in human platelets [29].

A remarkable result reported here is that the specific inhibition of Syk S297 phosphorylation by the PKC inhibitor GFX closely coincided with enhanced cvx/EB-stimulated Syk Y352/Y525/526 phosphorylation (indicating enhanced Syk activity), enhanced Syk substrate phosphorylation

(LAT, PLCγ2), and enhanced cvx/EB-induced InsP1/InsP3 increase. Considering the dynamic aspects of Syk activity by phosphorylation/dephosphorylation observed in other cells [22], the effects observed here are substantial. Interestingly, the PKC inhibitor GFX was reported to enhance GPVI-mediated Syk phosphorylation at Y525/526 without affecting the Y352 and Y323 sites in human platelets, effects not observed with murine platelets [46]. However, direct PKC substrates were not investigated. In our studies, platelet agonists, but also the PKC activator PDBu, stimulated not only Syk S297 phosphorylation, but also, used here as a control, the phosphorylation of MARCKS, a well-established PKC substrate [35], and both responses were prevented by the PKC inhibitor GFX.

There is another line of evidence for the conclusion that PKC inhibition enhances Syk and Syk substrate tyrosine phosphorylation with the possible involvement of Syk S297 phosphorylation. We recently showed that both cAMP- and cGMP-elevating platelet inhibitors (iloprost, riociguat), enhance Syk tyrosine phosphorylation/Syk activity [28]. Now, we show that iloprost (PKA pathway) strongly inhibits cvx/EB-stimulated Syk S297 phosphorylation. It is likely that, under our conditions, iloprost inhibits the platelet effects of ADP and TxA2, which, when released during cvx/EB stimulation, are important secondary messengers and (as we clearly showed for ADP) are able to stimulate PKC-mediated Syk S297 phosphorylation.

Our study demonstrates the regulation of Syk S297 phosphorylation in a primary human cell, the platelet, in response to several agonists such as ADP, cvx, and EB (Figure 10). Increased S297 phosphorylation is mediated by PKC and negatively affects the tyrosine phosphorylation state of Syk at both the interdomain-B (Y352) and the kinase domain (Y525/526). The S297 serine site of Syk is located at a crucial site of Syk regulation and activation, the interdomain-B, just at the border of the c-terminal SH2-domain [18,20] (Figure 10).

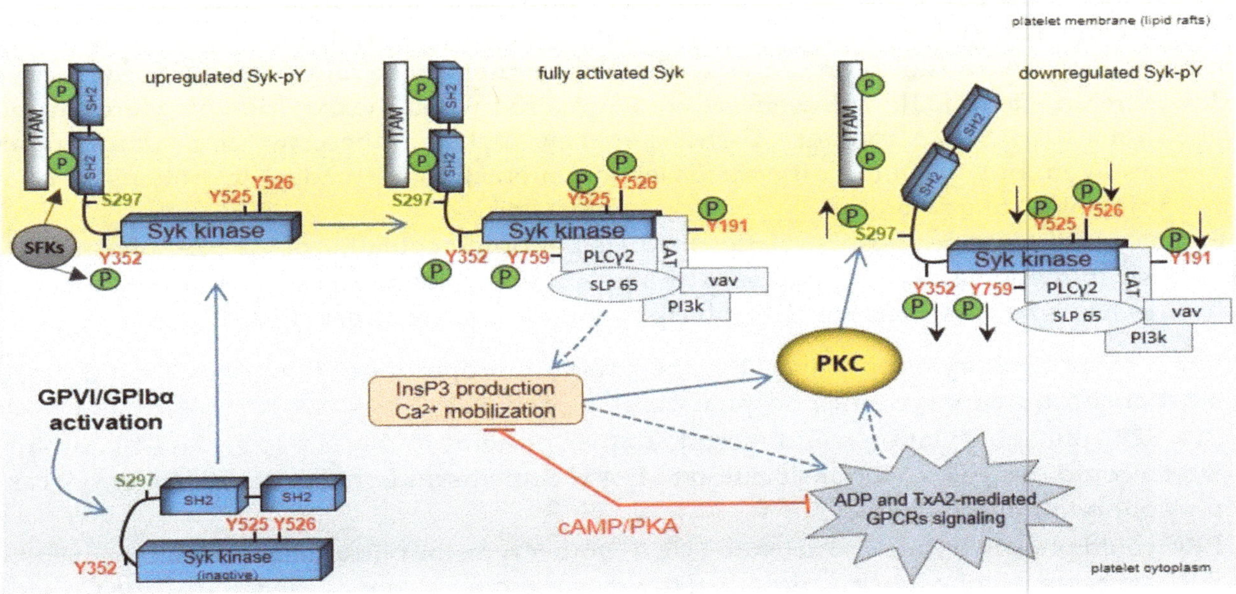

Figure 10. Model showing feedback regulation of Syk upon GPVI and GPIbα activation in human platelets. Clustering of GPVI and GPIbα upon agonist binding activates platelets initiated by src family kinases (SFKs) and the spleen tyrosine kinase (Syk). SFK activation leads to dual phosphorylation of the "immunoreceptor tyrosine-based activation motifs" (ITAMs), which represents a docking site for the two src homology 2 (SH2) domains of Syk. The interaction between Syk-(SH2)^2 domains and dually tyrosine-phosphorylated ITAMs recruits and translocates Syk to lipid rafts of the platelet cell membrane. Syk tyrosine phosphorylation of Y525/526 (kinase domain) and Y352 (interdomain-B) by autophosphorylation and SFKs leads to a fully activated kinase. Consequently, specific Syk substrates and regulatory proteins, such as LAT, PLCγ2, SLP65, vav, and phosphoinositide-3 kinase (PI3k) are

tyrosine-phosphorylated and recruited to a "LAT signalosome" complex. Activation of the Syk substrate PLCγ2 produces the release of ADP (from platelet δ-granules) and of TxA2 via elevation of InsP3 and Ca^{2+}. These secondary messengers enhance platelet activation via specific GPCRs, which also involves activation of distinct PKC isoforms. GPVI/GPIbα not only induce Syk tyrosine phosphorylation but also Syk phosphorylation on S297, which is located within the interdomain-B. GPVI/GPIbα-induced Syk S297 phosphorylation is mediated by PKC and decreases Syk tyrosine phosphorylation/activation. The cAMP/PKA pathway reduces the PKC response. It is proposed that Syk S297 phosphorylation represents an important mechanism for Syk feedback inhibition.

ITAM-and/or tyrosine phosphorylation-dependent Syk activation occurs at the membrane and alters the interaction of the various Syk domains, releasing the autoinhibition and producing an active kinase. Although there is functional and structural information that the interdomain-B with its phosphorylation and various interacting proteins is essential for the transition of inactive to activated Syk, the exact molecular mechanism(s) for this transition inside a cell are unknown. A major part of the crucial interdomain-B (amino acids 262–337) was unfortunately not defined in the otherwise impressive crystal structure of full-length inactive Syk [20], perhaps due to the flexible nature of interdomain-B.

The role of Syk phosphorylation including serine 297 (S291 in murine Syk) was extensively investigated by two groups in the chicken B-cell model system DT40, but discrepant functional data were obtained [23,24]. With murine Syk, PKC-mediated Syk S291 phosphorylation enhanced the ability of Syk to couple and activate the antigen receptor complex, perhaps involving the interaction of the interlinker domain B with the chaperone prohibitin [24]. These murine Syk effects were impaired when the murine S291A Syk mutant was tested. The other investigation with human Syk reported the presence of multiple Syk phosphorylation sites and also multiple Syk-interacting proteins. The major Syk phosphosite S297 (when phosphorylated) was reported to recruit the adapter protein 14-3-3γ, which prevented the translocation of Syk to the plasma membrane and subsequent Syk activation and downstream effects [23]. These effects were impaired when the experiments were carried out with the human Syk S297A mutant. The discrepancies between these two important studies are yet to be resolved, but a striking difference was the interaction with adapter proteins recruited to the phosphorylated interdomain-B.

Enhanced Syk tyrosine phosphorylation mediated by downregulation of Syk S297 phosphorylation through PKC inhibition is very likely due to decreased tyrosine dephosphorylation. There are several possible mechanisms via which Syk S297 phosphorylation could regulate this.

(a) Syk S297 phosphorylation could alter the structure/properties of Syk, which could make Syk a better substrate for tyrosine phosphatases (e.g., TULA-2).
(b) Syk S297 phosphorylation could recruit adapter proteins to this site (e.g., 14-4-3γ, prohibitin), which could decrease Syk translocation to ITAM proteins and, therefore, diminish Syk tyrosine phosphorylation/activation.
(c) PKC could phosphorylate and activate TULA-2/other tyrosine phosphatases resulting in enhanced Syk tyrosine dephosphorylation. However, there is no evidence for PKC-mediated phosphorylation and activation of TULA-2 [43].
(d) PKC could regulate other proteins, which affect Syk phosphorylation/dephosphorylation.

It is also possible that there are cell-specific mechanisms of Syk regulation, perhaps due to different binding/adapter proteins. These aspects should be investigated with platelets in the future. Our data with human platelets suggest that ITAM-dependent activation of Syk tyrosine phosphorylation is limited by Syk-dependent, PKC-mediated increased S297 phosphorylation, which can be prevented by PKC inhibition. This may represent a feedback inhibition important to prevent hyperactivation of Syk. Feedback inhibition is often observed with receptor-dependent signaling including tyrosine kinases. For example, PKC-stimulated phosphorylation of the epidermal growth factor receptor (EGFR) was reported to represent negative feedback inhibition of this signaling by converting the active EGFR dimer back to the inactive monomer [47]. Of special interest with respect to soluble tyrosine kinases, it was reported that PKCβ-mediated phosphorylation of the Bruton-tyrosine kinase BTK on S180 prevented membrane association and further tyrosine phosphorylation/activation of BTK in B and mast cells [48].

Considering the central role of Syk in immune cells and platelets, its established role in many disease processes including myeloid malignancies [15,49,50], and its increasingly recognized role as a therapeutic target [51,52], it will be now very important to elucidate the precise mechanism via which Syk is controlled by S297 phosphorylation.

Clinically used Syk inhibitors not only target Syk but also other tyrosine protein kinases. A clear direction of further development would be to find more specific Syk inhibitors, perhaps inhibitors, which do not directly target the kinase domain but a Syk-specific regulatory/allosteric domain such as the interdomain-B. Another problem with clinically used protein kinase inhibitors and also Syk inhibitors is the emergence of resistance, which was very recently addressed in a large genomic screen in acute myloid leukemia (AML) [50]. This screen identified activation of the RAS/MAPK/ERK pathway as one major mechanism of resistance to Syk inhibitors, which was further validated in AML cell lines with innate and acquired resistance to Syk inhibitors. The authors discussed their data, whereby resistance to tyrosine kinase inhibitors during targeted therapies may be due to secondary mutations of the target kinase, activation of other upstream stimulators or downstream effectors of the pathway, or the activation of parallel pathways. These very recent clinical findings highlight that improvement of our understanding of Syk regulation at the tyrosine kinase level directly, but also at the level of both upstream and downstream of this crucial component of platelet, immune, and myeloid cell activation, is essential for further advances in the area of Syk pathophysiology and inhibitors. Our present study on the feedback inhibition of Syk is perhaps one small step in this direction and also provides a human-relevant model of Syk signaling.

4. Materials and Methods

4.1. Ethics Approval

All healthy donors gave their informed consent for inclusion before they participated in the study. The study was conducted in accordance with the Declaration of Helsinki, and the protocol was approved by the local Ethics Committee of the University Medical Center Mainz (Study No. 837.302.12; 25.07.12; FF109/2015).

4.2. Preparation and Stimulation of Human Platelets

Whole blood was collected in citrate tubes from healthy volunteers who did not take any medication at least 10 days before the experiments. Platelet washing and the isolation procedure were performed as previously described [28]. A modified washing procedure was used for the experiments with ADP stimulation. PRP was diluted 1:1 with CGS buffer (120 mM NaCl, 12.9 mM Tri-Na-citrate, 30 mM glucose, pH 6.5) then centrifuged at 400× g for 10 min at room temperature (RT). The platelet pellet was resuspended directly with HEPES buffer (150 mM NaCl, 5 mM KCl, 1 mM $MgCl_2$, 10 mM glucose, 10 mM HEPES, pH 7.4). Human washed platelets were adjusted to 3×10^8 platelets/mL and kept at 37 °C prior to stimulation under stirring conditions with the corresponding platelet agonist or activator in the presence or absence of vehicle control or effectors.

4.3. Platelet Aggregometry

Aggregation of washed human platelets was monitored by light transmission aggregometry (LTA) using an Apact4S Plus aggregometer (DiaSys Greiner, Flacht, Germany). Platelets were stimulated by ADP (Sigma-Aldrich, Saint Louis, MO, USA), convulxin (Enzo life sciences, Lausen, Switzerland), echicetin beads (EB) (prepared as previously described [28,53]), or PDBu (Sigma-Aldrich, Saint Louis, MO, USA) under stirring conditions (1000 s^{-1}) at 37 °C. Platelets were pre-treated with different effectors prior to stimulation: Src family kinase inhibitor PP2 (Abcam, Cambridge, UK), Syk inhibitors OXSI-2 (Merck, Darmstadt, Germany) and PRT-060318 (Sellckem, Houston, TX, USA), PKC pan-inhibitor GF109203X (Enzo life sciences, Lausen, Switzerland), or cAMP-elevating agent iloprost (Bayer AG, Leverkusen, Germany).

4.4. SDS-PAGE and Western Blot Analysis

Platelet samples for Western blot analysis were prepared by adding 3× Lämmli buffer (200 mM tris/HCl, 15% (*v*/*v*) glycerol, 6% (*w*/*v*) SDS, 0.06% (*w*/*v*) bromphenol blue, 1:10 β-mercaptoethanol) directly in the aggregation cuvettes to stop platelet responses, then boiled at 95 °C for 10 min under gentle shaking. Platelet proteins were separated by electrophoresis using 8% SDS-polyacrylamide gels followed by immunoblotting as previously described [28].

Phospho-antibodies against Syk S297, Syk Y525/526, Syk Y352, LAT Y191, PLCγ2 Y759, and MARCKS S159/163 (Cell Signaling Technologies, Danvers, MA, USA) were used diluted 1:1000 in 5% BSA or 1:700 for MARCKS. Blots probed with phospho-antibodies were stripped and reprobed with a corresponding antibody detecting the total protein, anti-Syk, or anti-PLCγ2 (Santa Cruz Biotechnology, Santa Cruz, CA, USA), or other loading controls anti-β-actin or anti-α-actinin (Cell Signaling Technologies, Danvers, MA, USA) diluted 1:1000 in 5% BSA. After incubation with the appropriate secondary antibody, horseradish peroxidase (HRP)-conjugated goat anti-rabbit and anti-mouse IgG (BioRad Laboratories, Hercules, CA, USA) enhanced chemiluminescence (ECL) detection was performed using Fusion FX7 (Vilber Loumat GmbH, Eberhardzell, Germany).

4.5. Zn^{2+}-Phos-Tag^{tm}-SDS-PAGE and Western Blot Analysis

The phos-tag method was developed and described by the group of Koike [31,32] and used here for platelet Syk. Platelets were lysed using 3× Laemmli buffer for phos-tag (200 mM tris, 15% (*v*/*v*) glycerol, 6% (*w*/*v*) SDS, 2% (*w*/*v*) bromphenol blue, 1:10 β-mercaptoethanol) in the absence of EDTA. Samples were boiled at 95 °C for 10 min under gentle shaking. Platelet proteins were separated by electrophoresis using gels containing 6% *w*/*v* polyacrylamide, 35 μM phos-tag^{TM} compound, and 69 μM $ZnCl_2$ in the separating gel. Then, 1× phos-tag running buffer was used for gel electrophoresis (0.1 M tris, 0.1 M MOPS, 0.1% (*w*/*v*) SDS, 5 mM sodium bisulfite, pH 7.8). Prior to protein membrane transfer, gels were washed twice for 10 min with 1× transfer buffer containing 1 mM EDTA to remove the Zn^{2+} ions. A third washing step was performed with 1× transfer buffer without EDTA before the proteins were transferred to polyvinylidene difluoride (PVDF) membranes using a transfer buffer for phos-tag (25 mM tris, 192 mM glycine, 10% (*v*/*v*) methanol, 5% (*w*/*v*) SDS, pH 8.4).

Immunoblotting was performed as previously described [28]. Mouse monoclonal Syk antibody (Santa Cruz Biotechnology, Santa Cruz, CA, USA) was used as primary antibody (1:1000 in 5% BSA) and HRP-conjugated mouse antibody was used as secondary antibody (1:5000 in 5% BSA).

4.6. Inositol Monophosphate (InsP1) Measurement

Accumulated inositol monophosphate (InsP1), reflecting the produced inositol triphosphate (InsP3) upon platelet stimulation, was measured by using the IP-One ELISA kit (Cisbio, Codolet, France) in the presence of LiCl (1 mM), which prevents the degradation of InsP1 into myo-inositol. Washed platelets (3×10^8 platelets/mL) were stimulated with convulxin or EB in the absence or presence of effectors at 37 °C under stirring conditions, lysed after 5 min of activation, and centrifuged at 16,000× *g* for 10 min at 4 °C. InsP1–HRP conjugate and anti-InsP1 monoclonal antibody were pre-incubated with platelet lysates for 3 h prior to measurement, and InsP1 was quantified according to the manufacturer's instructions.

4.7. Intracellular Ca^{2+}-Release Measurement

Washed platelets (3×10^8 platelets/mL) were loaded for 30 min at 37 °C with 5 μM Fluo-3 acetoxymethyl (AM) esters (Life Technologies, Carlsbad, CA, USA), a Ca^{2+} indicator dye with excitation/emission of 506 nm/526 nm for the Ca^{2+}-bound form. Intracellular Ca^{2+} was monitored as mean fluorescence intensity for 2 min detected by flow cytometry (BD FACSCANTO II, BD Biosciences, Heidelberg, Germany) after stimulation with cvx or EB, without supplementation of extracellular Ca^{2+} and in the absence or presence of effectors. Data were analyzed by FACSDiva software v6.1.3 (BD Biosciences, Heidelberg, Germany) expressing the ratio of mean fluorescence intensity over time of treated platelets vs. basal platelet condition.

4.8. Statistical Analysis

Each dataset represents at least three different experiments from at least three different healthy volunteers when data are expressed as means ± standard deviation (SD). Statistical analysis was performed using GraphPad Prism 8 (GraphPad Software, San Diego, CA, USA). One-way or two-way ANOVA, followed by Tukey's multiple comparison test for comparison of more than two groups, or the two-tailed Student's t-test for comparison of two groups were used. A p-value < 0.05 was considered as significant.

Author Contributions: Conceptualization, K.J. and U.W.; methodology, S.G., S.M., and S.D.; data curation, S.M. and S.D.; formal analysis, S.M. and K.J.; writing—original draft preparation, S.M. and U.W.; writing—review and editing, S.M., S.D., S.G., U.W., and K.J.; supervision, K.J. All authors have read and agreed to the published version of the manuscript.

Acknowledgments: We are grateful to Ulrich Grädler (Merck KGaA, Darmstadt, Germany) for his fruitful discussions on the Syk structure.

Abbreviations

ACK1	Activated CDC42 kinase 1
ADP	Adenosine diphosphate
BCR	B-cell antigen receptor
BTK	Bruton-tyrosine kinase
cAMP	Cyclic adenosine monophosphate
cGMP	Cyclic guanosine monophosphate
cvx	Convulxin
EB	Echicetin beads
GFX	2-[1-(3-Dimethylaminopropyl)indol-3-yl]-3-(indol-3-yl) maleimide
HRP	Horseradish peroxidase
InsP1	Inositol monophosphate
InsP3	Inositol triphosphate
ITAM	Immunoreceptor tyrosine-based activation motif
JAK	Janus kinase
LAT	Linker adaptor for T cells
MARCKS	Myristoylated alanine-rich C- kinase substrate
PDBu	Phorbol 12, 13-dibutyrate
PI3k	Phosphoinositide- 3 kinase

PKA	Protein kinase A
PKC	Protein kinase C
PLCγ2	Phospholipase Cγ2
S	Serine
PAGE	Polyacrylamide gel-electrophoresis
SH2	Src homology 2
Syk	Spleen tyrosine kinase
T	Threonine
TxA2	Thromboxane A2
vWF	Von Willebrand factor
Y	Tyrosine

References

1. Jurk, K.; Kehrel, B.E. Platelets: Physiology and biochemistry. *Semin. Thromb. Hemost.* **2005**, *31*, 381–392. [CrossRef] [PubMed]
2. Versteeg, H.H.; Heemskerk, J.W.M.; Levi, M.; Reitsma, P.H. New fundamentals in hemostasis. *Physiol. Rev.* **2013**, *93*, 327–358. [CrossRef] [PubMed]
3. Jurk, K. Analysis of platelet function and dysfunction. *Hamostaseologie* **2015**, *35*, 60–72. [CrossRef] [PubMed]
4. Offermanns, S. Activation of platelet function through G protein-coupled receptors. *Circ. Res.* **2006**, *99*, 1293–1304. [CrossRef]
5. Brass, L.F.; Ma, P.; Tomaiuolo, M.; Diamond, S.L.; Stalker, T.J. A Systems Approach to the Platelet Signaling Network and the Hemostatic Response to Injury. In *Platelets in Thrombotic and Non-Thrombotic Disorders*; Springer: Berlin/Heidelberg, Germany, 2017; pp. 367–378.
6. Gardiner, E.E.; Andrews, R.K. Platelet Adhesion. In *Platelets in Thrombotic and Non-Thrombotic Disorders: Pathophysiology, Pharmacology and Therapeutics: An Update*; Gresele, P., Kleiman, N.S., Lopez, J.A., Page, C.P., Eds.; Springer: Cham, Switzerland, 2017; pp. 309–319. [CrossRef]
7. Rayes, J.; Watson, S.P.; Nieswandt, B. Functional significance of the platelet immune receptors GPVI and CLEC-2. *J. Clin. Investig.* **2019**, *129*, 12–23. [CrossRef]
8. Senis, Y.A.; Mazharian, A.; Mori, J. Src family kinases: At the forefront of platelet activation. *Blood* **2014**, *124*, 2013–2024. [CrossRef]
9. Ozaki, Y.; Suzuki-Inoue, K.; Inoue, O. Platelet receptors activated via mulitmerization: Glycoprotein VI, GPIb-IX-V, and CLEC-2. *J. Thromb. Haemost.* **2013**, *11* (Suppl. S1), 330–339. [CrossRef]
10. Falati, S.; Edmead, C.E.; Poole, A.W. Glycoprotein Ib-V-IX, a receptor for von Willebrand factor, couples physically and functionally to the Fc receptor gamma-chain, Fyn, and Lyn to activate human platelets. *Blood* **1999**, *94*, 1648–1656. [CrossRef]
11. López, J.A. The Platelet Glycoprotein Ib-IX-V Complex. In *Platelets in Thrombotic and Non-Thrombotic Disorders: Pathophysiology, Pharmacology and Therapeutics: An Update*; Gresele, P., Kleiman, N.S., Lopez, J.A., Page, C.P., Eds.; Springer: Cham, Switzerland, 2017; pp. 85–97. [CrossRef]
12. Stegner, D.; Haining, E.J.; Nieswandt, B. Targeting Glycoprotein VI and the Immunoreceptor Tyrosine-Based Activation Motif Signaling Pathway. *Arterioscler. Thromb. Vasc. Biol.* **2014**, *34*, 1615–1620. [CrossRef]
13. Zeiler, M.; Moser, M.; Mann, M. Copy Number Analysis of the Murine Platelet Proteome Spanning the Complete Abundance Range. *Mol. Cell. Proteom.* **2014**, *13*, 3435–3445. [CrossRef]
14. Poole, A.; Gibbins, J.M.; Turner, M.; van Vugt, M.J.; van de Winkel, J.G.J.; Saito, T.; Tybulewicz, V.L.J.; Watson, S.P. The Fc receptor gamma-chain and the tyrosine kinase Syk are essential for activation of mouse platelets by collagen. *EMBO J.* **1997**, *16*, 2333–2341. [CrossRef] [PubMed]
15. Mocsai, A.; Ruland, J.; Tybulewicz, V.L.J. The SYK tyrosine kinase: A crucial player in diverse biological functions. *Nat. Rev. Immunol.* **2010**, *10*, 387–402. [CrossRef] [PubMed]
16. Cheng, A.M.; Rowley, B.; Pao, W.; Hayday, A.; Bolen, J.B.; Pawson, T. Syk tyrosine kinase required for mouse viability and b-cell development. *Nature* **1995**, *378*, 303–306. [CrossRef] [PubMed]

17. van Eeuwijk, J.M.; Stegner, D.; Lamb, D.J.; Kraft, P.; Beck, S.; Thielmann, I.; Kiefer, F.; Walzog, B.; Stoll, G.; Nieswandt, B. The Novel Oral Syk Inhibitor, Bl1002494, Protects Mice From Arterial Thrombosis and Thromboinflammatory Brain Infarction. *Arterioscler. Thromb. Vasc. Biol.* **2016**, *36*, 1247–1253. [CrossRef] [PubMed]
18. Bradshaw, J.M. The Src, Syk, and Tec family kinases: Distinct types of molecular switches. *Cell. Signal.* **2010**, *22*, 1175–1184. [CrossRef]
19. Tsang, E.; Giannetti, A.M.; Shaw, D.; Dinh, M.; Tse, J.K.Y.; Gandhi, S.; Ho, H.D.; Wang, S.; Papp, E.; Bradshaw, J.M. Molecular Mechanism of the Syk Activation Switch. *J. Biol. Chem.* **2008**, *283*, 32650–32659. [CrossRef]
20. Gradler, U.; Schwarz, D.; Dresing, V.; Musil, D.; Bomke, J.; Frech, M.; Greiner, H.; Jakel, S.; Rysiok, T.; Muller-Pompalla, D.; et al. Structural and Biophysical Characterization of the Syk Activation Switch. *J. Mol. Biol.* **2013**, *425*, 309–333. [CrossRef]
21. Sada, K.; Takano, T.; Yanagi, S.; Yamamura, H. Structure and function of Syk protein-tyrosine kinase. *J. Biochem.* **2001**, *130*, 177–186. [CrossRef]
22. Geahlen, R.L. Syk and pTyr'd: Signaling through the B cell antigen receptor. *Biochim. Biophys. Acta-Mol. Cell Res.* **2009**, *1793*, 1115–1127. [CrossRef]
23. Bohnenberger, H.; Oellerich, T.; Engelke, M.; Hsiao, H.H.; Urlaub, H.; Wienands, J. Complex phosphorylation dynamics control the composition of the Syk interactome in B cells. *Eur. J. Immunol.* **2011**, *41*, 1550–1562. [CrossRef]
24. Paris, L.L.; Hu, J.J.; Galan, J.; Ong, S.S.; Martin, V.A.; Ma, H.Y.; Tao, W.A.; Harrison, M.L.; Geahlen, R.L. Regulation of Syk by Phosphorylation on Serine in the Linker Insert. *J. Biol. Chem.* **2010**, *285*, 39844–39854. [CrossRef] [PubMed]
25. Beck, F.; Geiger, J.; Gambaryan, S.; Veit, J.; Vaudel, M.; Nollau, P.; Kohlbacher, O.; Martens, L.; Walter, U.; Sickmann, A.; et al. Time-resolved characterization of cAMP/PKA-dependent signaling reveals that platelet inhibition is a concerted process involving multiple signaling pathways. *Blood* **2014**, *123*, e1–e10. [CrossRef] [PubMed]
26. Beck, F.; Geiger, J.; Gambaryan, S.; Solari, F.A.; Dell'Aica, M.; Loroch, S.; Mattheij, N.J.; Mindukshev, I.; Potz, O.; Jurk, K.; et al. Temporal quantitative phosphoproteomics of ADP stimulation reveals novel central nodes in platelet activation and inhibition. *Blood* **2017**, *129*, E1–E12. [CrossRef] [PubMed]
27. Trabold, K.; Makhoul, S.; Gambaryan, S.; van Ryn, J.; Walter, U.; Jurk, K. The Direct Thrombin Inhibitors Dabigatran and Lepirudin Inhibit GPIbalpha-Mediated Platelet Aggregation. *Thromb. Haemost.* **2019**, *119*, 916–929. [PubMed]
28. Makhoul, S.; Trabold, K.; Gambaryan, S.; Tenzer, S.; Pillitteri, D.; Walter, U.; Jurk, K. cAMP- and cGMP-elevating agents inhibit GPIb alpha-mediated aggregation but not GPIb alpha-stimulated Syk activation in human platelets. *Cell Commun. Signal.* **2019**, *17*, 1–10. [CrossRef] [PubMed]
29. Jurk, K.; Walter, U. New Insights into Platelet Signalling Pathways by Functional and Proteomic Approaches. *Hamostaseologie* **2019**, *39*, 140–151. [CrossRef]
30. Nagy, Z.; Smolenski, A. Cyclic nucleotide-dependent inhibitory signaling interweaves with activating pathways to determine platelet responses. *Res. Pract. Thromb. Haemost.* **2018**, *2*, 558–571. [CrossRef]
31. Kinoshita, E.; Kinoshita-Kikuta, E.; Kubota, Y.; Takekawa, M.; Koike, T. A Phos-tag SDS-PAGE method that effectively uses phosphoproteomic data for profiling the phosphorylation dynamics of MEK1. *Proteomics* **2016**, *16*, 1825–1836. [CrossRef]
32. Kinoshita, E.; Kinoshita-Kikuta, E.; Koike, T. Advances in Phos-tag-based methodologies for separation and detection of the phosphoproteome. *Biochim. Biophys. Acta-Proteins Proteom.* **2015**, *1854*, 601–608. [CrossRef]
33. Dangelmaier, C.A.; Quinter, P.G.; Jin, J.G.; Tsygankov, A.Y.; Kunapuli, S.P.; Daniel, J.L. Rapid ubiquitination of Syk following GPVI activation in platelets. *Blood* **2005**, *105*, 3918–3924. [CrossRef]
34. Thelen, M.; Rosen, A.; Nairn, A.C.; Aderem, A. Regulation by phosphorylation of reversible association of a myristoylated protein kinase C substrate with the plasma membrane. *Nature* **1991**, *351*, 320–322. [CrossRef] [PubMed]
35. Albert, K.A.; Nairn, A.C.; Greengard, P. The 87-KDA protein, a major specific substrate for protein-kinase-C—purification from bovine brain and characterization. *Proc. Natl. Acad. Sci. USA* **1987**, *84*, 7046–7050. [CrossRef] [PubMed]
36. Gilio, K.; Harper, M.T.; Cosemans, J.M.; Konopatskaya, O.; Munnix, I.C.; Prinzen, L.; Leitges, M.; Liu, Q.; Molkentin, J.D.; Heemskerk, J.W.; et al. Functional divergence of platelet protein kinase C (PKC) isoforms in thrombus formation on collagen. *J. Biol. Chem.* **2010**, *285*, 23410–23419. [PubMed]

37. Makhoul, S.; Walter, E.; Pagel, O.; Walter, U.; Sickmann, A.; Gambaryan, S.; Smolenski, A.; Zahedi, R.P.; Jurk, K. Effects of the NO/soluble guanylate cyclase/cGMP system on the functions of human platelets. *Nitric Oxide Biol. Chem.* **2018**, *76*, 71–80. [CrossRef]
38. Burkhart, J.M.; Vaudel, M.; Gambaryan, S.; Radau, S.; Walter, U.; Martens, L.; Geiger, J.; Sickmann, A.; Zahedi, R.P. The first comprehensive and quantitative analysis of human platelet protein composition allows the comparative analysis of structural and functional pathways. *Blood* **2012**, *120*, e73–e82.
39. Kunapuli, S.P.; Bhavanasi, D.; Kostyak, J.C.; Manne, B.K. Platelet Signaling: Protein Phosphorylation. In *Platelets in Thrombotic and Non-Thrombotic Disorders: Pathophysiology, Pharmacology and Therapeutics: An Update*; Gresele, P., Kleiman, N.S., Lopez, J.A., Page, C.P., Eds.; Springer: Cham, Switzerland, 2017; pp. 297–308. [CrossRef]
40. Pula, G.; Crosby, D.; Baker, J.; Poole, A.W. Functional interaction of protein kinase C alpha with the tyrosine kinases Syk and Src in human platelets. *J. Biol. Chem.* **2005**, *280*, 7194–7205. [CrossRef]
41. Bijli, K.M.; Fazal, F.; Minhajuddin, M.; Rahman, A. Activation of Syk by protein kinase C-delta regulates thrombin-induced intercellular adhesion molecule-1 expression in endothelial cells via tyrosine phosphorylation of RelA/p65. *J. Biol. Chem.* **2008**, *283*, 14674–14684. [CrossRef]
42. Oh, H.; Ozkirimli, E.; Shah, K.; Harrison, M.L.; Geahlen, R.L. Generation of an analog-sensitive syk tyrosine kinase for the study of signaling dynamics from the B cell antigen receptor. *J. Biol. Chem.* **2007**, *282*, 33760–33768. [CrossRef]
43. Tsygankov, A.Y. TULA proteins as signaling regulators. *Cell. Signal.* **2019**, *65*, 109424. [CrossRef]
44. Thomas, D.H.; Getz, T.M.; Newman, T.N.; Dangelmaier, C.A.; Carpino, N.; Kunapuli, S.P.; Tsygankov, A.Y.; Daniel, J.L. A novel histidine tyrosine phosphatase, TULA-2, associates with Syk and negatively regulates GPVI signaling in platelets. *Blood* **2010**, *116*, 2570–2578. [CrossRef]
45. Reppschlager, K.; Gosselin, J.; Dangelmaier, C.A.; Thomas, D.H.; Carpino, N.; McKenzie, S.E.; Kunapuli, S.P.; Tsygankov, A.Y. TULA-2 Protein Phosphatase Suppresses Activation of Syk through the GPVI Platelet Receptor for Collagen by Dephosphorylating Tyr(P)(346), a Regulatory Site of Syk. *J. Biol. Chem.* **2016**, *291*, 22427–22441. [CrossRef] [PubMed]
46. Buitrago, L.; Bhavanasi, D.; Dangelmaier, C.; Manne, B.K.; Badolia, R.; Borgognone, A.; Tsygankov, A.Y.; McKenzie, S.E.; Kunapuli, S.P. Tyrosine phosphorylation on spleen tyrosine kinase (Syk) is differentially regulated in human and murine platelets by protein kinase C isoforms. *J. Biol. Chem.* **2013**, *288*, 29160–29169. [CrossRef] [PubMed]
47. Kluba, M.; Engelborghs, Y.; Hofkens, J.; Mizuno, H. Inhibition of Receptor Dimerization as a Novel Negative Feedback Mechanism of EGFR Signaling. *PLoS ONE* **2015**, *10*, e0139971.
48. Kang, S.W.; Wahl, M.I.; Chu, J.; Kitaura, J.; Kawakami, Y.; Kato, R.M.; Tabuchi, R.; Tarakhovsky, A.; Kawakami, T.; Turck, C.W.; et al. PKCbeta modulates antigen receptor signaling via regulation of Btk membrane localization. *EMBO J.* **2001**, *20*, 5692–5702. [CrossRef]
49. Geahlen, R.L. Getting Syk: Spleen tyrosine kinase as a therapeutic target. *Trends Pharmacol. Sci.* **2014**, *35*, 414–422. [CrossRef]
50. Cremer, A.; Ellegast, J.M.; Alexe, G.; Frank, E.S.; Ross, L.; Chu, S.H.; Pikman, Y.; Robichaud, A.; Goodale, A.; Haupl, B.; et al. Resistance mechanisms to SYK inhibition in acute myeloid leukemia. *Cancer Discov.* **2019**. [CrossRef]
51. Fueyo, J.; Alonso, M.M.; Kerrigan, B.C.P.; Gomez-Manzano, C. Linking inflammation and cancer: The unexpected SYK world. *Neuro-Oncology* **2018**, *20*, 582–583. [CrossRef]
52. Szilveszter, K.P.; Nemeth, M.; Mocsai, A. Tyrosine Kinases in Autoimmune and Inflammatory Skin Diseases. *Front. Immunol.* **2019**, *10*, 1862. [CrossRef]
53. Navdaev, A.; Subramanian, H.; Petunin, A.; Clemetson, K.J.; Gambaryan, S.; Walter, U. Echicetin coated polystyrene beads: A novel tool to investigate GPIb-specific platelet activation and aggregation. *PLoS ONE* **2014**, *9*, e93569.

Permissions

All chapters in this book were first published by MDPI; hereby published with permission under the Creative Commons Attribution License or equivalent. Every chapter published in this book has been scrutinized by our experts. Their significance has been extensively debated. The topics covered herein carry significant findings which will fuel the growth of the discipline. They may even be implemented as practical applications or may be referred to as a beginning point for another development.

The contributors of this book come from diverse backgrounds, making this book a truly international effort. This book will bring forth new frontiers with its revolutionizing research information and detailed analysis of the nascent developments around the world.

We would like to thank all the contributing authors for lending their expertise to make the book truly unique. They have played a crucial role in the development of this book. Without their invaluable contributions this book wouldn't have been possible. They have made vital efforts to compile up to date information on the varied aspects of this subject to make this book a valuable addition to the collection of many professionals and students.

This book was conceptualized with the vision of imparting up-to-date information and advanced data in this field. To ensure the same, a matchless editorial board was set up. Every individual on the board went through rigorous rounds of assessment to prove their worth. After which they invested a large part of their time researching and compiling the most relevant data for our readers.

The editorial board has been involved in producing this book since its inception. They have spent rigorous hours researching and exploring the diverse topics which have resulted in the successful publishing of this book. They have passed on their knowledge of decades through this book. To expedite this challenging task, the publisher supported the team at every step. A small team of assistant editors was also appointed to further simplify the editing procedure and attain best results for the readers.

Apart from the editorial board, the designing team has also invested a significant amount of their time in understanding the subject and creating the most relevant covers. They scrutinized every image to scout for the most suitable representation of the subject and create an appropriate cover for the book.

The publishing team has been an ardent support to the editorial, designing and production team. Their endless efforts to recruit the best for this project, has resulted in the accomplishment of this book. They are a veteran in the field of academics and their pool of knowledge is as vast as their experience in printing. Their expertise and guidance has proved useful at every step. Their uncompromising quality standards have made this book an exceptional effort. Their encouragement from time to time has been an inspiration for everyone.

The publisher and the editorial board hope that this book will prove to be a valuable piece of knowledge for researchers, students, practitioners and scholars across the globe.

List of Contributors

Shreepa J. Chauhan, Anita Thyagarajan, Yanfang Chen and Ravi P. Sahu
Department of Pharmacology and Toxicology, Boonshoft School of Medicine Wright State University, Dayton, OH 45345, USA

Jeffrey B. Travers
Department of Pharmacology and Toxicology, Boonshoft School of Mesdicine Wright State University, Dayton, OH 45345, USA
Department of Dermatology, Wright State Physicians, Wright State University, Dayton, OH 45345, USA

Madhumita Chatterjee
Department of Cardiology and Angiology, Universitätsklinikum Tübingen, Medizinische Klinik III, 72076 Tübingen, Germany

Agnes Ehrenberg, Laura Mara Toska, Lisa Maria Metz, Meike Klier, Irena Krueger, Friedrich Reusswig and Margitta Elvers
Department of Vascular and Endovascular Surgery, Experimental Vascular Medicine, Heinrich-Heine University Medical Center, 40225 Düsseldorf, Germany

Preeti Kumari Chaudhary, Sanggu Kim, Seung-Hun Lee, Kyung-Mee Park and Soochong Kim
College of Veterinary Medicine, Chungbuk National University, Cheongju 28644, Korea

Youngheun Jee
College of Veterinary Medicine and Veterinary Medical Research Institute, Jeju National University, Jeju 63243, Korea

Lia Pulsatelli
Laboratorio di Immunoreumatologia e rigenerazione tissutale, IRCCS Istituto Ortopedico Rizzoli, Via di Barbiano 1/10, 40136 Bologna, Italy

Erminia Mariani
Laboratorio di Immunoreumatologia e rigenerazione tissutale, IRCCS Istituto Ortopedico Rizzoli, Via di Barbiano 1/10, 40136 Bologna, Italy
Dipartimento di Scienze Mediche e Chirurgiche, Alma Mater Studiorum-University of Bologna, Via Massarenti 9, 40138 Bologna, Italy

Bjoern F Kraemer
Medizinische Klinik und Poliklinik I, Klinikum der Universität München, Marchioninistrasse 15, 81377 Munich, Germany

Katharina Grundler Groterhorst
Medizinische Klinik und Poliklinik I, Klinikum der Universität München, Marchioninistrasse 15, 81377 Munich, Germany
Walter Brendel Centre of Experimental Medicine, University Hospital, Ludwig-Maximilians-University, Marchioninistr. 27, 81377 Munich, Germany

Hanna Mannell
Walter Brendel Centre of Experimental Medicine, University Hospital, Ludwig-Maximilians-University, Marchioninistr. 27, 81377 Munich, Germany
Biomedical Center, Ludwig-Maximilians-University, Großhaderner Str. 9, 82152 Planegg, Germany

Joachim Pircher
Medizinische Klinik und Poliklinik I, Klinikum der Universität München, Marchioninistrasse 15, 81377 Munich, Germany
DZHK (German Center for Cardiovascular Research), Partner Site Munich Heart Alliance, 80802 Munich, Germany

Yana Roka-Moiia and Marvin J. Slepian
Departments of Medicine and Biomedical Engineering, College of Medicine, University of Arizona, Tucson, AZ 85724, USA

Silvia Bozzi, Chiara Ferrari, Gabriele Mantica, Annalisa Dimasi, Marco Rasponi, Andrea Santoleri and Alberto Redaelli
Department of Electronics, Information and Bioengineering, Politecnico di Milano, Milan 20133, Italy

Mariangela Scavone and Marco Cattaneo
Divisione di Medicina, Ospedale San Paolo
Dipartimento di Scienze della Salute, Università degli Studi di Milano, Milan 20122, Italy

Filippo Consolo
Università Vita Salute, San Raffaele, Milan 20132, Italy

Mariangela Scavone, Gianmarco Podda and Marco Cattaneo
Unità di Medicina 2, ASST Santi Paolo e Carlo Dipartimento di Scienze della Salute, Università degli Studi di Milano, 20142 Milan, Italy

Tatiana Mencarini and Alberto Redaelli
Department of Electronics, Information and Bioengineering, Politecnico di Milano, 20133 Milan, Italy

Maria Valeria Catani, Isabella Savini, Valentina Tullio and Valeria Gasperi
Department of Experimental Medicine, Tor Vergata University of Rome, 00133 Rome, Italy

Bjoern F. Kraemer
Medizinische Klinik und Poliklinik I, Klinikum der Universität München, Marchioninistrasse 15, 81377 Munich, Germany

Tobias Lamkemeyer
Cluster of Excellence Cologne (CEDAD), Mass Spectrometry Facility at the Institute for Genetics, University of Köln, Josef-Stelzmann-Str. 26, 50931 Köln, Germany

Mirita Franz-Wachtel
Proteasome Center Tuebingen, University of Tuebingen, Auf der Morgenstelle 15, 72076 Tübingen, Germany

Stephan Lindemann
FB 20–Medizin, Philipps Universität Marburg, Baldingerstraße, 35032 Marburg, Germany
Medizinische Klinik II, Klinikum Warburg, Hüffertstr. 50, 34414 Warburg, Germany
Medizinische Klinik und Poliklinik III, Universitätsklinikum Tübingen, Otfried-Muller-Str. 10, 72076 Tübingen, Germany

Francesca Salamanna, Melania Maglio, Maria Sartori, Matilde Tschon and Milena Fini
Laboratory of Preclinical and Surgical Studies, IRCCS Istituto Ortopedico Rizzoli, Via di Barbiano 1/10, 40136 Bologna, Italy

Stephanie Makhoul, Stephanie Dorschel, Ulrich Walter and Kerstin Jurk
Center for Thrombosis and Hemostasis (CTH), University Medical Center Mainz of the Johannes Gutenberg University Mainz, 55131 Mainz, Germany

Index

www.ingramcontent.com/pod-product-compliance
Lightning Source LLC
LaVergne TN
LVHW060602210926
848485LV00005B/1358